GOLDEN YEARS OF MOSCOW MATHEMATICS

SECOND EDITION

HISTORY OF MATHEMATICS • VOLUME 6

GOLDEN YEARS OF MOSCOW MATHEMATICS

SECOND EDITION

Smilka Zdravkovska
Peter L. Duren
Editors

American Mathematical Society
London Mathematical Society

2000 *Mathematics Subject Classification.* Primary 01–06, 01A60, 01A65, 01A70.

For additional information and updates on this book, visit
www.ams.org/bookpages/hmath-6

Library of Congress Cataloging-in-Publication Data

Golden years of Moscow mathematics / Smilka Zdravkovska, Peter L. Duren, editors. — 2nd ed.
p. cm. — (History of mathematics ; v. 6)
ISBN-13: 978-0-8218-4261-4 (alk. paper)
ISBN-10: 0-8218-4261-7 (alk. paper)
1. Mathematics—Russia (Federation)—Moscow—History—20th century. I. Zdravkovska, Smilka, 1947– II. Duren, Peter L., 1935–

QA27.R8G65 2007
510.947′310904—dc22 2006052223

10 9 8 7 6 5 4 3 2 1 12 11 10 09 08 07

Contents

Preface to the Second Edition

It is a truism to state that much has changed for the Moscow mathematical community in the almost twenty years since the idea for this book was conceived and we approached about forty mathematicians in the Soviet Union for contributions. Some accepted, others did not reply, and some declined with an apology that times were too uncertain for them to be completely open about airing their views.

In the meantime, many have emigrated, others remained, but the contacts between the first and second groups are more numerous than anyone could have anticipated in the early 1990's: some of those who emigrated return regularly to Moscow for seminars, teaching and conducting research, and some of those who remained go to conferences, sabbaticals and other visits abroad. This intermingling of Russian mathematicians with those from outside of Russia has had several consequences for mathematics, both within and without Russia. Thus, to mention but one example, we see in the U.S. (say, at UC Davis, or Penn State) the organization by Russian mathematicians of "mathematical circles" as a method for training high school and undergraduate students by their slightly older peers, as was common in Russia (see footnote on p. 143).

When we were asked to prepare a preface for this second edition, it seemed natural to solicit an update by a Moscow mathematician. V. M. Tikhomirov graciously agreed and contributed on very short notice the beautiful article "On Moscow Mathematics—Then and Now", which is much more than an update, giving us an overview of 20th-century Moscow mathematics.

Reading this account was like a trip down memory lane. We hope it will provoke the same reaction from those who have experienced Moscow mathematics in that period, and, for others, that it will provide some flavor of what it was like to be there.

Tikhomirov sent us his article just before the 2006 International Congress of Mathematicians in Madrid began. So he had no way of knowing that out of the four Fields medals in Madrid, two would be awarded to mathematicians born and trained in Russia, one of them in Moscow. One of them has remained in Russia, and the other lives and works in the U.S. but has kept close ties with mathematicians in the old country.

So we join our hope to that of Tikhomirov and many others that the new century will be as productive for Moscow and Russian mathematics in general as the preceding one was.

In addition to the new article by Tikhomirov, this edition contains an index of names and a short list of errata. We are grateful to V. I. Arnold, who sent a whole page of corrections as soon as the first edition appeared.

Smilka Zdravkovska and Peter Duren
Ann Arbor, September 2006

Preface

The idea to put together a volume like this one occurred while I was reading an article on A. A. Lyapunov in the Russian magazine *Priroda* (Nature). Even though it was written with a somewhat heavy (for Western taste) dose of turgid sentimentality, the stimulating atmosphere of the Moscow intellectual circles was almost palpable. I mentioned this to Bill LeVeque, who was then Executive Director of the American Mathematical Society; he immediately liked the idea that the AMS publish such a volume, and he contacted Peter Duren, the Chairman of the AMS History Committee, about it. For various reasons, the project has dragged on for almost five years. Originally, it was to be coedited by Allen L. Shields and myself: his encyclopaedic knowledge, which included—but was certainly not limited to—Russian history and culture, and his global interest in mathematics, which is what one seems to mean when talking about "Russian flavour" in mathematics, made him an ideal person for the undertaking. Unfortunately, Allen died on 16 September 1989. At that point, for me, most things seemed irrelevant. Fortunately, Peter Duren had by then become a coeditor, for which I am deeply grateful. Without his crucial contribution, this volume would not have appeared.

The collection was first intended to have wider scope and something like "Soviet Mathematics; Recollections" as a title. But as the articles actually accepted all focused on Moscow, the current title seemed more appropriate.

The revolutionary changes in the past few years have affected this volume in several ways. I will mention just a minor consequence: whereas originally we were going to use the Mathematical Reviews transliteration of Russian names, this became impossible, with many of the mathematicians travelling abroad and using Western versions of their names and some insisting on using specific spellings for various people mentioned in their articles. So we had to resign ourselves to inconsistent spellings of names. Another inconsistency is in the translation of the Russian words *fakultet* (department), and *kafedra* (section in a department), which variably occur as Faculty, School or Department for the first, and Section, Chair or Division for the second. Finally, let us mention here two terms that often appear in this volume: "Candidate's degree" which corresponds to a Ph.D. from a university in the United States, and "Doctorate" which is much more difficult to obtain.

I was very fortunate to be an undergraduate student at the Mathematics Department (Mekh-Mat) of Moscow State University in the 1960s. I will never forget this exciting *milieu*, where one would learn at least as much from fellow students and those slightly older or younger than oneself as one would from the professors. After the first two years, one had to choose a *kafedra* and an undergraduate advisor under whose guidance one would write the undergraduate (research) thesis in the fifth year. One could pick for credit from among literally dozens and dozens of courses and seminars, most taught or conducted by first-rate mathematicians. One passed the knowledge on to (and learned from) the bright high-school children in the mathematical *kruzhoks* (circles). And of course mathematics was but one of the many interests (though maybe the major) of the groups bound by close friendships.

A marked chill in the atmosphere was felt in 1968, after 99 mathematicians signed a letter protesting the treatment of Esenin-Volpin (this is explained in several of the articles in this volume; see, for example, the articles by Fuchs (p. 220) and Sossinsky (p. 235)). Obtaining permission from all the cosignatories to publish their names here was too daunting a task, so this tempting idea had to be dropped. This letter with its signatures is published in *Sobranie Dokumentov Samizdata* (Collection of Samizdat Documents), Vol. 1, AS No. 20 (available at the University of Michigan Library, for example), so the interested reader might want to look it up. Any mathematician will readily recognize most of the names.

It seems that the decline of Mekh-Mat has been constant since then (see an article by S. P. Novikov about it in *Moscow News*, February 18, 1990). That, along with the possibility for travel and emigration of the top mathematicians, has threatened the very existence of Moscow mathematics as described here. So it was with joy that I learned of the creation by a group of enthusiasts[1] of the Independent University of Moscow (IUM). It started in the Fall of 1991, comprising just a mathematics and a physics departments. The mathematics department has fifty students (all in the freshman year; each year an extra year is planned to be added), hand-picked by leading mathematicians after a competitive examination.

The IUM is currently housed in a high school which lets the IUM use its building in the evening, after the regular students have left. Evening classes are a necessity also because the IUM students for the time being must also be enrolled in some other educational institution in order to obtain draft postponement. Three courses were taught this past year: Algebra (by Rudakov), Analysis (by Kirillov and Vassiliev) and Mathematical English (by Sossinsky). All this is done on a practically voluntary basis.

[1]The Scientific Council of the Mathematics Department of IUM consists of: V. I. Arnold, A. A. Beilinson, R. L. Dobrushin, B. A. Dubrovin, L. D. Faddeev, A. A. Kirillov, A. G. Khovanskiĭ, S. P. Novikov, A. N. Rudakov, Ya. G. Sinai, M. A. Shubin, and V. M. Tikhomirov, N. Konstantinov serves as Acting Dean.

The founding of IMU is a very important first step, in Arnold's words, in the attempt to solve the problem of conserving and regenerating the cultural potential of Russia. It should lead to more golden years of Moscow Mathematics.

Smilka Zdravkovska
Ann Arbor, 6 June 1992

ACKNOWLEDGMENT. The editors would like to thank all the authors of the articles for their contributions. In addition, many people have helped us in various ways. We would particularly like to thank Tatiana Belokrinitskaya, Maryse Brouwers, Igor Dolgachev, Viktor Havin, Dmitry Khavinson, Askold Khovanskiĭ, Nikolai Nikolskiĭ, Abe Shenitzer, Andrej Urumov, and Bojana Urumova.

Encounters with Mathematicians

A. P. YUSHKEVICH

Close to 70 years have passed since the fall of 1923 when I entered Moscow University. During that time I met, and worked with, many outstanding mathematicians. The basis of these encounters was invariably the interest of these scholars in the history of mathematics, a subject that has been my specialty since my student days.

The attitudes of mathematicians, both pure and applied, toward the history of their subject vary. They depend on the *Zeitgeist*, on the dominant research interests at a given time, on their individual professional activities, and so on. Some have no taste whatsoever for the history of science and regard its study as the lot of noncreative specialists, others are amused by biographical anecdotes, others value highly its educational potential, and still others use it to elaborate broad methodological conceptions. Finally, there are scholars interested in the history of mathematics for its own sake. For them it is a subject to which they devote some of their creative endeavors. Here we might mention foreign mathematicians such as Chasles, Zeuthen, F. Klein, Hardy, A. Weil, and Dieudonné, and Russian mathematicians such as P. S. Aleksandrov, Delone, Kolmogorov, Markushevich, Luzin, and V. I. Smirnov. This list could be extended by mentioning Vasil′ev, V. F. Kagan, and such now flourishing mathematicians as A. D. Aleksandrov and Gnedenko.

In the 1920s and 1930s many Moscow mathematicians had very positive attitudes toward the history of mathematics and studied it to a greater or lesser extent. This was largely due to animated discussions of fundamental concepts and methods of mathematics provoked by the logico-philosophical problems of set theory and the theory of functions. The idea of the infinite has been fraught with methodological and logical difficulties virtually from its inception, and, quite naturally, attempts to resolve these difficulties relied on the historical analysis of the question. The first quarter of this century was marked by the emergence of rival schools of formalism, logicism, intuitionism, and constructivism, and by the rapid growth of mathematical logic. The flourishing of the school of the theory of functions of a real variable at

Moscow University was, quite naturally, accompanied by a growth of interest in the philosophy of mathematics and thus in its history. Another significant contributing factor was an interest in the history of culture in general and in our country in particular. Of course, the interest in the history of mathematics grew in other scientific centers as well. Historical sections appeared in a variety of textbooks such as V. V. Stepanov's book on differential geometry, A. I. Markushevich's book on the theory of functions of a complex variable, the three-volume course on differential and integral calculus of the Leningrad professor G. M. Fikhtengol′ts, in A. F. Bermant's course of mathematical analysis for higher technical schools (!), and in other texts. One should also mention the contributions of mathematicians to books on the history of universities and learned societies, and papers connected with the celebration of anniversaries such as those of Newton, Lagrange, Euler, and Chebyshev, as well as those of our contemporaries.

I was more or less acquainted with many of the Soviet mathematicians just listed and I cooperated with some of them. Here I will write about N. N. Luzin, P. S. Aleksandrov, A. N. Kolmogorov, and V. I. Smirnov.

1. N. N. Luzin

Shortly after entering the university I attended a course on higher algebra given by Nikolai Nikolaevich Luzin (1883–1950). His inspiring lectures fascinated the audience; I have said more about this elsewhere. From 1930 on, I lectured at the Moscow Higher Technical School (MVTU) and for many years Luzin may be said to have disappeared from my field of vision. In the 1930s, my contact with Moscow State University (MGU) was limited to participation in the seminar on the history of mathematics that came into being in 1933 and was at first run by M. Ya. Vygodskiĭ and S. Ya. Yanovskaya. Echoes reached me of the slanderous public campaign, launched in 1936, which described Luzin as "an enemy in a Soviet mask". When I realized that he was being subjected to a harsh "working over" in print and at various meetings of scientific workers submissive to Stalin's regime, I felt deeply sorry for him. To express my sympathy, I wrote a warm dedication on a copy of the just-published second edition of L. Carnot's *Reflections on the metaphysics of infinitesimals*, translated under my editorship, and took it to his home. Luzin responded with a letter, which, I am sorry to say, is no longer in my possession. I do recall that the letter included a few phrases about the importance of the history of mathematics, and his view that an historical analysis might perhaps make it possible to prove Riemann's hypothesis on the distribution of the zeroes of the zeta function.

Between 1946 and 1947 I had the opportunity of working closely with Luzin in the following connection. S. I. Vavilov, the then president of the Soviet Academy of Sciences (AN SSSR), a physicist and a historian of science, proposed the publication of a Russian translation of the correspondence

N. N. LUZIN

between Euler and Goldbach, published in part in 1842 in the original languages by the academician P. N. Fuss. My father-in-law V. S. Gokhman, a physicist by training, was to translate the text, Luzin was to write an introduction and serve as general editor, and I was to prepare the notes. That is why I got together and talked with Luzin a number of times, visited him in his new apartment (he moved from Arbat, #25, to Sretenskiĭ Boulevard, #6), and travelled with him to see the head of the central state archive (a certain Dzincheradze), who was in charge of the publication. At the time Luzin reflected on the properties of the set of natural numbers and its subsets, and some of our conversations were of a scientific nature. Luzin devoted to this topic two small papers published in 1943 and 1947, respectively. These papers are reproduced in vol. 1 of his *Collected works* (Moscow, 1953). Luzin always took a lively interest in philosophical problems of set theory and the theory of functions. One day he called me and asked me to lend him my copy of the Russian translation of Cavalieri's *Geometry of indivisibles*, a book that came up in one of our conversations. A driver brought back the book and the following letter (I gave the original to the archive of the Academy of Sciences):

> Dear Adolf Pavlovich,
>
> It is with deep gratitude that I return to you your copy of Bonaventura Cavalieri's splendid book. The day after you were kind enough to bring it to me in person I became very seriously ill (a case of kidney complications following the flu) and could not possibly use it. But the sense of certainty that I had such an important original source played a vital and beneficial role in my consciousness. I did not think that I would be ill for three months.
>
> I am returning your book and wish to express the hope that we will meet again, and not just in connection with Euler. I have no doubt that we mathematicians are at the threshold of a change of consciousness comparable to that which Lobachevskiĭ was fated to bring about in his time. But this time it will pertain to the natural numbers rather than to space.
>
> The nineteenth century was exceptionally quiet and sluggish. At present consciousness moves quickly and what now seems paradoxical will, within 2–3 scientific generations, enter firmly the general consciousness and become a triviality.
>
> Once more, thanks for the book.
>
> Yours,
>
> N. Luzin
>
> Euler is excellent, especially when he experiments with the summing of infinite series. We have no sure sense of him, and writers at the beginning of the century have thoroughly distorted him.

This letter illustrates Luzin's unfailing courtesy in personal relations. Its substance is closely related to the papers on subsets of the natural numbers, the second of which, dated 2 June 1947, ends with a remark about impending changes in arithmetic similar to those carried out by Lobachevskiĭ in geometry. These papers link Luzin's ideas about the need to explore the descriptive classification of countable sequences to the subsequently developed theory of algorithms.

As for the reference to Euler's experiments with the summing of infinite series, Luzin may have had in mind his generalized summing of divergent series as well as the "interpolation" of sequences $u_1, u_2, \dots, u_n \dots,$ that is, the finding of expressions suitable for the definition of u_n for fractional values of n. Incidentally, in Euler's work, generalized summing and "interpolation" are often combined.

The translation of the Euler-Goldbach correspondence was carried out in two stages. The first one was to publish only Euler's letters. When their translation was completed, Luzin wrote an introduction in which there was no mention of Goldbach's letters. This was at the beginning of 1948; I am certain of the date because Luzin wrote me a note, dated 6 January 1948, in which he asked me to send him the Euler materials which I needed to write up my as yet unfinished notes. Then, following Luzin's suggestion, Goldbach's letters were translated. But Luzin did not think it necessary to supplement his introduction, the importance of Goldbach's letters for the study of Euler's creative contribution notwithstanding. In his role as editor-in-chief, Luzin signed the manuscript as ready to be printed. But the translation of this remarkable correspondence, the most extensive and longest of all exchanges between Euler and his many correspondents (196 letters written between 1729 and 1764, the year of Goldbach's death), never saw the light of day. On 28 February 1950 Luzin died suddenly of a heart attack, and on 25 January 1951 Vavilov also died of the same cause. Luzin had just turned 67 and Vavilov was less than 60. There is little doubt that Luzin's health was undermined by his 1936 ordeal, and Vavilov's by the loss of his brother, the eminent botanist N. I. Vavilov who perished in the torture chambers of the GPU, and by his work in 1945 as president of the Soviet Academy of Sciences (AN SSSR) under moral and political conditions which he must have found well nigh unendurable. Without "support from above" the publication of the Euler-Goldbach correspondence failed to materialize and the manuscript of the translation is still in my possession. I wish to note that in a letter to Gokhman, dated 21 August 1946 and pertaining to the publication of the Euler-Goldbach correspondence, Vavilov wrote: "The means to pay for all this work will be found in the Academy" (IMI, vols. XXXII–XXXIII, 1990, p. 527). After Vavilov's death such means could not, apparently, be found. All I managed to do was publish Luzin's introduction in issue XVI of IMI (1965). Of special interest here are the remarks about Euler's set-theoretic ideas. Of course, this paper is not included in the three-volume *Collected works* of Luzin published between 1953 and 1959. I wish to add that, together with the late Berlin academician E. Winter and a number of German and Soviet collaborators, I was able to publish the full Euler-Goldbach correspondence with numerous German and Latin commentaries—the two languages of the original (Berlin, 1965).

Luzin's introduction to the Euler-Goldbach correspondence is not his only work on the history of mathematics. He was very much interested in the history and philosophy of mathematics. In many of his major works there are historical and methodological digressions. Also, Luzin published a number of papers on the history of mathematics; see [**2**].

2. P. S. Aleksandrov

I had known Pavel Sergeevich Aleksandrov (1896–1982), as well as Luzin, since my student days. I did not attend his lectures. However, he was my examiner in an integral calculus course. To this day I remember this examination with a measure of embarrassment. At the time, the spirit of the "Luzitania"—the jocular name for the group of Luzin's students—dominated the department of mathematics. We were interested in the theory of functions, and, when it came to the integral calculus, in existence theorems and generalizations of the idea of the integral; the whole group, including myself, participated actively in the seminar "The Denjoy integral", run by D. E. Men′shov, one of Luzin's senior students. This being so, I prepared for the integral calculus examination using Russian translations of the relevant works of L. Bieberbach and of de la Vallée-Poussin. I ignored the tutorial sessions devoted to the evaluation of indefinite integrals and to the geometric and other applications. Aleksandrov, then a docent, asked me first to prove the existence of the integral of a continuous function. I answered "à la Bieberbach". Then he asked me to "please compute the volume of a torus". I was baffled. I did not know the word "torus" and asked what figure it denoted. Rather than explain, Aleksandrov asked me to evaluate some indefinite integral. He looked at the page of my computations and let me go in peace.

Many years passed since this first encounter. I saw Aleksandrov a number of times at meetings of the Moscow Mathematical Society (MMO). During his presidential tenure (1932–1964) the meetings often took the form of lectures on the history of mathematics. Nevertheless, for a long time I had no occasion to associate with him directly. True, when the first issue of "Historical-mathematical investigations" (IMI), which I began to publish together with G. F. Ribkin, was in preparation, Aleksandrov, Gnedenko, and Smirnov submitted the paper "Mathematics at Moscow University in the twentieth century up to 1940". this issue, published in 1948, contained a number of articles on mathematics at Moscow State University (MGU), written in part before World War II in honor of the 185th anniversary of the university, which fell in 1940. (For some reason, Stalin decided that it was desirable to solemnly commemorate this odd-dated anniversary.) I do

P. S. ALEKSANDROV

not recall any conversations with Aleksandrov in connection with this article. A few years later Aleksandrov wrote a brilliant paper on the same topic. It appeared in 1955, in issue of VIII of IMI devoted to the 200th anniversary of MGU. The paper contains characterizations of all of the important scientific schools and their results in the first half of the twentieth century. In particular, it includes a discussion of investigations in the history of mathematics. (Here Aleksandrov was probably helped by S. A. Yanovskaya, whose name is mentioned on the first page of the paper.) The paper included photographs of many of the professors of that time. The innovative pedagogical activities of B. K. Mlodzeevskiĭ and D. F. Egorov are carefully elucidated. There is a suitably detailed description of Egorov's famous seminar, where participants learned of the latest scientific developments in Western Europe. But Egorov's photograph could not be included. The memory of the struggle against "Egorovshchina" was still too fresh. As a result of this struggle Egorov was removed from the position of director of Scientific research institute of mathematics and mechanics of MGU and from the post of president of the Moscow Mathematical Society (MMO) and sent to Kazan. Soon thereafter he died in a hospital. I should add that the same issue included a splendid article on mechanics at MGU during the same period, written by V. V. Golubev. Golubev died in 1954 at the celebration of his 70th anniversary.

My sporadic meetings with Aleksandrov began in the 1950s. Of these I recall only isolated episodes. In 1951 my assistant V. V. Gussov defended his Candidate's dissertation on the history of cylindrical functions in Russia and in the USSR before the Learned Council of the Faculty of mechanics and mathematics (Mekh-Mat). The mathematicians liked the dissertation because it was rich in concrete material and the subject was discussed within the large framework of the development of analysis as a whole. After the defense Aleksandrov came up to me, spoke in flattering terms about the dissertation, and added: "*Allaverdy* to you." * . Gussov's paper on the history of the gamma function and of cylindrical functions appeared in issues V (1952) and V1 (1953) of IMI. Gussov soon left Moscow and devoted himself to teaching at the Far Eastern Polytechnical Institute (in Vladivostok).

In 1955 there began intensive preparations in the USSR and in the German Democratic Republic (GDR) for the celebration of the 250th anniversary of Euler's birth (15 April 1707). The academies of sciences of both countries decided to cooperate in this enterprise. We formed an anniversary committee headed by M. A. Lavrent′ev, then vice-president of the Academy of Sciences (AN). I was appointed scientific secretary.

In the spring of 1956 I went to Berlin to help coordinate the anniversary programs. This was my first trip abroad after my family's half-year stay in France, in the spring and summer of 1914 (at which time I was just eight

---

* *Editors' note*: In Russian, this is a rather recondite form of praise.

years old).[1] In Berlin I had discussions with the geophysicist G. Ertel, vice-president of the GDR Academy of Sciences, the mathematician (academician) K. Schröder, the representative of the Euler commission, the historian (academician) E. Winter (mentioned above), and the mathematician Dr. K. R. Bierman, the scientific secretary of the anniversary committee of the GDR Academy of Sciences. Dr. Bierman and I became, and are, close friends. We fixed dates and programs of celebrations and sketched plans for various publications. All the plans were realized. The celebrations stimulated not only further study of Euler's life and works but also the study of the history of both academies, where Euler worked for nearly 60 years. The jubilee sessions took place on 21–23 March in Berlin and on 15–18 April in Leningrad. The participants included scholars from the USSR, the GDR, and other countries. The Soviet delegation that went to Berlin was headed by Aleksandrov (a corresponding member of the Berlin academy since 1950) and included the corresponding member of the Soviet Academy of Sciences B. N. Delone, A. T. Grigor′yan, A. G. Postnikov, and me. We went to Berlin by train. During the journey Aleksandrov kept to himself. Most likely, he thought about the lecture which he was to give in the name of the Soviet Academy at the meeting of the Academy of Sciences of the GDR. He had no books with him—he relied entirely on his memory. He spoke German as fluently as Russian and his historical-mathematical education was truly impressive. He was a very good speaker. His half-hour talk on the first day of the session was remarkable. When the representative of the Berlin Academy asked him for the text of his talk for the purpose of publication, Aleksandrov told him that he had no written text and that he had no time to write one down. My talk, and the talks of Delone and Postnikov, were included in a collection published by the Berlin academy. As soon as the meeting ended Aleksandrov returned to Moscow. The remaining members of the delegation took part in an interesting trip to the "witches' mountain", the Brocken, located between the two Germanys. On the way we visited picturesque Vernigorod as well as Kvedlinburg, a town whose medieval character is largely preserved. Aleksandrov took no active part in the Euler session in Leningrad; it the seems that he did not even come to Leningrad at the time. A week before the Leningrad meeting, on Aleksandrov's initiative, there took place a festive meeting of the Moscow Mathematical Society and the Learned Council of the Faculty devoted to the same jubilee. On this occasion I gave a lecture titled "The life and mathematical activity of Leonhard Euler" published in the Uspekhi (UMN), XII, 4, (1957). For more detailed information on the Euler

[1] In Paris we stayed in Mr. Shomar's hotel "Parisiana" (at 4, rue Tournafort). Luzin stayed in this hotel until May 1914. Some of Luzin's students also stayed there. All of them left endearing memories. When I was in Paris 60 years later I visited the hotel and talked to its two owners, the daughters of Mr. Shomar. They remembered fondly Luzin, "so well-mannered and pious", Men′shov, and "happy and tall Michel", that is Lavrent′ev. Both sisters are somewhat older than I.

anniversary celebrations in Berlin and Leningrad, see *Vestnik Akad. Nauk SSSR*, 1967, #6, pp. 121–123.

As a result of working with Lavrent′ev I learned of his remarkable business qualities, the speed with which he came to necessary decisions, and of his directness which sometimes shaded into offhandedness. The Berlin trip also gave me an opportunity to get to know Delone better. He too was very direct. He often talked ironically, at times even cynically, about our common acquaintances. Many years later we published a joint paper titled "Academician Leonhard Euler" (*Priroda*, #701).

Following as much as possible the chronological order, I wish to mention that on one occasion I "swapped" reviews with Aleksandrov. My *History of mathematics in Russia up to 1917* was published in 1968. One year later the book was reviewed in vol. 24 of UMN by Aleksandrov and Bashmakova. I think that the review was largely, and perhaps entirely, written by Bashmakova but Aleksandrov was familiar with it and with the book. When we met he said: "Of course, I read your book." Most likely, he looked it through and read carefully pages 574–575 devoted to his and Suslin's work on descriptive function theory. He also forgave me a lapse which was very unpleasant for him; I had written that Suslin called the A-sets discovered by Aleksandrov analytic sets. Actually, the term was introduced by Luzin, and this irked Aleksandrov (see below). In 1976 there appeared Lobachevskiĭ's *Scientific-pedagogical Nachlass*, edited by Aleksandrov and Laptev. The book included a brilliant paper on Lobachevskiĭ's educational ideas presented by Aleksandrov at MGU in March of 1967. Following Aleksandrov's request, Professor Oleĭnik, assistant editor of UMN, suggested that I write a review of the book. I did so with pleasure. The review appeared in UMN, vol. 33, issue 3 (1978).

Another encounter with Aleksandrov, related to the study of Euler's works, took place in 1967, in connection with the arrival in Moscow of representatives of the Euler Commission of the Swiss Academy of Science (or the Society of Naturalists). They proposed to the Institute of History of Science and Technology of the Soviet Academy joint publication of the fourth series of Euler's collected works, including his scientific correspondence. The group included the Basel mathematician W. Habicht, son-in-law of the eminent algebraist and historian of mathematics B. L. van der Waerden. Habicht was aware that, while in Göttingen, Aleksandrov had known van der Waerden very well, so he asked to meet him. At the time, Aleksandrov was head of the department of mathematics of Mekh-Mat. The visit took place in his office and lasted for about an hour. Habicht listened with interest to Aleksandrov's description of his encounters with van der Waerden and to various details of mathematical life in Göttingen. When we left, the Basel professor said that Aleksandrov's German was better than his own—and he (Habicht) had spoken German since childhood.

Let me set aside personal recollections for a while and say a few words about the importance of the trips to Göttingen by Aleksandrov and other Soviet mathematicians, trips that began in 1923 and ended as a result of the Nazi takeover in 1933. In those years, the actual organizer of mathematical life in Göttingen was Courant, an applied mathematician and a student and friend of Hilbert's. In her biography of Courant, Constance Reid notes that he always promoted Aleksandrov's yearly lectures on topology. Basing herself on a conversation with Busemann (b. 1905), who studied in Munich, Paris, and Göttingen and went to the USA in 1936, Reid writes that the arrival of Russian mathematicians—of Aleksandrov as well as of Uryson, Kolmogorov, Lysternik, Gelfond, Schnirelman, and Pontryagin—"filled a gap, because they knew well certain areas insufficiently represented in Göttingen" [**3**, p. 125]. Earlier, in 1910–1912, Luzin had stayed in Göttingen.

Incidentally, I wish to note a circumstance that is essential for a biography of Luzin. At a meeting of the commission set up by the President of the Academy of Sciences to investigate the charges laid against him in 1936, Luzin stressed the importance of his stay in Göttingen; namely, it was there that, at Landau's insistence, he wrote his first publishable paper, titled "On a certain power series" (Rendiconti Circ. Palermo). Incidentally, this was his only paper in German. Luzin's spoken German was not as good as his French, but good enough to converse. In Göttingen, Luzin met not only Landau but also Klein and other local professors.

In the later 1970s I began to visit Aleksandrov in his room in the new building of MGU. The first visits were prompted by an archival discovery made by the Colombian historian of mathematics Louis K. Arboleda. In 1979 Arboleda investigated in Paris the archive of Fréchet and found in it 48 letters by Aleksandrov and Uryson, the co-founder, with Aleksandrov, of the Moscow School of Topology. The letters were dated 1923–1933. (Seven of these letters were written jointly by Aleksandrov and Uryson and 41 by Aleksandrov; Uryson drowned in a storm in the Atlantic in 1924.) The study of these letters prompted Arboleda to write a paper on the rise and early stage of the Moscow School of Topology. Arboleda told me this during our meetings in Paris. The fact that no one abroad had researched this topic made the project all the more interesting. Arboleda asked me to find out whether Fréchet's replies were in existence. (They were not, said Aleksandrov.) I told the young Colombian scholar that the right thing to do was to show the discovered materials to Aleksandrov and to obtain his approval. I offered to serve as a go-between. When I returned to Moscow I called Aleksandrov and he asked me to read Arboleda's materials to him (by then Aleksandrov was almost completely blind). The reading of the text took a number of evenings. The text mentioned various publications of Aleksandrov and Uryson. I was amazed by Aleksandrov's memory; he correctly stated the place and time of the appearance of the respective publications without waiting for me to read Arboleda's references. Aleksandrov approved Arboleda's work and a Russian translation of his paper (prepared by F. A. Medvedev), titled "The birth of the Russian school of topology. Remarks on the letters of P. S. Aleksandrov and P. S. Uryson to Maurice Fréchet", was published in issue XXV of IMI (1980). At that time, following my suggestion, Aleksandrov wrote a letter to our embassy in Paris in which he asked about the condition of Uryson's grave in the small town of Batz. An employee of the embassy was sent to Batz and reported that there is a tombstone over the grave with a sculpted likeness of Uryson in profile, and that the mayor's office looks after the tombstone. The embassy sent Aleksandrov photographs of the tombstone and the portrait, one of which he gave to me; it is still in my possession.

In 1979 there appeared in UMN, vol. 34, issue 6, the first part of Aleksandrov's "Pages of an autobiography". The second part appeared 1980 (UMN, vol. 35, issue 3). Aleksandrov gave me a reprint of his memoirs with a dedication written with a trembling hand; as mentioned earlier; he was almost blind at the time. I mention the date 30 May 1980 because it reminds me of a certain conversation I had with him. I will say something about this conversation below.

After our trip to Berlin to attend the Euler jubilee, my relations with Aleksandrov became largely confidential. On a number of occasions I asked him about the early history of "Luzitania" and his relations with Luzin. On the whole, his replies were close to what he had written in his memoirs. But there were also certain additions (the same is true of Aleksandrov's appearances before the commission that dealt with the "case" of Luzin). Aleksandrov always had a high regard for the merits of his former teacher as a scientist and as a leader of "Luzitania" but their relations clouded over from the time when Aleksandrov and Uryson, also a member of "Luzitania", gave up the study of function-theoretic issues—Luzin's key interest—and embarked in earnest on the study of topology. Their first joint paper dealt with compact topological spaces and was published in 1923, and for some time their topological studies alternated with the study of descriptive function theory. When they first went to Göttingen the two friends had with them Luzin's letter of recommendation addressed to Klein, dean of the Göttingen mathematicians. (This letter is in German and is dated 14 May 1924. It is kept in the manuscript division of the library of Göttingen University. When I visited Göttingen in October 1981 I obtained a photocopy of this letter and published a Russian translation of it; see IMI, XXIII, 1983.)[2] Soon, however, topological themes became dominant in the papers of Aleksandrov and his friend and this provoked Luzin's displeasure. Other Luzin students also organized independent, but in a sense filial, schools of probability, number theory, and so on. At the meeting of the previously mentioned commission that took place on 7 July 1936, Aleksandrov stated that when he went over

[2] Aleksandrov does not mention this letter when describing in his memoirs his first visit to Göttingen and his visits to Klein and Hilbert (UMN, vol. 34, issue 6, p. 245).

to topology Luzin told him: "As long as you study topology we can have no contact."

The profound deterioration of the relations between Aleksandrov and Luzin was caused neither by Aleksandrov's "betrayal" of descriptive function theory nor by Luzin's jealousy over Aleksandrov's new interests but by the story of the discovery of the so-called A-sets. In the first part of his "Pages of reminiscenses" Aleksandrov has this to say: In the summer of 1915, following Luzin's suggestion, he, Aleksandrov, began to study the question of cardinality of B-sets and in October of that year delivered a lecture in the student circle (*kruzhok*) on a result he obtained. In addition to the students, one of whom was M. Ya. Suslin, the listeners included professors Mlodzeevskiĭ, Egorov, and, of course, Luzin. Present was also the young Polish mathematician W. Sierpinski, then living in Moscow, who later headed in Poland a very active school of the theory of sets and functions. Aleksandrov's result, to the effect that every uncountable B-set contains a perfect kernel, was important and Luzin sent a French text of his note to Paris where it was published in 1916 (Comptes Rendus Acad. Sci. Paris, vol. 117).[3] In the proof of his result Aleksandrov used a certain set-theoretic operation which Suslin, who in the meantime had got to know Aleksandrov very well, proposed to call in the latter's honor the A-operation. Following Luzin's suggestion, Suslin studied one of Lebesgue's papers, found an error in it, and in analyzing the error discovered a class of sets different from B-sets and called them A-sets. Suslin's result was published in 1917 and, in a sense, created a sensation. The term A-sets was important to Aleksandrov, for it had to do with his first result, the one which was "for this reason, possibly, the dearest to me". Aleksandrov wrote with great bitterness that much later Luzin called A-sets analytic sets, and, according to Aleksandrov, this term masked the significance of his A-operation in the construction of A-sets.[4] "But by that time," he added, "my personal relations with Luzin, once profound and heartfelt, were essentially forfeited" (op. cit., p. 235). He tells laconically that in 1918, because of rationing and difficulties of everyday life, Luzin, Khinchin, Menshov, and Suslin took jobs at the Polytechnic Institute in Ivanovo-Voznesensk, now Ivanovo (the Institute was moved from Riga, then under German occupation). Luzin travelled frequently to Moscow where he continued to work with his other students. V. V. Golubev and I. I. Privalov settled down at the University of Saratov. "Suslin didn't get on in Ivanovo and soon lost his job there." He wanted to transfer to Saratov. "One expected Luzin to give him

---

[3] The same result was published in the same year by the German mathematician F. Hausdorff in *Mathematische Annalen* for 1916. In view of the state of war between Russia and Germany, Aleksandrov could not know this. While Aleksandrov did not mention Hausdorff's results in his memoirs, Hausdorff included Aleksandrov's paper, as well as his own, in the bibliography of his work on set theory (see footnote 5).

[4] *Editors' note.* For a mathematical description of the A-operation and A-sets, see the article in this volume by V. M. Tikhomirov, "A. N. Kolmogorov".

a recommendation. But Luzin failed to give it," and so Suslin went to his parents who lived in the village of Krasavka in the Saratov district. Shortly thereafter he contracted spotted fever and died in 1919 (op. cit. p. 241).

After reading this incomplete account I asked Aleksandrov to supply as many of the missing details as possible. At the end of 1979 or at the beginning of 1980 Aleksandrov called me up and said that I could come to him the next day. The door of his apartment would not be locked and, at the specified time, he would sit on a chair opposite the door. When I arrived we went into the apartment. We were alone. Then I learned the details of Aleksandrov's view of the relations between Luzin and Suslin. I went home, typed Aleksandrov's story, paid him another visit shortly thereafter, and read to him my version of his story. Aleksandrov confirmed the correctness of my narration. I could not bring myself to ask him to certify this with his signature—after all, he was not able to read the text. Then Aleksandrov said that this account should not be published during his lifetime.

After Aleksandrov's death I asked his closest friend, Kolmogorov, how to proceed. After reading my account Kolmogorov said that it was too early to publish it. After Kolmogorov's death I showed my account to professor V. M. Tikhomirov who had known both of them quite well. Tikhomirov said that Aleksandrov had told him the same story he had told me. I decided to put my account in a sealed envelope with the inscription "obtained from A. P. Yushkevich. To be opened five years after his death" and to deposit it in the Archive of the Soviet Academy of Sciences.

Ten years have passed since Aleksandrov's death and I think it proper to describe in a few words the essence of Aleksandrov's story. Firstly, by introducing the term "analytic sets" Luzin had consciously damaged Suslin's claim to priority; he could have kept the term "ensemble (A)" proposed by Suslin in his note of 1917. Secondly, Aleksandrov regarded Luzin as morally responsible for Suslin's death and thought that he (Luzin) was conscience-stricken for many years; witness the fact that Suslin's portrait was always on his desk.

An extra justification for my disclosure of the essence of the story Aleksandrov had told me is that he had said much the same things at meetings of the commission in the Luzin "case" in the presence of many of its members, and that the verbatim record of this commission's meetings is about to be published.[5]

At the meeting of the commission on 7 July 1936 Aleksandrov highly praised Luzin's contribution to the creation of the Moscow school of function theory and to the shaping of his, Aleksandrov's, mathematical personality. While he did not accuse Luzin of "formal plagiarism" he stated that in his publications Luzin "cited Suslin's lesser theorems but said nothing about who established the concept of that set [A-set]". On 11 July Luzin was presented

[5] These reports are kept in the Archive of the Soviet Academy.

with a full written list of his transgressions, approved by the commission, which included the charge that he "appropriated Suslin's fundamental ideas and systematically persecuted him".

A perusal of Luzin's publications shows no trace of appropriation of Suslin's ideas.

Suslin's first note on A-sets was published in a French translation by Luzin (as Suslin did not himself write in French), who presented it to the Paris Academy of Science, vol. 164, pp. 88–90. It was followed by Luzin's note on pp. 91–93. The latter contained a few new theorems on A-sets but it began as follows: "In this paper I wish to state some consequences of Mr. Suslin's results." I am quoting from the Russian translation in [**4**, p. 270] in which (as in other cases) the word "Mister" is left out. Here the term "analytic set" is not yet used; Luzin introduced it later. It is possible that Luzin wanted to distance himself from Suslin. It is also possible that he regarded himself as a co-discoverer of A-sets because of some preliminary conversations with Suslin—the human heart is a mystery. In his note Luzin mentioned Sierpinski.

Luzin's next, fuller note, co-authored with Sierpinski, dated 8 April 1918 and published in French in the Bulletin of the Cracow Academy of Science, began with the words: "In the note...Suslin introduced an important class of sets which he called" A-sets ([**4**, p. 273]). It is clear that it makes no sense to talk of Luzin's appropriation of Suslin's "fundamental ideas".

Some time later Luzin introduced a new class of sets which he called projective. Now he thought it proper to put next to Suslin's name the name of Sierpinski as well as his own, and to introduce the term "analytic sets" suggested to him by Lebesgue. In the large "Memoir on analytic and projective spaces," published in French in Matematicheskiĭ Sbornik, vol. 33, issue 3, 1926, he says: "In 1916–1918, Suslin, Sierpinski and I...studied a new class of sets", and further: "(Following Lebesgue's suggestion) we called these sets analytic" ([**4**, p. 320]).[6] Also in 1926, Luzin published a detailed memoir in French titled "On analytic sets" in Sierpinski's journal *Fundamenta Mathematicae* in which he gave detailed reasons for his terminology. At the very beginning, [**4**, p. 384], he referred to the respective 1917 notes of Suslin and himself in which only the term "A-sets" was used. Luzin presented three lemmas and two theorems of Suslin and included at the end a special remark, with yet another reference to both notes of 1917, and added the following statement: "Suslin's premature death and the difficult international relations prevented the publication of a detailed account of the theory of analytic sets. Sierpinski obtained independently the proofs of all propositions of this theory and published them in a series of papers in his journal *Fundamenta Mathematicae* [**4**, p. 451]." The same reference to the 1917 notes is found

[6] Aleksandrov called them "Suslin sets". Following his advice, Hausdorff used this term in the second edition of his famous monograph *Die Mengenlehre* (1927). Both terms are in use today.

at the beginning of the third chapter of Luzin's fundamental *Lectures on analytic sets and their applications*, published in French in Paris in 1930. It is characteristic that a Russian translation of this work, abbreviated in the bargain, appeared only in 1953—that is, three years after the author's death. Also, Lebesgue's interesting introduction was left out. More than thirty years later, professor V. A. Uspenskiĭ published a translation of this introduction with very valuable comments (UMN, vol. 40, issue 3, 1985, pp. 9–14). A complete version of Luzin's monograph (without Lebesgue's introduction) is found in [**4**, pp. 69–219].

I repeat: It is quite clear that the charge that Luzin appropriated Suslin's fundamental ideas was baseless. Luzin was right when he defended himself against the accusations of the commission that handled his "case" by saying, on 11 July 1936, that everything he did later in the theory of analytic sets "helped Suslin and brought out with ever greater clarity what was fundamental in what he had done".

Luzin never mentioned Suslin's use of Aleksandrov's A-operation in the construction of A-sets. True, Luzin had no need for this operation in his own work. However, while he had assigned the proper place to the Aleksandrov-Hausdorff theorem in his 1926 paper mentioned above, and had noted the importance of the property of B-sets discovered by them, he had said nothing about the role of the A-operation in Suslin's results. This was a reflection of the poor relations between Luzin and Aleksandrov that had begun in previous years. In this case there were grounds for Aleksandrov's resentment which, obviously, he could not express without discomfort.

As for Suslin's "persecution", openly discussed at the meetings of the commission and hinted at in Aleksandrov's memoir, certain facts and statements made by Luzin before the commission add up to the following: As mentioned earlier, due to the dislocations of the first years of the civil war, Luzin took a job in 1918 at the Polytechnic Institute in Ivanovo-Voznesensk and took Suslin with him. His feelings about priority in the discovery of A-sets notwithstanding, Luzin recommended Suslin for a position at the Polytechnic Institute. Also, in spite of the fact that Suslin had graduated just a year earlier, that is in 1917, and thus had no teaching experience whatever, and in spite of the fact that he had published just one, admittedly excellent, paper, he was given the title of professor at Luzin's request. All Luzin stipulated was that Suslin should pass his Master's examination and that he should be active in research. However, Suslin's scientific activity in Ivanovo-Voznesensk was insubstantial, he did not take his Master's examination, and over a period of two years wrote only one minor paper (published posthumously in vol. IV of *Fundamenta Mathematicae*, 1923). This resulted in serious friction between Luzin and Suslin.

In view of the above, one could hardly call Luzin's refusal to recommend Suslin for a professional position at the University of Saratov "persecution". In spite of the fact that Luzin could not have foretold that Suslin would contract spotted fever and could not be blamed for the untimely death of his talented student, there is no doubt that Luzin felt that he was somehow implicated. And when he recalled on 11 July the circumstances of Suslin's death at a meeting of the academic commission, he said: "And then a catastrophe occurred that has weighed heavily on all my life." Maybe that is why Luzin kept Suslin's portrait on his desk.

In the summer of 1980 I met twice with Aleksandrov. At that time it was proposed to publish a Russian translation of Cantor's works on set theory with an introduction or epilogue by Aleksandrov. A more detailed account of this publication can be found in my recollections of Kolmogorov. Here I mention that I asked Aleksandrov to write such an essay. He dictated a first version, dated 14 August 1980. Its fate is discussed below. At that time Aleksandrov's health deteriorated to such an extent that it was no longer proper to call on him.

Aleksandrov published a number of works on the history of mathematics. Some were mentioned above. I tried to characterize his contribution to the historical-mathematical literature in [**5**]. Except for a small paper titled "A. I. Markushevich" and signed by Aleksandrov, Kolmogorov, and myself and published in English in *Historia Mathematica* 8 (1981), pp. 125–132, I did not co-author works with Aleksandrov. I wish to add that I remember with gratitude that Aleksandrov stood up for me at a difficult moment in my life. The details are found in the essay on Kolmogorov.

3. A. N. Kolmogorov

I was neither Andreĭ Nikolaevich Kolmogorov's personal friend nor one of his close collaborators. We had some infrequent business meetings and we co-authored a few papers. In other contexts, I could only watch, as it were, from the sidelines. Nevertheless there are recollections preserved in my memory, and these may be of some interest.

I first saw Kolmogorov in a corridor of the mechanics–mathematics department, in the so-called "old" building of Moscow University at 11 Mokhovaya Street (later the Karl Marx Boulevard). A simply-dressed young man, his head somewhat forward, walked along the corridor. My friends whispered that this young man was one of the best, if not the best, mathematics student in the department. This was in 1924 or 1925. (I entered Moscow State University in the fall of 1923; Kolmogorov completed his studies there in 1925.)

Somewhat later I attended one of his presentations. At that time Moscow mathematicians enjoyed discussing the foundations of analysis, and especially the intuitionism of L. E. J. Brouwer, whose position influenced that of H. Weyl. Weyl opposed D. Hilbert's formalism and B. Russell's logicism. Not only I, but also qualified mathematicians, found it difficult to grasp the idea that the law of the excluded middle may not apply to infinite sets and processes—that, for example, it is not possible to claim a priori that the only possibilities for the sequence of digits $1, 2, \ldots, 9$ are that it does, or it does not, occur once or an arbitrary number of times in the decimal expansion of π. This is not the place to elaborate the issue. What counts is that, quite naturally, the mathematicians of Moscow, where there was a strong school in the theory of functions, took a lively interest in all questions pertaining to the foundations of analysis. As part of the activities of the mathematical section of the Natural Science Department of the Communist Academy, there were regular meetings of a seminar on methodology. The scientific workers in the section were L. A. Lyusternik and L. M. Likhtenbaum. The Department was headed by Otto Yul′evich Schmidt, and the seminar was run by Professor Aleksandr Yakovlen′ch Khinchin. Many hundreds of listeners attended the seminar sessions, at which various philosophical and historical questions of mathematics were discussed. Lectures were given, and were followed by discussions. I remember the heated arguments that arose in connection with R. von Mises' frequency theory of the foundations of probability.

In 1924–1926, I and A. O. Gel′fond, then a close friend of mine, paid frequent visits to Lyusternik and Likhtenbaum. They were not in the same class as mathematicians, but they were both very sociable and, in my view, charming. Lyusternik read to us his excellent satirical verses in which he poked fun at some contemporaries such as the physics professor A. K. Timiryazev who stubbornly rejected the theory of relativity but was otherwise a very nice and widely educated person. These new friends introduced us to Khinchin's seminar, where we got to know I. V. Arnol′d, V. I. Glivenko, and other young scientists interested in the foundations of analysis. S. A. Yanovskaya, whom I knew earlier, tried to solve questions in the foundations of mathematics and to criticize inimical conceptions from a viewpoint which she regarded as Marxist.

I first heard Kolmogorov in Khinchin's seminar and was much impressed by him. The discussion pertained to intuitionist logic—a subject to which Kolmogorov devoted a number of important papers between 1925 and 1932. (Yanovskaya's survey of these papers is found on pp. 28–30 of her article in "Mathematics in the USSR during the thirty-year period 1917–1947", Gos. Moscow-Leningrad, 1948.) The presentation I have in mind followed Khinchin's lecture on intuitionism, subsequently published in the Bulletin of the Communist Academy, No. 16, 1926. The lecture was discussed by Arnol′d, Glivenko, Kolmogorov, and some others. I found it all very interesting, listened carefully, and took notes. Unfortunately, the notes got lost, and I can only convey a very general impression. Khinchin was an excellent lecturer. He always spoke very clearly and beautifully. This lecture was no exception. Still, it left no deep impression on me; it seemed somewhat trivial. Glivenko's presentation was more vivid, fresher. But Kolmogorov's short and oratorically unsophisticated presentation was especially striking. His interpretation of intuitionistic logic was highly original and persuasive.

It anticipated K. Gödel's thesis that intuitionistic mathematics only appears to be narrower than classical mathematics. Kolmogorov's paper was published in 1925.

I was enthusiastic about the new themes. Confident of success, I undertook the translation of Weyl's generalizing papers. These papers were intelligible to broad mathematical circles and brimmed with historical comments (the 1926 survey paper "Die neutige Erkenntnislage in der Mathematik" was brilliant). I also translated a number of highly specialized papers by Brouwer. My translation of Weyl's collection of papers, titled *On the philosophy of mathematics*, was published in 1934. (It was decided not to publish the translations of Brouwer's rather specialized papers.) In the introduction, I noted some important results of Gödel and A. Heyting. I also made some passing quasi-Marxist observations on intuitionism, which strike me now as quite immature.

This work proved that I was a competent translator. As a result, when Kolmogorov wanted to translate into Russian Heyting's book *Mathematische Grundlagenforschung. Intuitionismus. Beweistheorie* he recommended me for the job. The translation was published in 1936. Kolmogorov added a substantial introduction at the end where he noted that the paper was very difficult to translate but that I proved equal to the task.

My relations with Kolmogorov were strengthened through the publication of the second edition of the Great Soviet Encyclopaedia (GSE). At the beginning of 1949, Kolmogorov replaced Professor V. F. Kagan as editor of the mathematics section. At Kagan's suggestion, I had written articles on Integral calculus, Logarithms, and Figures, for the first edition of the GSE. My role in the second edition was more active. For the most part, I communicated with Kolmogorov through V. I. Bityutskov, who is still with the Soviet Encyclopaedia.

My involvement with the second edition of the GSE began by chance, and it was then that I had my first long conversation with Kolmogorov. In 1950, or more likely in 1951, a meeting of the Moscow Mathematical Society was convened to discuss the manuscripts for the next volumes of the second edition of the GSE. I pointed out some elementary inaccuracies in the manuscript of "Differential calculus" (in the formulation of Rolle's theorem and elsewhere). At the end of the meeting, Kolmogorov approached me and suggested that I write a new version of "Differential calculus". I agreed. My paper was accepted (it appeared in 1952, in vol. 14 of the GSE).

My next paper for the GSE was titled "Mathematical symbols". Its first version, by I. G. Bashmakova and me, attracted Kolmogorov's special attention. He supplied substantial additions, and the paper, with all three of us listed as authors, appeared in the GSE, vol. 17, 1952. After that, I became a permanent author of papers on the history of mathematics for the second edition of the GSE. My "Integral calculus" was reprinted with some additions. (A German translation of this paper and of my paper on differential calculus appeared in 1956.) My last paper for the second edition of the GSE was a biography of Euler. It appeared in 1957, in vol. 48.

My papers were reprinted, sometimes with minor editorial changes, in the third edition of GSE, but always without my name as their author. This was a time when the consequences of the struggle against so-called "cosmopolitanism" were still operative, and, since I was a Jew, the new editorial board preferred not to mention me. I wish to add that Kolmogorov replaced my short article in the second edition titled "Method of exhaustion", with his own piece.

Tempora mutantur, et nos mutamur in illis. When in 1988 Yu. V. Prokhorov, assisted by Bityutskov, undertook to publish the Russian Encyclopaedic Dictionary of Mathematics I was made a member of the editorial board that supervised the historical sections, and my name was put back at the end of each of my articles. This large volume opens with Kolmogorov's splendid and well-known article titled "Mathematics", first published in (1938) the first edition of the GSE. At the request of the editorial board I made some minor corrections and historical additions to the article. (This was after Kolmogorov's death.) The second version of this article was enlarged.

In the article "Mathematics", Kolmogorov proposed a global periodization of mathematics. In this connection, I talked to him in the 1980s, and offered some critical observations. Speaking of European mathematics, Kolmogorov called the period from Greek antiquity to the beginning of the seventeenth century the period of elementary mathematics, and the seventeenth and eighteenth centuries, the period of creation of the mathematics of variable magnitudes. I begged to differ. I said that, granted the notions of variable magnitude and function had not been formulated by the ancient Greek mathematicians, their mathematics could not be described in its entirety as "elementary", whether in the pedagogical sense of the word or in any other sense. I gave as examples the infinitesimal methods used not only by Archimedes but also by some of his Greek predecessors and followers, the geometry of Apollonius, the mathematical-philosophical debates dealing with the infinite, and, in particular, with Zeno's paradoxes, and so on. To distinguish between the Greek epoch on the one hand and the Egyptian and Babylonian epochs on the other, I suggested the term "period of the formation of mathematics as a system of deductive disciplines". I also pointed to approaches to the mathematics of variable magnitudes in the Oxford and Paris schools in the fourteenth century. This lengthy conversation took place when Kolmogorov's illness (Parkinson's disease) made it difficult for him to speak, so that I did most of the talking. Kolmogorov listened with attention and, possibly, with forbearance, but said only "yes, all this is interesting, and calls for thought".

I gratefully acknowledge the help offered by Kolmogorov in the following connection. Before the second world war, anti-Semitism in the USSR was mostly in a latent state. It began to grow during the first years of the war—

with Stalin's help, and, possibly, on Stalin's initiative, and it took different forms depending on the time and the place. Since 1932, I had been employed in the Moscow Higher Technical School (MVTU), now a university, and in 1941 I became head of the department of mathematics. My relations with the administration were always normal. Beginning at the winter of 1945, I had a half-time position at the Institute for History of Science of the Soviet Academy of Sciences. (In 1953 its name was changed to "Institute for the History of Science and Technology".) The first gentle sign of trouble came from the MVTU: I was dismissed from the admissions committee. The excuse was concern for my summer vacations. The real reason was the need to get rid of an inconvenient witness to the "regularization" of the national makeup of newly admitted students. In the winter of 1951, soon after I was awarded the Order of the Red Banner of Labor, the administration ordered me to prepare a report of my activities and to present it at the meeting of the Scientific Council. There were many speeches. The leading professors, including I. I. Kukolevskiĭ, the most influential and honest of them, praised the work of my department. When the administration proposed a resolution censuring the department for poor organization of teaching and research, there were expressions of bewilderment. The administration withdrew its resolution for the ostensible purpose of reworking it and of presenting the new text during the next meeting of the Scientific council. Time passed without further developments. When I expressed puzzlement at this state of affairs, I was politely informed that there was no need for a new resolution, that all was well with the department without it. This was pure deception. As I later found out, already in the winter of 1952 the Bauman district committee of the Communist Party had decided to dismiss me. Three days before the end of the academic year they decided to settle the matter in one fell swoop. MVTU was under the jurisdiction of a division of the Ministry of Higher Education (KVSh) headed by Professor Arzhannikov. The latter sent for me, offered me a "Kazcek" cigarette, politely praised the work of my department, which, he said, I had organized so well that any other professor could easily continue to run it, and...offered me a new department in the newly organized institute in the town of Izhevsk. Following my categorical rejection of the offer, Arzhannikov changed his tone and told me that the solution had been accepted by the "management". At the time, Arzhannikov's deputy was a former student of mine. On his advice, and while Arzhannikov was on some business trip, I handed in an application to the administration of the MVTU in which I asked to be released from working in the KVSh system. At the same time I agreed to work full time in the Academic Institute—a step I was advised to take much earlier. At the MVTU they gave me a splendid testimonial, including a statement that I was leaving at my own request. This testimonial is still in my possession. I hoped that this was the end of my troubles, but I was mistaken.

Soon reports appeared in print about "The murderers in white coats"—

doctors, mostly Jewish, who allegedly murdered high officials in their care. In January of 1953, a meeting of council members and of higher executives of the Academy of Sciences was convened to investigate all employees, and the "purge" began. I was dismissed on February 1, my topic was eliminated from the [research] plan, and my membership in the Scientific Council was revoked. At a meeting of that Council it was declared that now the political atmosphere at the Institute had been purified.

I decided to appeal to A. N. Nesmeyanov, the president of the Academy of Sciences. In this Kolmogorov helped me with his authority. He demonstrated indubitable courage in coming to the defense of a pariah. Here is the full text of Kolmogorov's letter to Nesmeyanov. Kolmogorov gave it to me, and I transmitted it to Nesmeyanov through a secretary whom I had known since the earlier presidency of S. I. Vavilov.

> Dear Aleksandr Nikolaevich!
>
> I know Adolf Pavlovich Yushkevich well, as do all mathematicians interested in the history of mathematics. He belongs to the very small group of leading Soviet historians of mathematics and occupies a prominent position among them by virtue of his erudition and extensive literary activity. I personally had dealings with A. P. Yushkevich in connection with his work on the GSE, where he is the author of a large number of papers, many of which could have been assigned to no one but him. In the Soviet Academy of Sciences he carried out a major task of writing for the History of Sciences in the USSR.
>
> I think that it would be difficult to find in the USSR a scholar in the history of mathematics of comparable erudition and capacity for work. This being so, I doubt very much that the Institute of the History of Science is acting correctly in depriving itself of so highly qualified a scientific worker.
>
> I therefore take the liberty of asking you to receive A. P. Yushkevich and to consider the correctness of his dismissal.
>
> Respectfully,
>
> A. Kolmogorov, 11 February 1953.

The president received me immediately. He listened patiently to my explanations without uttering a word. By way of concluding my visit he said that he would hand over the matter for consideration to one of his secretaries, Professor Kiselev. On a number of occasions, the latter gave me evasive answers; it was clear that he was playing a waiting game. Stalin died on March 5. Rumors spread about the impending review of the "Jewish question". At the time, our Institute was part of the Division of History and Philosophy of the Academy of Sciences, headed by a historian, A. M. Pankratova, a comrade in the Odessa underground communist organization (VKP(b)) of my friend S. A. Yanovskaya. At Yanovskaya's request, Pankratova agreed to receive me. I came to her with a letter signed by P. S. Aleksandrov, president of the Moscow Mathematical Society, and Professors A. G. Kurosh and Yanovskaya. It was dated April 19, 1953. The letter opened with the statement that I was dismissed for "fortuitous motives", gave a rather detailed analysis of some of my works, and ended with a request to consider the "vexing misunderstanding" that happened to me. I have only copies of the two quoted letters. The originals are probably in the archives of the Soviet Academy of Sciences.

Pankratova received me very politely. She told me that I need not worry, that the general question of the "doctor issue", and so on, was being considered by the Central Committee of the Soviet Communist Party (she was either a member or a candidate of the Central Committee), and that all would be satisfactorily resolved in less than a week. Indeed, Pravda soon published a front-page article in which the "doctors' plot" was described as a criminal fabrication. As for my personal situation, what changed was that I was again invited to various meetings of the Institute, and all acted as if nothing had happened. But it took a special meeting of the administration, attended by Pankratova, to reverse my dismissal order and to preserve my uninterrupted work record.

I am most grateful to Kolmogorov, Aleksandrov, and others, for their support at a difficult period. At the time, this called for a great deal of civic courage.

There was an analogous case when Kolmogorov was unable to overcome the chauvinist stance of the management of the mechanical-mathematical department of Moscow State University. In 1952–53, my son Aleksandr wrote a diploma paper under Kolmogorov's supervision on the differentiability of transition probabilities of Markov processes with a countable number of states. Kolmogorov approved for publication this paper and my son's short paper on Markov's notion of entropy. Both were published. In one of them, the notion of a strong Markov process was introduced. On Kolmogorov's suggestion, the postgraduate department of probability recommended that the author be appointed a postgraduate fellow at Moscow State University. But all this came to nought. When I turned to Kolmogorov, he told me that he was unable to help, and that it was only with great difficulty that he had managed, a year earlier, to overcome the objections to a similar appointment of his talented student R. L. Dobrushin. In the end, my son acquired as supervisor E. B. Dynkin, now a professor at Cornell University and a member of the U. S. Academy of Sciences. Later they jointly elaborated the general theory of strong Markov processes.

Kolmogorov was always tactful. Here is an example of his tact. In D. E. Menshov's *Recollections*, based on my taped interview with him, there appeared the statement that "Kolmogorov was not one of N. N. Luzin's immediate students, but did study with V. V. Stepanov" (Istoriko-matematicheskie Issledovaniya (IMI), XXVII, 1984, p. 328). P. L. Ulyanov, a corresponding member of the Soviet Academy of Sciences, noticed this and expressed justified surprise. I told this to Kolmogorov who sent a letter to the editors of IMI in which he explained that he had indeed been "an immediate student of Luzin", and that his "conversations with him...were extremely important and continued until...1925." At the same time he noted the [important] role of his "studies with Stepanov and Menshov" in the formation of his interest (IMI, XXVII, 1985, pp. 337–338). Also, the IMI editors explained that the confusion was due to an error that occurred in the process of transcribing the tape. This ended the awkward incident.

On one occasion Kolmogorov abandoned his usual tact. This happened in the fall of 1946, when a number of new mathematical academicians were to be appointed. One of the candidates was P. S. Aleksandrov, Kolmogorov's closest friend and corresponding member of the Academy since 1929. What I am about to say was told to me by Kolmogorov. There are other, quite similar accounts. During the initial discussion of the candidates, Luzin noted Aleksandrov's great contributions to topology and supported his candidacy. But a few hours later, just before the vote in the mathematics section, he stated that what the country needed at the moment was an "applied" profile. Since Aleksandrov was interested in a group of very abstract problems, his candidacy was inappropriate. The usually restrained Kolmogorov lost his temper and slapped Luzin (very slightly) in the face. This incident became immediately known, throughout the Academy and beyond. S. I. Vavilov, president of the Academy of Sciences and a man of the noblest character, called Kolmogorov and told him "Well, old man, you sure started something; the Academy has seen no such thing since Lomonosov's time." Vavilov told Stalin what had happened. Stalin said: "Well, such things occur with us as well." That was the end of it. I am sure that Kolmogorov took the incident very much to heart. Later he invariably wrote about Luzin with the highest regard. One of his utterances is mentioned above.

During the 1946 elections the following "applied" mathematicians were appointed to the Academy: M. V. Keldysh, M. A. Lavrenti′ev, A. N. Dinnik, and I. G. Petrovskiĭ. Aleksandrov was made an academician in 1953.

This incident took place before the pernicious Parkinson's disease began to hinder Kolmogorov's speech. It became progressively more difficult to converse with him, but our cooperation lasted a few more years. Kolmogorov took a lively and active interest in the history of mathematics, and in a number of cases gave me substantial support. Above all, he agreed to co-edit with me a series of books titled *Nineteenth-Century Mathematics*, three of which appeared in 1978–1987. He put together the program of the series, selected authors, and wrote jointly with me a general introduction containing key evaluations of the tendencies of mathematics in the nineteenth century and at the beginning of the twentieth. The authors were mathematicians and historians of mathematics with the required qualifications. The essays published thus far deal with the following branches of mathematics: mathematical logic, algebra, number theory, probability, the theory of analytic functions, geometry, ordinary differential equations, the calculus of variations, the theory of finite differences, and the Chebyshev approach to the theory of functions. The part of the fourth volume devoted to the history of partial differential equations is finished. Still to come are the parts dealing with numerical analysis and the social history of mathematics. I hope that all this will be completed in a year or two. I wish to mention Kolmogorov's involvement in a detailed joint review (15 pp.) of *Abrégé d'histoire des mathématiques*, t. 1–2, Paris, 1978, under the editorship of J. Dieudonné. Kolmogorov wrote part of the review dealing with functional analysis, probability and mathematical logic. A. I. Markushevich wrote about analytic functions, and elliptic functions together with Abelian integrals, Bashmakova dealt with algebra and number theory, and I wrote about the eighteenth century and the foundations of analysis. We rated highly the introduction and the whole book. We expected Aleksandrov to write about the topological chapter of the book. However, after getting acquainted with the work, he refused, and limited himself to a footnote in which he noted the completely inadequate coverage of the early Moscow school of topology. The reproach was justified. When I conveyed it to the author of the chapter (G. Hirsch), he told me that the original text contained an analysis of the results of that school but it was eliminated by Dieudonné. Dieudonné thought that, compared with the subsequent progress in toplogy, the results in question were not very significant. Also, the chapter was very large anyway. Our review appeared in Russian and in French.

Kolmogorov and I considered the preparation of one or two volumes on twentieth-century mathematics. His modesty was striking. He was well aware of his creative powers, but had a realistic estimate of the bounds of his erudition. He said that while he was very much in favor of writing such a work, its upper time limit would have to be the time of WWII. In other words, it would be necessary to limit oneself, by and large, to the first half of the century, and outline in very general terms the directions of the subsequent development of mathematics. "I can no longer follow the mathematics of recent decades", Kolmogorov added by way of an explanation. To our sorrow, Kolmogorov did not live long enough to take part even in determining the program of the proposed work; its editorial board is now headed by Yu. I. Manin.

I recall an earlier episode that also demonstrates Kolmogorov's modesty. When, during the second world war the Soviet Union and its allies made a decisive breakthrough, many of the scientists evacuated in the fall of 1941 from the capital began to return. The Moscow Mathematical Society renewed its activity. An early meeting was devoted to a debate about the Society's program of activities. Aleksandrov, the president, asked those present what percentage of lectures they followed. Most mentioned figures of 10 to 15 percent. Kolmogorov thought for a while, and said that, essentially, he understood, that is, could evaluate, half the lectures. Only Stepanov, who had a phenomenal memory, said that he understood 100 percent of the lectures. One could readily believe him.

Kolmogorov was of great help in connection with the Russian edition of Georg Cantor's *Works on set theory* (Moscow 1985). F. A. Medvedev prepared the translation, commentaries, and Cantor's biography (use was also made of old translations prepared by my father). The difficulty was that our Institute does not publish translations. I sought Kolmogorov's advice. Kolmogorov found an ideal solution in three minutes' time. He called Academician P. L. Kapitsa, editor-in-chief of the series *Classics of science*, explained the difficulty, and Kapitsa immediately included Cantor's book in his series.

There was another difficulty, or rather awkwardness. Aleksandrov was to write an afterword, which unfortunately turned out to be too concise and entirely inadequate. I brought the text to Kolmogorov and he advised that it be published as a separate contribution in a volume of our *Historical-mathematical investigations* (IMI). This is where the article appeared soon after the author's death (in vol. XXVII, 1983, pp. 290–292). A new afterword was prepared jointly by Kolmogorov and by me. It saw the light of day while Kolmogorov was still alive.

In one case, however, Kolmogorov's support was, at the time, without effect. Shortly before his death, Lyusternik sent me some recollections on physics at MGU in the early 1920s in the form of a small poem, as well as jocular verses about O. Yu. Schmidt's "liberation", on Stalin's orders, from the duties of vice-president of the Soviet Academy of Sciences in March 1942 when the Academy was in Kazan. Lyusternik asked me to publish this material. The poem and the verses were sophisticated, apt, and stylistically superb. Memoirs of scholars, including those in the form of poems, belong to the history of science. I wanted to publish the manuscript in IMI but encountered categorical opposition from the administration of IMET. In their view, Lyusternik's verses did not constitute scientific work. Following my request, Kolmogorov supported me, but his intervention was of no avail. But times change. This material was published in issue XXI–XXIII of IMI, vol. XXII–XXIII (1991), ten years after the author's death. I am sure that many readers will find it interesting.

In the last years of his life Kolmogorov could not speak as a result of the ravages of Parkinson's disease. The drugs that helped for a number of years and were brought from abroad by many friends and acquaintances (including myself) lost their effectiveness. Some of his students maintained round-the-clock attendance at the patient's bedside. Of course, I cut back on my visits.

My relatively few encounters with Kolmogorov have left powerful impressions in my memory. This is especially true of our conversations in his home, in a room graced with an excellent copy of the well-known painting by K. S. Petrova-Vodkina "The bathing of a red horse" which was obviously Kolmogorov's favorite painting.

4. V. I. Smirnov

Vladimir Ivanovich Smirnov (1887–1974) was a professor at the University of Leningrad and an academician. Just as in the case of other "mathematical friendships", my encounters with Smirnov had to do with his very active interest in the history of mathematics. Like his teachers Steklov and Gyunter, Smirnov belonged to the Petersburg-Petrograd school of mathematics and for 25 years was the dean of Leningrad mathematicians. But he differed greatly from the previous generations of the Chebyshev school. Here Chebyshev had formulated the main ideological directions. In his research he was spiritually and thematically very close to Euler and to the Paris École Polytechnique and did not admire certain important areas of mathematical thought that developed in the second half of the nineteenth century such as the theory of analytic functions, group theory, and set theory. His closest students and successors also avoided many new developments. Thus Lyapunov did not share Riemann's geometric views, Steklov failed to see the true worth of functional analysis (and the theory of integral equations), and Uspenskiĭ, who was younger than they, underestimated the theory of functions of a real variable. Smirnov's generation, including J. L. Tamarkin, A. S. Besicovitch, and the prematurely deceased A. A. Fridman, was far more open to new trends, even the newest. Smirnov told me that their intellectual development took place at the university as well as outside; they read and discussed the newest mathematical literature. As an example, he mentioned their study of the Riesz-Fischer theorem of metric function theory, published independently by the two scholars in 1907, three years before Smirnov's graduation. Smirnov and his friends did not discuss these extracurricular activities with their teachers. It was only gradually that these areas of modern mathematics came to be cultivated at Leningrad University. Some of the people who promoted this process were G. M. Fikhtengolts, who accepted a position at the university, Luzin and some of his students who came to Leningrad to celebrate the 100th anniversary of Chebyshev's birth (1921), and last but not least, Smirnov himself who became a professor in 1915.

As a scientist and teacher Smirnov developed under the new conditions the traditions of his predecessors at the university and at the Academy of Science, traditions that in many respects derived from Euler, the forefather of the Petersburg school. In the previously mentioned paper by Delone and myself, the former suggested that this school be called the Euler-Chebyshev school. Not only Chebyshev (and, of course, his older colleagues Ostrogradskiĭ and Bunyakovskiĭ) but also academician A. A. Markov were familiar with Euler's works. Smirnov told me that during an examination Markov asked him who was the author of a certain formula he had written on the board while answering a question. Smirnov pleaded ignorance. Markov told him that the formula was introduced by Euler, stated the relevant work and date, and lectured him on the need to know Euler's works. Markov was well

V. I. SMIRNOV

acquainted with the works of many other eighteenth-century mathematicians. One indication of this is the bibliography at the end of his *Calculus of finite differences*, the second edition of which appeared in 1910. The interest of Chebyshev and his students in Euler's works was due not only to their interest in history but also to the fact that they frequently found in his works the starting points for their researches. Whatever the reasons, they were thoroughly familiar with the works of their intellectual predecessors. The same is true of Smirnov.

For a long time Smirnov turned to history of mathematics only on particular occasions. Thus, beginning in 1921, he published papers in memory of Chebyshev and other Russian scientists, and, after the premature death of his student I. A. Lappo-Danilevskiĭ, he edited his works. His systematic study of history begin in 1945. His survey paper, titled "Russian mathematics in the 18th and 19th centuries" and published in the third issue of "Priroda" (Nature), may be said to have determined Smirnov's subsequent historical studies. He liked and was well-read in history, and applied this general interest to the physical and mathematical sciences. His virtually exclusive interest was in the development of science in Russia and in the USSR, and above all in the Academy of Sciences and at his university. Here I will just mention his splendid work on Daniel Bernoulli (1959) and his many projects (beginning in 1948) devoted to A. M. Lyapunov, such as the editing of the unpublished works of this great scientist, of his correspondence, and of his most important monographs, to which he added an analysis of his creative endeavors. Smirnov also contributed significantly to the organization of studies in the history of knowledge. He did this as a member of the commission on the history of the physical and mathematical sciences established in 1949. After Vavilov's death in 1951, Smirnov took his place as chairman. He also became chairman of the Learned Council of the Leningrad branch of the Archive of the Soviet Academy of Sciences, and chairman of the Euler commission of our academy. Under his chairmanship the latter commission did a great deal of work involving the study, description, and partial publication of Euler's manuscript Nachlass.

At Vavilov's suggestion I took part, now and then, in the meetings of this commission. Some of its other members were the famous expert in optics, and corresponding member of the Academy of Sciences, T. P. Kravets, and the astronomer professor I. I. Idel′son. Its very active learned secretary was I. M. Radovskiĭ. It was then that I got to know Smirnov and soon visited him at home.

Smirnov was direct and jovial. He had none of the snobbishness of many highly-placed scientists. He regarded everyone as his equal. He readily and gladly shared recollections and personal impressions and never maligned people.

I have already mentioned Smirnov's stories of his student years. To his story about A. A. Markov, Senior, I will add a characteristic sidelight.

Smirnov attended Markov's course on probability, which he began with the words: "I understand that the Kazan Mathematical Society, of which I am a member, has suggested the following topic: The axiomatic basis of probability. Well, let's go to work."

The conversations I had with Smirnov in his home dealt with a great variety of topics. At the time he lived on Kirov Street in the corner house #31. He stood at the window that faced the street, looked at the opposite building and said pensively: "Well, I spent my childhood in that house and at the end of my life I live across the street from it." At one time the building belonged to the department of churches. Smirnov's father was a priest and he himself was Orthodox and deeply religious. There was an icon in his room. However, he never touched on religious topics and completely tolerated other religions as well as atheism. In addition to history he was very much interested in philosophy. He once pointed to an open book on the table and said confidentially: "I am retyping this." It was Josephus' *The Jewish war*. He kept up with our philosophical literature. Some time later I stayed in the holiday home in Komarovka. After settling down, I went over to Smirnov's dacha in the neighboring academic settlement to borrow some light reading matter. I asked him for some crime story. Smirnov pointed to a pile of books on a shelf at the door and said: "Choose. Or perhaps I'll suggest to you a more interesting item" and handed me the fifth volume of *The encyclopedia of philosophy* given to him by an acquaintance. "Take it," he added, "it contains extremely interesting articles." Indeed, the work contained articles such as scholasticism, existentialism, and others, written in an up-to-date style, free of the then common malevolent "criticism" made worse by references to recent foreign Russian editions unavailable to most readers. These and other articles were indeed more interesting than crime stories. The piquancy of the fifth volume was due to the fact that its editors were famous academicians-philosophers such as P. N. Fedoseev and P. F. Yudina. Most likely, the editors did not bother to read the articles. On the other hand, given the stature of this editorial committee, the reviewers found it awkward to publish critical reviews.

Here I wish to mention that Smirnov neither opposed the ideological flow nor went with it or made advantageous compromises. He told me that he was visited in Komarov by a representative of Leningrad State University and asked to sign a collective letter that evaluated the works of Solzhenitsyn. Smirnov asked to see the works in question. Naturally, the messenger had none with him. Then Smirnov told him that never in his life had he expressed an opinion about unknown works and said: "Bring the books, I'll familiarize myself with them, and then we'll see." That was the end of the story.

Smirnov was a great music lover. He played the piano, including works for four hands. As he put it, he was a regular visitor of the Leningrad symphony, and, at the ebb of life, the oldest. He was catholic in his musical tastes. Once, as soon as I came in, he told me with joy that Metner's second symphony, long out of favor, was about to be performed. Smirnov had an excellent memory. He knew by heart the scores of all Tchaikovsky operas, the Beethoven symphonies, and so on.

As chairman of the Euler commission of the Academy of Science, Smirnov played a very active part in the preparation for publication of Euler's archival Nachlass. For many years, a group of scholars studied the great scientist's vast correspondence. We were the joint editors of the annotated index of this correspondence that included some 2900 letters and business notes (about 70 percent of his letters have survived). This was far from a simple job. In addition to the two of us, the participants included T. N. Klado, Yu. Kh. Kopelevich, T. A. Lukina, I. G. Mel′nikov, and K. R. Bierman and F. G. Lange of the Academy of Science of the GDR. Later, a somewhat supplemented German translation of this index was published in Basel as volume 1 of the fourth series of Euler's collected works. One of the participating editors of this volume was Professor Habicht, mentioned earlier.

On 18 April 1957 our academy organized a festive reception as a conclusion of the 1957 Euler jubilee. It was marred by an unpleasant incident. The many foreign guests included the academician M. Fréchet, who delivered an address of welcome.[7] Smirnov's response was addressed to Fréchet. It emphasized the importance of Fréchet's work in general and its positive effect on the progress of mathematics in our country in particular. Some official of the administration of the Presidium of the Academy of Science thought that Smirnov was guilty of "worship of foreigners". He went up to Smirnov during the speech and quite distinctly advised him to be brief. Smirnov's address collapsed. But the rudeness of the official did not surprise me.

At about this time Smirnov took an active part in the preparation of the three-volume *History of the Soviet Academy of Sciences.* He was one of the editors of the first volume (1958) and a co-author of the chapters on mathematics and mechanics in the second volume (1964). He and E. P. Ozhigova wrote the mathematical chapter for the third volume, but this chapter never saw the light of day. This was so because chief editors and editors in charge of the volume kept on changing and failed to live up to their responsibilities.

I took part in writing the mathematical chapters in the second volume. This gave me an opportunity to work closely with Smirnov. I wrote at home preliminary versions of three essays on the development of mathematics in the Academy of Science between 1803 and 1917. Then Smirnov suggested that I stay for a few days at his dacha in Komarovka so that the two of us could complete the work. This was in the summer of 1963. Then I could fully appreciate the friendliness of Smirnov and his wife. We worked as follows. I sat at a writing-table and read slowly my text. Smirnov walked slowly in the room, listened, and, whenever necessary, proposed more precise expressions

[7] Fréchet's brilliant and substantial address dealt with French-Russian relations in mathematics from Euler to Aleksandrov and Uryson. It was published in Historical-mathematical investigations (IMI), issue XXVII, 1983.

and dictated additions. I already mentioned his excellent memory which extended to many areas of mathematics and, especially, to various areas of analysis. I heard that Smirnov had dictated many chapters of his *Course of higher mathematics*, that was reprinted many times and consisted in 1947 of five hefty volumes, to the typist and would tell her how much space to leave in various places for symbols and formulas. Now I could witness his tremendous mathematical erudition. He seldom consulted the books on the shelves. When we worked on N. Ya Sonin's results on the theory of cylindrical functions he picked up the volume of *Historical-mathematical investigations* containing Gussov's paper mentioned earlier and having located the necessary information praised the paper. "Such historical works are useful," he said.

While he seldom consulted sources for information in analysis, he did this frequently in geometry and in number theory. In dealing with specialized issues in these areas he checked the correctness of formulations and kept on saying: "I am not a geometer," or "I am not a specialist in arithmetic." It was pleasant and instructive to work with him, for he could make relatively complex ideas and concepts intelligible. We worked two "shifts" a day, a few hours at a time.

Late in the evening Smirnov liked to take a walk and I was glad to accompany him. We talked about many things. I have already mentioned some interesting details. I will add that Smirnov was highly responsive to the difficulties of his friends and did what he could for them. For example, he helped the corresponding member (of the Academy of Sciences) N. S. Koshlyakov when he was subject to repression. Subsequently, Koshlyakov was fully rehabilitated.

Smirnov was a highly cultured person. He had a special love for Russian culture. In his old age he made a trip by car to see Vladimir, Suzdal, and other towns of the Golden Ring.

When I learned of his death I wrote one of the first obituaries [**9**], in which I tried to honor his human qualities and his contribution to our knowledge of the history of mathematics.

References

1. A. P. Yushkevich, *From student recollections* (1923–1929), Matematika, Sb. nauchn.-metod. stateĭ, Moscow, Vysshaya Shkola, 1976, No. 6, pp. 99–112.

2. ——, *The history of mathematics in the works of N. N. Luzin*, Voprosy Istorii Estestvenozn. i Tekhniki, 1984, No. 1, pp. 98–106.

3. C. Reid, *Hilbert*, Springer, New York, 1970.

4. N. N. Luzin, *Collected Works. Descriptive theory of functions*, Akad. Nauk SSSR, Moscow, 1952.

5. A. P. Yushkevich, *On the works of P. S. Aleksandrov on the history of mathematics*, Istoriko-Matematicheskie issledovaniya, 1985, vol. 29, pp. 125–157.

6. *Matematika v SSSR za tridtsat' let*, 1917–1947, Moscow, Gos. izd-vo tekh.-teoret. lit-ry, 1948.

7. K. Marx and F. Engels, *Collected works*, vol. 20.

8. A. P. Yushkevich, *A. N. Kolmogorov on mathematics and its history*, Voprosy Istorii Estestvenozn. i Tekhniki, 1983, No. 3, pp. 67–74.

9. ——, *Academician V. I. Smirnov*, Voprosy Istorii Estestvenozn. i Tekhniki, 1975, No. 1, pp. 98–100.

Moscow

Translated by ABE SHENITZER
with the editorial assistance of HARDY GRANT

The Moscow School of the Theory of Functions in the 1930s

S. S. DEMIDOV (MOSCOW)

The Moscow School of the theory of functions of a real variable was officially set up in 1911 when D. F. Egorov (1869–1931), professor at the Moscow University, published in the *Comptes Rendus* of the Paris Academy of Sciences an article **[A1]** stating the theorem that any convergent sequence of measurable functions can be transformed into an uniformly convergent one by ignoring a set of arbitrary small measure, the theorem which later became classic.

In the next year *Comptes Rendus* published an article **[A2]** by Egorov's pupil N. N. Lusin [Luzin] (1883–1950). It described a theorem on the C-property of measurable functions which later grew famous: each measurable function can be converted into a continuous one, if its values within a set of an arbitrary small measure are suitably altered.

Those two works are the very first in the history of a most notable mathematical school of the twentieth century. Its appearance was preceded by a comparatively short period of acquaintance with the ideas of the French School of the theory of functions of a real variable in Moscow, which proved very receptive to them. A number of publications (see **[B1–B16]**) are dedicated to that prehistory and its principal figures (N. A. Bugaev, B. K. Mlodzeyevsky, P. A. Florensky, and especially D. F. Egorov and N. N. Lusin), including the recent publications based upon new archives files **[B10–B16]**.

The prerevolutionary years of the school are marked first of all by the extraordinary creative activity of Lusin himself, whose work was attentively supervised by Egorov (one can learn some facts about it from recently published letters of Egorov to Lusin **[A3]**). In 1915 the famous work of Lusin, *Integral and trigonometric series*, was published **[A4]**. Results of one of Lusin's first pupils A. Ya. Khinchin (1894–1959), a student at the time, were included in it. In 1916 Khinchin published in *Comptes Rendus* an article **[A5]** on the theory of the integral which is known now as the Denjoy–Khinchin integral.

Another pupil of Lusin, D. E. Menshov (1892–1988), published in *Comptes*

D. F. EGOROV

Rendus [**A6**] his well-known construction of a trigonometric series converging to the zero function outside a perfect set of measure zero. In *Matematicheskiĭ Sbornik* [**A7**] he published a construction of a function integrable in the sense of Dirichlet but not in the sense of Borel.

At the same time *Comptes Rendus* presented an article of another of Lusin's pupils, P. S. Aleksandrov (1896–1982) [**A8**]. An important theorem on the power of Borel sets was proved, and even more important was the introduction of operations which later got the name of "*A*-operations". This work was done by Aleksandrov in 1915 when he still was a student. And only a year later another Lusin student, M. Y. Suslin (1894–1919), used those operations to create the famous *A*-sets which are sometimes called Suslin sets and sometimes analytical sets. An article about them was published in 1917 [**A9**]. The *A*-set theory was later developed by Lusin himself and his pupils, and also by F. Hausdorff and others. In the opinion of one of the representatives of the school and its historian L. A. Lyusternik, it was the *A*-sets theory which marked the milestone when the Moscow School of Theory of Functions stopped "research within the limits of the traditional French school and took up its own topics".

At the same time there were attempts by Moscovites to apply the theory of functions of a real variable to different fields: Lusin himself, his friend V. V. Golubev (1884–1954), and Egorov's pupil, I. I. Privalov (1891–1941) used results of the metric theory of functions in examining boundary properties of analytic functions—see, for example, Golubev's magistral thesis [**A10**] and Privalov's book [**A11**].

Just before the revolution of 1917 a group of gifted mathematicians formed around Egorov and Lusin (Menshov, Khinchin, V. S. Fedorov (1893–198?), Suslin, Aleksandrov, V. I. Veniaminov (1895–1932), V. V. Stepanov (1889–195?), Privalov). They all actively developed problems of the theory of functions.

The years of revolution, Civil War and profound economic devastation badly affected the activity of the school. Harsh living conditions made many researchers move to other towns in order to survive. Lusin himself and a group of his pupils, including Khinchin, Fedorov, Menshov, Suslin, found themselves in Ivanovo-Voznesensk; Golubev and Privalov in Saratov. In 1919 Suslin died during an epidemic of typhoid fever. Nevertheless, the school survived and moreover grew stronger.

This was due to Egorov, staunch guardian of academic traditions, who had high authority among scientists of Moscow and "shielded" the school, and due to the indefatigable efforts of Lusin, who was a great teacher, able to attract talented youngsters coming to the University from all parts of the country shaken by the revolution.

The school became especially active in 1920, when Lusin moved back to Moscow. A circle of the most devoted pupils and collaborators formed up around him, and it was the basis of a "fraternity" known as "Lusitania" in

N. N. Luzin

A. YA. KHINCHIN

the history of mathematics. It included Professors Stepanov, Aleksandrov, Veniaminov, P. S. Uryson (1898–1924); and students N. K. Bari (1901–1961), U. A. Rojanskaya (1901–1967), V. I. Glivenko (1896–1940), N. A. Selivanov, and L. G. Shnirelman (1905–1938). A. N. Kolmogorov (1903–1987) entered the University in 1920 and joined Lusitania, and at the end of 1921 M. A. Lavrentiev (1900–1980) joined them, and in 1922 L. V. Keldysh (1904–1976), E. A. Leontovich (1905–19??), P. S. Novikov (1950–1975), I. N. Khlodovsky (1903–1951), and G. A. Seliverstov (1905–1944). In the same year some "old-timers" returned to Moscow and rejoined Lusitania; those were Privalov, Menshov, and Khinchin.

In 1923–1924 Bari, Kolmogorov, Seliverstov, and Lavrentiev presented their first works on the theory of functions. Moscow became one of the acknowledged European mathematical centers. During the elections to the Administrative Board of the Moscow Mathematical Society in 1923, representatives of the new school won with an overwhelming majority: Egorov became President, Lusin Vice-President, Privalov Secretary.

At the same time, 1923 was the last year of existence of Lusitania in its traditional form. Its disintegration had begun. In 1922 the "topological" school of Aleksandrov and Uryson succeeded from it. In 1923 Khinchin began working in the theory of probability and number theory; in 1924–1925 Stepanov, Lyusternik, and Lavrentiev took up differential equations and calculus of variations, and Stepanov also examined almost-periodic functions. Aleksandrov, Khinchin, and Privalov got their own post-graduate students.

Lusin's disciples were maturing. Former students were becoming distinguished mathematicians with creative aspirations and tasks of their own. They started to feel too cramped within the limited framework of the master's ideas. The master himself was aware of the crisis in the themes of the school and was desperately looking for a way out. This is confirmed by his recently published letter to O. Yu. Schmidt dated February 24, 1926[1] [**A12**].

The decline and disintegration of Lusitania was prompted by Lusin's long absence from Moscow—his long scientific missions in 1925 and in 1926 and finally the one from the summer of 1928 to the autumn of 1930. Various subjective reasons also played a big role. His relations with Egorov deteriorated: they both were aspirants to the title of Academician at the elections in 1929. (In the long run, it turned out that they were not even rivals—Egorov was elected Honorary Academician, and Lusin became Academician at the Department of Philosophy. S. N. Bernstein (1880–1962), I. M. Vinogradov (1891–1983), and N. M. Krylov (1863–1945) were elected Academicians of Mathematics.)

His relations with his pupils worsened also because of his ambitious character. His emotionalism, bordering on psychopathy, attracted the attention

[1] "...the exhaustion of the creative soil, in which, after a lot of digging out, I managed to find only imperfect creative potential and some petrified debris of previous periods–made me suffer" [**A12**, p. 281].

of creative young people, but this very emotionalism was a factor in destabilizing human relations within the limited circle of a scientific school. Lusin's idiosyncrasy left the only possibility for its organization: talented pupils in the role of sons who look up in awe and admiration to their revered father and master, who exercises autocratic control of the school's life with unquestionable authority. This organization could not possibly remain stable: by the end of the 1920s his personal relations with many disciples had seriously deteriorated and he was even on hostile terms with P. S. Aleksandrov. To a large extent, this was further promoted by the above-mentioned election campaign to the Academy of Sciences in 1929 when many pupils voted against their teacher.

When Lusin got back to Moscow in the autumn of 1930, there was not a vestige of Lusitania left. The atmosphere in the University was so stifling for him (we should also take into account the overall ideological atmosphere after the campaign against Egorov, but we will talk about it later) that he preferred to leave it. The focus of his activities shifted to the Academy of Sciences, where he filled the post of the head of the Department of the Theory of Functions at the Steklov Institute, and then he took charge of the Mathematical Group of the Academy of Sciences.

After Lusin had left the University his active teaching work stopped. His school disintegrated, bringing to life a number of new schools, some of which had a considerable impact on the development of the mathematical science of the twentieth century.

The history of the origin, the heyday and the disintegration of the Moscow School of Function Theory is typical of a scientific school. The old school dies, giving birth to a number of new schools. And during the process the relations of the rebellious disciples with their masters and between themselves go sour. It is as old as the hills, and this process has repeated itself over and over again for centuries. It repeated itself in our short history of the Moscow School, too. When the theory of functions was gaining ground in Moscow in the years 1918–1920, the key posts of the Moscow Mathematical Society were still held by the representatives of the old school: N. E. Zhukovsky, Mlodzeyevsky, etc. Young people were trying to make the best of the situation and to bring about a changeover. In the late 1920s and early 1930s the situation was similar: the young people were fighting for power, although this parallel holds true only to a certain extent. There was a considerable difference between the two situations. In the first case, the process was going on in a natural way: in the normal academic environment, within the framework of established and time-honored academic traditions. In the second case, the process was utterly unnatural—it took place in an atmosphere of heavy ideological pressure on the scientists and academic institutions. As a result, this natural though painful process of generation change took on very ugly forms.

The above-said remains within the standard picture traditional for Soviet historical science. It did not present false facts but it was essentially false because it was based sometimes quite consciously on ignoring a number of facts and, above all, on ignoring the ideological atmosphere of the Moscow mathematical circles in the 1920s and 1930s.

Right after the 1917 October Revolution, one of the principal goals of the Soviet power became breaking the old university system and establishing a new one which would train scientific personnel devoted to completely different ideology. The simplest way of solving this problem was to change the social structure of the students. And that is what the government decrees dated 2 August 1918 were aimed at, stating that all those who wanted to enter higher educational establishments could do so, whether or not they had a secondary-school diploma, but workers and peasants by birth were to be given preference (see [**B17**]).

During the first few years after the revolution the social structure of students did change considerably. Those of doubtful social origin and unacceptable ideology were ousted. For example, a purge of students took place in 1924. On the other hand, youngsters of proletarian and peasant origin were encouraged to enter. The average level of students declined drastically. But these new students became principal adherents of the proletarian ideology, influencing the activity of the department and even the process of training.

Although the structure of students could be changed quite rapidly, the effective reshuffling of professors was a much longer process. Those few professors who had proclaimed their leftist convictions long ago and had actively supported establishing the new power were mostly relied upon.

One of them was O. Yu. Schmidt, an expert in algebra, who began working at the University in 1926. He was an old communist, notable organizer of science, important official, and one of the most influential mathematicians of Moscow. Beginning in 1931–1932 he was head of the Mathematical Institute of Moscow University.

S. A. Yanovskaya began teaching at Moscow University in 1925, still studying at the Institute of Red Professors, which was founded in 1921 in order to "train Marxist personnel for research and scientific work and teaching in higher educational establishments and communist universities" (Small Soviet Encyclopedia, 1930, V.3.M.: OGIZ.RSFSR, p. 471). She acquired her mathematical education before the revolution at the Higher Courses for Women in Odessa. She had been a party member since 1918 and was a participant in the Civil War. At Moscow University she monitored a seminar on methodology of mathematics and natural sciences and taught mathematics beginning in 1930. She was one of the leading Communist Party ideologists among Moscow mathematicians.

M. Y. Vygodsky started working at the University in 1931. He was an old-time communist, he took part in the Civil War, and he worked at the Y. M. Sverdlov Communist University and the Institute of Red Professors.

O. YU. SCHMIDT

Together with Yanovskaya he organized a seminar on the history of mathematics functioning to the present time.

Unlike Schmidt, Yanovskaya and Vygodsky, who took part in the activity of the University, Ernest Kolman exercised ideological influence from the outside, participating now and then in various meetings and publishing articles. Czech by birth, he graduated from the Karlov University in Prague in 1913, and as a serviceman of the Czech corps found himself in Russia during World War I. After joining the Communist Party, he took an active part in the Civil War and the creation of the Soviet state. He held important posts in the Moscow committee and the Central committee of the Communist Party during the 1920s and 1930s, and even headed the Department of Science of the Moscow Party committee in 1936–1937 (see [**A19**]).

E. Kolman was one of the most sinister personalities of the science vs. ideology "dichotomy" of the 1920s and 1930s. All tragic events in the history of mathematics of the period were accompanied by his exposing addresses and appeals to enhance witchhunting.

At the end of the 1920s, state policy concerning scientific personnel entered a new phase. The year 1928 was marked by the "Shakhtinskoe" case, the end of 1930 by the "Prompartia" process (industrial party). The sabotage of specialists became the most popular topic of mass media and all-pervasive meetings. Saboteurs were looked for everywhere. And although until that time those in power had tolerated the fact that all high offices had been held by "reactionary professors" hostile to the new ideology, there could be no coexistence any more. An attack was launched. Proletarian students and the "progressive group" of teachers triggered a ruthless campaign against the old type managers, resulting in the latter being driven from their posts and very often arrested by the OGPU (predecessor to KGB).

Kolman wrote in his article "Sabotage in science", published in "Bolshevik" in 1931 [**A13**], that "The substitution by liberalism of the Bolshevist policy in science, of the party character of science is made even more felonious by the fact that reactionary theories are confessed to by such eminent professors as a disciple of Mach Frenkel in physics, vitalists Gurevich and Berg in biology, Savich in psychology, Koltsov in eugenics, Vernadsky in geology, Egorov and Bogomolov in mathematics. They all 'draw out' the most reactionary social theories in their respective domains of science."

It is not accidental that the name of Egorov is in this notorious list of "saboteurs"; Kolman and Co. had long singled him out. The patriarch of Moscow mathematicians, President of the Moscow Mathematical Society, Director of the Mathematical Institute of Moscow University, corresponding member of the Academy of Sciences of the USSR (honorary member of it since 1929), had never concealed his negative assessments of the new ways.

As I. Zaidenvar put it in 1930 [**A14**], proletarian students "declared war" on Professor Egorov in the second half of the 1920s. First Egorov had to leave the post of Chairman of the Subject Commission, then he was removed from the Mathematical Institute. Immediately after his dismissal, the Institute underwent reconstruction, more proletarian personnel began working, and the methodology was altered; for example, individual magistral examinations were abolished (as one can guess, they used to be very nasty for "proletarian" students—S.D.). In general (Zaidenvar rejoices—S.D.) the activity of the Institute "envigorated". A meeting of young scientists of Moscow University, which took place on 21 December 1929 [**A16**, p. 18] was marked by "post-graduate students and some active students strongly attacking Professor Egorov, criticizing his ossification, inertia, lack of political zeal in reforming and renovating pedagogical, research and scientific activity, and methodology".

Remaining firm, Egorov did not intend to repent. Speaking at the discussion of the program report of Schmidt, new director of the Mathematical Institute, he stated (I quote from Kolman's article [**A13**] that "it is the enforcement of a standard outlook upon scientists which is the real sabotage". Truly right were these words, but so untypical for that meeting! Egorov's role was very significant at the Kharkov First All-Union Congress of Mathematicians, which took place in June 1930. True to his convictions, he was among those few who rejected the proposal to sign a greeting to the seventeenth Communist Party Congress going on at that time. Soon afterwards he was arrested, and he died in exile in Kazan on 10 September 1931.

After his arrest the campaign against him went on. Kolman and the "proletarian students" (D. A. Raikov, G. V. Khvorostin) were rampant. They were echoed by "red professors" (O. Y. Schmidt, S. A. Yanovskaya, and others). Then came the turn of his disciples. I. Zaidenvar wrote in the above quoted article [**A14**, p. 74]: "It is characteristic that the Mathematical Society didn't react to Professor Egorov's arrest and planned a routine working meeting, ignoring the fact of the arrest." Such a response the instigators of the campaign could not tolerate. And surely the "disciples" were instructed. As a result "an intiative group of young Soviet mathematicians" was set up, and it raised the question of restructuring the Mathematical Society. It put forward a declaration which went like this [**A15**] (quoting Charles Ford's translation [**B11**]):

"The intensified class war in the USSR has pushed the right wing of the professorate into the camp of counter-revolution. The reactionary professorate has been at the head of all the recently uncovered wrecking organizations and counter-revolutionary parties. Thanks go to the brilliant efforts of the OGPU [predecessor to the KGB] for uncovering the crimes of a whole series of scientific bonzes who have known how to conceal themselves artfully behind various masks—from cold loyalty to a loudly advertised warm attachment to Soviet power. Active counter-revolutionaries have appeared even among mathematicians. Professor Egorov was arrested for participation in a counter-revolutionary organization. He is the acknowledged leader of the Moscow School of Mathematics, president of the Mathematical Society, former director of the Mathematical Institute and the candidate of Moscow mathematics in the Academy of Sciences. This same Egorov is the preserver of academic traditions, against which the proletarian student body had already undertaken struggle. Nearly unanimously the Moscow mathematicians came to his defense. There has been a full clarification of the role of academic traditions in our nation, traditions coming from pre-revolutionary Russia, in the promotion of counter-revolutionary and restorationist attitudes among scientists. By the preaching of 'pure science', by the renunciation of the class struggle among scientific workers, by the preservation of caste prejudices among scientists, the counter-revolutionaries have preserved for themselves the leadership positions in scientific organizations."

Of course, such a condemnation of Egorov was forced upon young mathematicians, who were led by seemingly reasonable considerations. The action, which could inflict no harm on the arrested Egorov, enabled them to preserve the Mathematical Society (the very existence of which was threatened, for many scientific societies had been closed in those years) and their own status. Moreover, their position was even strengthened. Lyusternik and Gelfond became respectively Editor-in-Chief and Secretary of the principal press organ of the Society, *Matematicheskiĭ Sbornik*. Aleksandrov was elected president of the Society in 1932. Of course, they paid a high price for it: they had to take up Kolman's game and show their willingness to support mean actions. They had to be more active in the future and as a matter of fact they were. For example in the notorious case of Lusin, who until recently had been their beloved teacher.

Having returned from a prolonged trip to Paris in the autumn of 1930, Lusin had prudently decided to leave the department and concentrate on the work at the Academy of Sciences, left nearly unhinged in the course of the notorious campaign. During Stalin's reconstruction of the 1930s the Academy strengthened its status and became a central agency, controlling all scientific research conducted in the country. Thus Lusin's authority, as a new chief of the Mathematical Group of the Academy, increased. For the young to advance meant a struggle for power. In so doing, the young triggered a campaign for Lusin's dismissal. This could not be achieved by usual academic means. Among them only Aleksandrov was a Corresponding member at that time (since 1929). L. G. Shnirelman got the same rank in 1933. There was only one very effective way of struggle left, i.e., ideological. Those who took part could not fail to understand what kind of consequences their victory could bring to Lusin, remembering the fate of Egorov. The attackers knew the path they had chosen. They had already taken a first step by participating in the campaign against Egorov (though being forced to and only in the last stage). They were now tempted to take a second much more serious step. And the tempter was the all-powerful system of Stalin.

Total subjugation of all spheres of culture, science included, to the state ideology was one of the goals of Stalin in the 1930s. In the late 1920s the aim was to remove outspoken dissidents from high offices, but in the middle 1930s it was to do away with those managers whose ideology was doubtful and establish an all-embracing ideological control upon scientific thought. Mathematics was no exception.

Such programs started with a few major scandalous cases setting an example to be followed all over the country. Lusin was a suitable target, being a notable mathematician well-known in Europe and highly ranked in the Academy of Sciences. Moreover, he represented a school which had long been known for its "idealistic" views. The tireless Kolman started the ball rolling.

In 1931, the Mathematical Department of the Communist Academy issued a digest called "Struggle for Materialistic Dialectics in Mathematics" [**A17**], Kolman being the editor. The preface, undoubtedly written by Kolman, states [**A17**, p. 6–7] (largely repeating his own article [**A17**, p. 10–19] published in the same digest): "After the collapse of capitalism the Moscow Mathematical School calmly keeps to its previous course, holding its serfowners' and clerical philosophy in disguise... Let the Soviet people not interfere with our mathematical activity, we are neutral, we are beyond politics... The Vasilievs, Bogomolovs, and Florenskies published their mystical, utterly idealistic meditations in Soviet editions; so do the Egorovs, concealing their deadly hatred for socialism and refraining from signing the appeal to foreign scientists about saboteurs." (Let me mention that a campaign against Lusin's friend, the great Russian philosopher P. A. Florensky, was started at that time. He was arrested in 1933 and killed in a labor camp in 1937. Kolman was among his most zealous attackers in the mass media—S.D.)

The final words in the above quotation refer to Lusin, whose name is directly mentioned later. Criticizing Lusin's "intuitional" notion of the natural numbers and stressing "the lavish commendments to Lusin by Henri Lebesgue" that eloquently expose Lusin's hostile ego, the author of the preface (i.e., Kolman himself) exclaims: "It's absolutely clear from the above-said that the fortress of idealism in mathematics should be completely destroyed."

In Kolman's book, *The subject and the method of modern mathematics* [**A18**], which was submitted for publication on April 2, 1936 (i.e., several months before Lusin's trial), we came across another charge of the same kind but a much harsher one. "This reactionary way of thinking (of the representatives of the Moscow Philosopho-Mathematical School) has been preserved intact by one of the 'pillars' of the school whose name is Lusin and who has rendered it a more modern pro-fascist coloring." [**A18**, p. 290]

So the assault on Lusin had been well planned beforehand. Some influential circles (it is quite possible that it was Kolman himself and his associates) started the "practical organization" of Lusin's case. The masterminds of the trial knew that they would find obedient, sometimes even zealous, and (what counted most of all) personally involved "performers" among his former disciples. To start a trial one needed to expose an act of sabotage.

The whole thing was carefully planned. First, Lusin was invited to a graduation exam in mathematics at a Moscow secondary school and was asked to contribute to the *Izvestia* newspaper an article covering his impressions. Those who masterminded the scheme knew beforehand that Lusin would abound in praise—they counted on his personal traits of character (this proves that the people who organized the farce were on good terms with Lusin). They were not wrong: the issue of *Izvestia* dated June 27 featured an item by Lusin entitled "Pleasant disappointment". The gist of it was that the academician expecting the usual low level of the students' academic performance was "pleasantly disappointed" at their competence. It turned out that there were no poor students. Another move was made five days after that: *Izvestia* printed the answer of the school's principal, who argued that Lusin was biased, that the students did not have mathematical skills and could not work with textbooks properly. The principal wrote that schools did not need hypocritical praise but constructive criticism. "Wasn't it your goal," said the organizers of the campaign through the principal, "to whitewash our shortcomings and thus to damage our school?"

Thus, the issue of sabotage was raised. The next day, July 3, *Pravda* carried an editorial entitled "Enemies hiding behind Soviet mask". The letter of the school's principal, the author of the article argued, had raised slightly the curtain hiding the hostile and subversive activities of the academician. In a day (!) *Pravda's* office had been inundated with letters from "scientists speaking for mathematics". The article argued that the analysis of Lusin's activities demonstrated that this hypocritical account was "but one ring in the long chain of Lusin's camouflage, very instructive as to the enemy's artful methods of disguise".

There are good reasons to conjecture that the author of this article was Kolman **[B18]**; but even if not, there can be no doubt that the author had first-hand knowledge of Kolman's report, because the arguments in Kolman's text **[A18]** and those of the *Pravda* article were much the same.

In the ensuing *Pravda* articles dated July 9, 10, 12, 13, 14 and 15, an all-out campaign against Lusin was unleashed (see **[B18, B19, B20]**). Meanwhile, at various research and academic institutions, meetings took place where Lusin was branded as an enemy of the Soviet state. At Moscow State University, Yanovskaya made a report denouncing Lusin; she was echoed by L. A. Lyusternik, N. N. Buchgolz, Aleksandrov, and Kolmogorov **[A19]**. Meetings also took place at the Steklov Institute, at the Energy Institute of the Academy of the USSR, at the Academy of the Byelorussian SSR, and at other institutions.

The main events took place at the session of a special committee set up by the Presidium of the USSR Academy of Sciences, on July 7. The chairman of the committee was the Vice-President of the USSR Academy of Sciences, G. M. Krzhizhanovsky, and the committee was composed of Academicians A. E. Fersman, S. N. Bernstein, O. Yu. Schmidt, I. M. Vinogradov, A. N. Bach, N. P. Gorbunov, Associate members Shnirelman, S. L. Sobolev,

Aleksandrov, and Khinchin. Also on the committee were A. N. Krylov, Gelfond, Lyusternik, and B. J. Segal. Lusin was invited to some of the committee's sessions.

We will not follow the discussion at these sessions (you can read about that in the article by A. P. Yushkevich on "Lusin's case" [**B19**]). It is sufficient to say that major charges against Lusin boiled down to the following: (1) he systematically gave undeserved benevolent reviews of the works of other mathematicians; (2) he published his best works abroad while only minor papers were published in the USSR; he was allegedly idolizing the West (this point should be considered in the context of the major change in the official nationalities policy of that time) [**B18**]; (3) he misappropriated results of his own disciples. All this was considered to be outright sabotage.

The tone of the ideological accusations against Lusin set up by the newspaper reports was ever-present at these sessions. The most aggressive mathematicians were Aleksandrov, Schmidt, Sobolev, Gelfond, and Lyusternik. (Khinchin, Kolmogorov, and Shnirelman demonstrated more reserve. Only two people—Bernstein and Krylov—tried to defend Lusin. Some of Bach's remarks were in Lusin's favor, although on other points he sided with the attackers.

In the chorus of denunciations and exposés, the voices of those who had the courage to defy this raging witchhunt were nearly drowned. The most daring attempt was made by P. S. Kapitsa, who on July 6 sent a letter to the Chairman of the Soviet Council of Ministers, Molotov, protesting the campaign again Lusin [**B21**, pp. 86–90]. Some scientists expressed their sympathy and support for Lusin. Among them were V. I. Vernadsky and S. A. Chaplygin. Many people expressed their attitude by refusing to take part in the investigation (even that required a lot of civic courage at the time).

The campaign was underway. Everyone was following the proceedings of the Academic Committee; its major direction did not leave any doubts as to the outcome—excluding Lusin from the Academy. It was quite easy to foresee the decision of the general meeting followed by the inevitable intrusion of the "competent" organs (the euphemism for security police).

But it did not turn out that way. The campaign was abruptly stopped by somebody high up. This happened sometime between July 11 and 13. The shorthand records reveal the abrupt change in the general tone of the speeches by the Committee members. The Presidium of the USSR Academy of Sciences on the basis of the Committee's conclusions ruled that "it is sufficient to reprimand and warn Lusin that in case he doesn't change his line of conduct the Presidium of the Academy will have to raise the question of excluding N. N. Lusin from the Academy" [**B19**, p. 113].

Why was Lusin's case terminated? We still cannot answer this. We can only join with A. P. Yushkevich and point out a possible reason for its termination. At that time, preparation for the trial of Zinoviev and Kamenev was well underway (it was to take place in August), and in relation to this formidable trial of Stalin's major political opponents and rivals, the case of Lusin seemed too unimportant.

Lusin got off lightly for those times. He withdrew from his mathematical school and concentrated on his own work. He managed to write a number of important works on the theory of differential equations, and on the history of mathematics. In 1938 he started research in differential geometry and gave a final solution to a problem of deformation of surfaces which had been a stumbling-block for Moscow mathematicians since the 1860s. In 1945 he gave a brilliant course of lectures under the title "Selected topics in the theory of functions of a complex variable". And still another group of young mathematicians started to work on the theory of functions of two real variables. He died of natural causes in 1950.

The "black-angel" of the Moscow mathematical circles, Kolman, soon disappeared from the political arena. In 1937 under very obscure circumstances[2] he was dismissed from the post of the official responsible for science in the Moscow Party committee and ceased to play any significant role in the life of the mathematical community of the USSR. He made a point not to mention Lusin in his autobiography **[A20]**, and when direct questions were posed to him he said he had nothing to do with the whole matter.

Other participants in the events also tried hard to forget about Lusin's case. It was useless to interrogate them about the details of this matter—they could not remember.

So, "the young" won. Most of them were talented and gifted people. Only one of them was a failure—Shnirelman—who committed suicide under obscure circumstances. Nevertheless, Lusin obstructed the election of Aleksandrov to the Academy till the end of his life—Aleksandrov was elected Member of the Academy of Sciences only in 1953 after Lusin's death. As for the others, Lusin did not try to meddle with their academic careers. At least, we do not have any clear evidence that he did.

In 1939 Kolmogorov was elected Member of the Academy of Sciences; Gelfond, Pontryagin (who became Member of the Academy in 1958), and Khinchin were elected Associate Members; in 1946 Lavrentiev became Academician, and Lyusternik was elected Associate Member.

As we see, this school had quite a satisfactory life story, and by our standards it was a lucky one, because everything could have been different—Lusin put to death, and his disciples might have followed him. There have been a lot of examples of this sort in the history of Soviet science. The best known is the complete destruction of the Soviet school of genetics. A similar outcome was quite possible for Lusin's school.

---

[2] In his autobiography **[A20]**, Kolman writes about his dismissal in a pretty calm way and, strange as it might seem, he "forgets" about his arrest in 1937 along with other officials of the Moscow Party committee. It was revealed only recently in the publication of the *Vechernaya Moskva* newspaper dated November 14, 1989 (see the interview of Khrushev's former aide I. P. Aleksakhin, "How the Moscow City Party committee was dissolved").

Why didn't it happen? It is quite useless trying to read some logic into it in a practical, be it cynical, way (like, say, the leadership of the country needed military industry, hence technology, hence physics, hence mathematicians might also come in handy). If all the country's highest-ranking officers were exterminated just before World War II, surely the leaders of the country did not display such a well-calculated foresight as to preserve mathematicians. Most probably it was just a matter of chance, although it is possible to voice some viewpoints on general factors that made the physical survival of the school possible.

Mathematics did not count much for the rulers of our country. It was politically unprofitable to have show trials in this field of science. But economics or, say, biology was quite different because these fields of science were not only theoretical but had practical spheres of application, so sabotage in biology could be quite real. Even in theory, biology was much more politicized than mathematics because biology involves Darwinism, and every schoolboy in Russia considered himself a Darwinist even in the nineteenth century. Biology was also more convenient, because it did not have a set of strict proofs. The proofs in biology can be interpreted in different ways, and at times the only proof is practice, and this takes a lot of time. In mathematics, any competent specialist can check everything rather fast. And it is clear why even such a hunted and shy man as Lusin could make fun of (!) the mathematics of Kolman when Lusin was being interrogated at the Academy; and Kolman held a very powerful position—his responsibility at the Moscow City Party committee was monitoring science.

Mathematics was the wrong science to start a witchhunt in. And one needed a strong leader—a leader with personality, confidence, talent of a certain kind, a leader who would defend some rather strong (not only in an exclusively scientific sense of the word) concept. Lysenko, the man who destroyed Soviet genetics, based his "science" on popular interpretations of Michurin's biological concepts. Lysenko had both the power and the charisma! Was there such a personality in mathematics? I think that the only person who could claim to be the one was Kolman. As we see, he had bad luck, and I think there were several reasons for it. One of the most important reasons for his failure was his own lack of mathematical talent. As a mathematician he was a joke to every competent specialist. Above all, he could not seem attractive to the Pillar of Sciences (Stalin), because Stalin was more inclined towards the Russian-peasant type of "scientists" who were guided not by tricky theoretical sophisms but by time-honored popular wisdom. But who knows what might have been? Kolman established ties with Lysenkovists in 1940 [**A21**, **B22**], and who knows, he might have gained the support of the so-called "people's academician" and then perhaps the whole Moscow school would have been persecuted. But providence preserved them.

REFERENCES

[A1] D. F. Egorov, *Sur les suites des fonctions mesurables*, C. R. Acad. Sci. Paris **152** (1911), 244–246.

[A2] N. N. Lusin, *Sur les propriétés des fonctions mesurables*, C. R. Acad. Sci. Paris **155** (1912), 580–582.

[A3] Letters from D. F. Egorov to N. N. Lusin, with preface by P. S. Aleksandrov; publication and notes by F. A. Medvedev with the collaboration of A. P. Yushkevich, Istor.-Mat. Issled. **25** (1981), 335–361. [Russian]

[A4] N. N. Lusin, *Integral and trigonometric series*, Moscow, 1915. [Russian]

[A5] A. Y. Khinchin, *Sur une extension de l'intégrale de M. Denjoy*, C. R. Acad. Sci. Paris **162** (1916), 287–289.

[A6] D. E. Menshov, *Sur l'unicité du développement trigonométrique*, C. R. Acad. Sci. Paris **163** (1916), 433–436.

[A7] ——, *Relations between the integrals of Borel and Denjoy*, Mat. Sb. **30** (1916), 288–295. [Russian]

[A8] P. S. Aleksandrov, *Sur la puissance des ensembles mesurables B*, C. R. Acad. Sci. Paris 162 (1916), 323–325.

[A9] M. Y. Suslin, *Sur une définition des ensembles mesurables B sans nombres transfinis*, C. R. Acad. Sci. Paris **164** (1917), 88–90.

[A10] V. V. Golubev, *Single-valued analytic functions with a perfect set of singular points*, Moscow, 1916. [Russian]

[A11] I. I. Privalov *The Cauchy integral*, Saratov, 1918. [Russian]

[A12] "A letter of N. N. Lusin to O. Yu. Schmidt", publication and notes by S. S. Demidov, Istor.-Mat. Issled. 28 (1985), 278–287. [in Russian]

[A13] E. Kolman, *Sabotage in science*, Bolshevik, 1931, no. 2, pp. 73–81. [Russian]

[A14] I. Zaidenvar, The October revolution in the Mathematical Society and in the Institute of Mathematics and Mechanics, Varnitso, 1930, no. 11–12, pp. 73–74. [Russian]

[A15] L. A. Lyusternik, L. G. Shnirelman, A. O. Gelfond, L. S. Pontryagin and P. A. Nekrasov, *Declaration of the initiative group for the reorganization of the Mathematical Society*, Nauchni Rabotnik, 1930, no. 11–12, pp. 67–71. [Russian]

[A16] I. G. Dolin, *At the meeting of post-graduate students*, Nauchni Rabotnik, 1930, no. 1, pp. 18–20. [Russian]

[A17] *Struggle for materialistic dialectics in mathematics*, GNTI (Gosudarstvennoe nauchno-tekhnicheskoe izdatel'stvo), Moscow-Leningrad, 1931. [Russian]

[A18] E. Kolman, *The subject and the method of modern mathematics*, Sotsekgiz., Moscow, 1936. [Russian]

[A19] *Against Lusin and Lusinists*, Front Nauki i Tekhniki **7** (1936), 123–125. [Russian]

[A20] E. Kolman, *We should not have lived that way*, Chalidze Publications, New York, 1982. [Russian]

[A21] ——, *Is it possible to give a statistical-mathematical proof or disproof of Mendelism?*, Dokl. Akad. Nauk SSSR **28** (1940), no. 9., [Russian]

B. Historical-Mathematical Investigations

[B1] A. P. Yushkevich, *History of mathematics in Russia up to the year* 1917, Nauka, Moscow, 1968. [Russian]

[B2] *History of native mathematics*, Vol. 2, Naukova Dumka, Kiev, 1967, pp. 437–443. [Russian]

[B3] P. S. Aleksandrov, *Mathematics at Moscow University in the first half of the 20th century*, Istor.-Mat. Issled. 8 (1955), 9–54. [Russian]

[B4] P. S. Aleksandrov, B. V. Gnedenko and V. V. Stepanov, *Mathematics at Moscow University in the 20th century* (*up to* 1940)", Istor.-Mat. Issled. **1** (1948), 9–42. [Russian]

[B5] F. A. Medvedev, *Set-theoretical training and functional-theory research in Russia*, Historical Sketches in Mathematics and Mechanics, Sbornik Stat., Izdat. Akad. Nauk SSSR, Moscow, 1963. [Russian]

[B6] P. I. Kuznetsov, *Dmitrii Fedorovich Egorov*, Uspehi Mat. Nauk **26** (1971), no. 5, pp. 169–206 [Russian] = Russian Math. Surveys **26** (1971), no. 5, pp. 125–164.

[B7] L. A. Lyusternik, *The early years of the Moscow mathematical school*, Uspehi Mat. Nauk **22** (1967), no. 1, pp. 137–161; no. 2, pp. 199–239; no. 4, pp. 147–185 [Russian] = Russian Math. Surveys **22** (1967), no. 1, pp. 133–157; no. 2, pp. 171–211; no. 4, pp. 55–91.

[B8] E. Phillips, *Nicolai Nicolaevich Luzin and the Moscow school of the theory of functions*, Historia Math. **5** (1978), 275–305.

[B9] A. Shields, *Years ago: Luzin and Egorov*, Math. Intelligencer **9** (1987), no. 4, pp. 24–27; Ibid. **11** (1989), no. 2, pp. 5–8.

[B10] S. S. Demidov, *N. V. Bugaev and the origin of the Moscow school of theory of functions of a real variable*, Istor.–Mat. Issled. **29** (1985), 113–124. [Russian]

[B11] Ch. Ford, *Dmitrii Egorov: Mathematics and religion in Moscow*, Math. Intelligencer **13** (1991), no. 2, pp. 24–30.

[B12] S. S. Demidov, *Bougaiev et la création de l'école de Moscou de la théorie des fonctions d'une variable réelle*, Mathematica, Festschrifte für Helmuth Gericke ("Boethius" Series, Vol. 12), Wiesbaden–Stuttgart, 1985, pp. 651–673.

[B13] ——, *On an early history of the Moscow School of theory of functions*, Philos. Math. (2) **3** (1988), no. 1, pp. 28–35.

[B14] ——, *Der philosophische Kontext der Herausbildung der Moskauer funktionentheoretischen Schule*, NTM Schr. Geschichte Natur. Tech. Medizin **25** (1988), no. 2, pp. 25–31.

[B15] S. M. Polovinkin, *On the student mathematical circle in the Moscow mathematical society in 1902–1903*, Istor.-Mat. Issled. **30** (1986), 148–158. [Russian]

[B16] S. S. Demidov, A. N. Parshin, and S. M. Polovinkin, S. M. *On the correspondence of N. N. Lusin with P. A. Florensky*, Istor.-Mat. Issled. **31** (1989), 125–190. [Russian]

[B17] A. F. Lapko and L. A. Lyusternik, *From the history of Soviet mathematics*, Uspehi Mat. Nauk 22 (1967), no. 6, pp. 13–140 [Russian] = Russian Math. Surveys **22** (1967), no. 6, pp. 11–138.

[B18] A. Levin, *Anatomy of a public campaign: Academician Luzin's case" in Soviet political history*, Slavic Review **49** (1990), no. 1.

[B19] A. P. Yushkevich, *The case of Academician N. N. Lusin*, Vestnik Akad. Nauk SSSR, 1989, no. 4, pp. 102–113. [Russian]

[B20] A. P. Youschkevitch and P. Dugac, *"L'affaire" de l'académicien Luzin de 1936*, Gazette des mathématiciens, 1988, no. 3, pp. 31–35.

[B21] P. L. Kapitsa, *Letters on science*, Moskovskii Rabochii, Moscow, 1989. [Russian]

[B22] A. Shields, *Years ago*, Math. Intelligencer **10** (1988), no. 3, pp. 7–11.

About Mathematics at Moscow State University in the late 1940s and early 1950s

E. M. LANDIS

This is an account about Moscow mathematicians whom I was able to observe over the last 45 years. I largely restrict myself to the older generation and the years from 1945 to 1953. Of course, these notes are highly subjective, as are all descriptions of events observed by individuals who have, to some extent, participated in them.

I propose to talk about mathematics in Moscow but the title refers to mathematics at Moscow State University (MSU). This discrepancy is apparent rather than real. Indeed, up to some 20 years ago, Moscow mathematical activity centered around the Faculty of Mathematics and Mechanics (Mekh-Mat)[1] at MSU. Even mathematicians who were not faculty members had very close ties to it; they lectured and conducted seminars at MSU. I will talk about the few exceptions below.

I entered the first year of Mekh-Mat at MSU in 1945. (Strictly speaking, I began my studies in 1939. But two months later I was called up and served in the army for six years, almost to the end of the war.) At that time Mekh-Mat was made up of the mathematical, mechanical, and astronomical sections. There was no specialization in the first two years. In addition to mathematics, mechanics, and astronomy, taught by faculty members, students in the first two years were taught physics in the School of Physics by members of the physics faculty. This arrangement was an echo of the "good old times", when the university lived up to the meaning of the term "university" and there were close ties among the faculties. Soon these ties broke, the general physics course was eliminated, and only certain chapters of theoretical physics were subsequently taught to advanced students on the Mekh-Mat premises. At that time astronomy was also eliminated from Mekh-Mat and became the responsibility of the School of Physics, and students were assigned to, and taught in, Mekh-Mat beginning with the first year.

[1] *Editor's note*: Faculty of Mathematics and Mechanics, or Mekh-Mat, is translated as the Department of Mechanics and Mathematics in some of the other articles in this issue.

During the first term of the first year I was not very clear as to who was who, so that the impressions I had of my teachers at the time were those of an unprejudiced student, and it is those impressions that I now share with the reader.

Analysis was taught by A. Ya. Khinchin. Khinchin lectured slowly, pedantically, and rather boringly. Analytic geometry was taught by B. N. Delone, who acted strangely. For example, when explaining symmetry, he drew an elephant on the blackboard, stood in front of the board on a chair, and pretended to be an elephant. To produce the right effect he moved his hand in front of his face in an imitation of an elephant trunk. He spouted anecdotes, some of doubtful propriety. For example, in connection with P. S. Aleksandrov he remarked: "Pavel Sergeevich is a good man and a fine mathematician but one doesn't know why Uryson drowned." (And to think that this was said to first-year students who tend to believe anything!) Algebra was taught by I. R. Shafarevich. Shafarevich was a *wunderkind*. When I attended this course he was 24 and I was his age. In that year he defended his doctoral dissertation[2]. He was B. N. Delone's student. Our knowledge of his defense of the dissertation was, of course, the knowledge of outsiders. Delone advertised the dissertation, explained that it dealt with the inverse theorem of Galois theory, and was remarkable in that it was just 24 pages long and very sophisticated. (It seems that this brevity record has not been broken to this day. Incidentally, the brevity record for candidate dissertations belongs to A. L. Lunts, whose dissertation is just four pages long.)

From Orlov, who lectured on astronomy, we learned just a few technical terms and many tales and anecdotes. What was most amusing about them was the serious tone in which he told them. It was from him that we learned of Tycho Brahe's silver nose that replaced the original lost in a duel.

Physics was taught by Kalashnikov, a good lecturer, and then by Ivernova. Her lectures were ladylike superficial and did not rise markedly above the high school level. The experiments that accompanied the lectures were good but rather antiquated. The result was that even I stopped attending the lectures in spite of the fact that I was a very conscientious student.

In addition, there were mathematical exercises, the study of a foreign language (one could choose French or German; English was regarded as unnecessary), physical education, and history of the [Communist] party.

I. R. Shafarevich conducted an algebra seminar which I attended. The topic was quaternions. The seminar struck me as formal and rather boring.

All of this changed for me beginning with the second term, when A. S. Kronrod organized a *kruzhok*[3] for the study of the theory of functions of a real variable. Kronrod was at the time a fifth-year student. He was unusual

[2] A candidate dissertation corresponds to a Ph.D. in the U.S., whereas a doctoral dissertation is of much higher rank.

[3] *Editor's note*: "Kruzhok" ("circle") is the Russian equivalent of "working group".

both as a person and as a mathematician. I became a mathematician when I joined his *kruzhok*. The way one worked in the Kronrod *kruzhok* was to solve some difficult textbook problems and then try to tackle new, unsolved problems. What made this approach possible was the nature of mathematics at the time and, in particular, the nature of the theory of functions of a real variable, where little background knowledge was needed to obtain new results. This method of training and involving people in scientific work went back to N. N. Luzin. At the time (before Bourbaki) the language of mathematics was extremely meager. Introduction of an unnecessary new term was regarded as bad form. This enabled one to work on new results without extensive preliminary knowledge.

I want to make a small detour and state my observations concerning methods of teaching mathematics (as well as other subjects). Throughout high school, what may be called the "intensity of education" is, in general, very low. New knowledge is imparted in microscopic doses. No effort is made to get students used to somewhat longer logical arguments. This continues at the university. I carried out the following experiments. Already before the war, Mekh-Mat operated mathematical *kruzhoks* for Moscow's elementary school students. In a *kruzhok* for seventh graders I tried to teach the differential and integral calculus through problems. The students learned the material after a minimal amount of training and learned it far better than high school students. On another occasion, in a seminar with a group of (admittedly gifted) second-year university students, we managed to cover a semester's work on analytic functions (14 two-hour lectures and 14 two-hour problem sessions) in two lessons. So much for teaching methods. End of detour.

True to the style of the Kronrod *kruzhok*, after solving some preparatory problems I became involved in scientific work. The first training problem I solved was to construct a function that takes on all real values on every interval without relying on the axiom of choice, and thus, in particular, on Zermelo's theorem. The second training problem was to reconstruct Kolmogorov's example of a function in L^1 whose Fourier series diverges everywhere. In the process of thinking about this problem one rapidly acquired all necessary knowledge of the theory of Fourier series.

After a few such problems one would handle independent scientific work. At the end of the first year I completed such an investigation in what is now called general topology.

Beginning with the second year, I attended only some of the required courses (I. M. Gel′fand's linear algebra, Gel′fond's number theory, Kolmogorov's functional analysis, Petrovskiĭ's partial differential equations, Khinchin's probability, and Rashevskiĭ's differential geometry) and a number of special courses (Men′shov's on series of orthogonal functions, P. S. Novikov's on descriptive function theory, Pontryagin's on topology, and Kurosh's on group theory and on the theory of algebras).

In addition to joining Kronrod's *kruzhok* I began to take part in student *kruzhoks* and seminars (Bari's on almost periodic functions, and Petrovskiĭ's and Kreines' on partial differential equations), and in research seminars (Men′shov's, Bari's, and P. S. Novikov's on the theory of functions of a real variable, and Petrovskiĭ's, Sobolev's, and Tikhonov's on equations of mathematical physics).

I write in such detail about myself for two reasons. One is that my road to science was typical of the road followed by capable university students then as well as later. The other is that, beginning with the late 1940's, I could consciously observe the life of Mekh-Mat. In the sequel I will write about mathematicians whose activities I began to observe as a student. My opinion about what most of them were like has not changed. But some of them have changed in time. I recall a conversation I had with I. M. Gel′fand many years ago. We fully agreed that people change so much under changing conditions that, in general, a man should not be held responsible for what he had done more than ten years earlier.

The core of the faculty's mathematics professors consisting of Luzin's students of the older, middle, and younger generations: Stepanov, P. S. Aleksandrov, Men′shov, M A. Lavrent′ev, Kolmogorov, Bari, Lyusternik, M. V. Keldysh, P. S. Novikov, L. V. Keldysh, Gel′fand, and others, and students of his students: Pontryagin, Gel′fand, Tikhonov, A. A. Lyapunov, and others. Not all of them were formally professors at MSU (for example, P. S. Novikov) but the essence of their activities was linked with Mekh-Mat.

In addition to Mekh-Mat there existed—and continues to exist—in Moscow the Steklov Mathematical Institute of the Academy of Sciences, popularly known as the "Steklovka". Many of the Mekh-Mat professors worked at the same time at the Steklov Institute.

During my days at MSU, Luzin practically never put in an appearance. I think that this was largely due to the nature of the personal relations between him and two of his most illustrious students, Kolmogorov and P. S. Aleksandrov. Of course, the fact that he was already ill at the time was a contributing factor. In 1947 he announced that he would give an advanced course on ordinary differential equations. He gave one lecture and then got ill.

The impression he made during that lecture was strange. He brought with him a number of hefty volumes containing just logical symbols, without a single word. He circulated these volumes in order to surprise the students. That is how the first hour passed. During the second hour he explained that the course would deal with differential equations determined by integral vortices and went over some examples. I, at least, never understood what he had in mind and what he wanted to say. This was probably also true of the others. I learned from Bari that at the time Luzin was preoccupied with the continuum hypothesis.

This was the one occasion when I saw Luzin. On the other hand, as editor of practically all of his books, I got to know his creative work very well indeed. The books in questions were: *The integral and the trigonometric series* (his doctoral dissertation, published as university notes in 1915, edited by me and supplemented by a number of his papers and extensive commentaries), *Lectures on the theory of analytic functions*, and the first volume of his collected works. All these were published after Luzin's death (he died in the beginning of 1950).

When describing the makeup of the mathematical section of Mekh-Mat I noted that it consisted largely of Luzin's students and students of his students. But there were exceptions.

I begin with Ya. I. Plessner. Professor Plessner fled from Germany to the USSR in 1933. He introduced Moscow's mathematicians to contemporary functional analysis. His lectures, talks, and his fundamental paper in the Uspekhi became the source of the Moscow school of functional analysis headed by I. M. Gel′fand. A close friend of Wiener's, Plessner introduced Moscow mathematicians to Wiener's theory of regularity of a boundary point in Dirichlet's problem. Before the war, as well as in the early postwar period, the works of American mathematicians were relatively unknown in Moscow. There were American journals in the huge combined library of Mekh-Mat and of the Moscow Mathematical Society but few Moscow mathematicians read them. It is worth noting that whereas Wiener's paper was published in 1924, in 1928 Lyusternik suggested to Petrovskiĭ a problem that was a special case of Wiener's problem. (Petrovskiĭ solved the proposed problem and, given his originality, went beyond it.) Ignorance of Wiener's result was not restricted to Moscow mathematicians. It seems that this result was properly appreciated in Europe only after M. V. Keldysh presented it simply and precisely in 1940.

In the late 1930s the Academy of Sciences moved from Leningrad to Moscow and with it came a number of mathematicians. The most eminent of them was S. N. Bernshteĭn. As early as 1904 Bernshteĭn solved Hilbert's nineteenth problem and continued to work on a large mathematical front in a number of very different areas: probability theory, approximation theory, differential equations, and differential geometry. But all of his major works were done before the war.

Another major mathematician to come from Leningrad was I. M. Vinogradov. He was an expert in number theory. What made him famous was his solution of the Waring and Goldbach problems and the related major advances in analytic number theory. In my time he no longer did any serious scientific work. What he did do was reissue his very good book on number theory.

Neither of these two mathematicians took part in the work of Mekh-Mat in the postwar period and both came there infrequently. Vinogradov, a monarchist and a Jew-hater, was the permanent director of the "Steklovka" practically to his death in 1983 and was intensely concerned about the racial purity of its new members.

S. L. Sobolev, a student of V. I. Smirnov, came to Moscow at that time. He was famous for his theorems on the embedding of function spaces and became an academician at age 26. In my time he got interested in computational mathematics and supervised the division of mathematics in the Kurchatov Institute of Physics. He founded the computing division of Mekh-Mat and then continued his study of mathematical computing at Novosibirsk, where he published routine papers on cubature formulas.

In spite of the fact that he worked in computing for the rest of his long life, this was not Sobolev's true calling. He was at heart a mathematical physicist. In addition to his embedding theorems, he contributed to this area high-level works, including papers on the rotation of fluids. Sobolev's failure to recognize his true calling is not unique. It is well known that Glazunov thought that he was meant to be a great conductor, but when he conducted his "Raymonda" all he managed to do was confuse the ballerinas.

O. Yu. Schmidt, the famous polar explorer, was an algebraist. He did not belong to the Moscow school of mathematics (he was a student of Grave). I regret to say that all I know of his work in algebra is his book *Finite groups*. At the time he was director of the Institute of Earth Physics. He came to Mekh-Mat infrequently. He was deeply interested in a new cosmogonic theory. In particular, he discovered that it is possible for a two-body system to capture a third (an event regarded as impossible since the time of Laplace). I recall his announcement of this fact at a meeting of the Moscow Mathematical Society. (He suffered from cancer at the time and died shortly thereafter.) Schmidt's proof of his discovery was computational. It was rigorous in the sense that it included an exact error estimate. A beautiful analytic proof of Schmidt's result was given somewhat later by K. A. Sitnikov, a student of P. S. Aleksandrov.

A. N. Kolmogorov and P. S. Aleksandrov were close friends and some of Aleksandrov's students were to a large extent students of both of them. Kolmogorov generated ideas with great ease but presented them in terse, summary form. Even in his publications, he often did not more than annotate his results in the Proceedings (Doklady) of the Academy of Sciences. By contrast, Aleksandrov's presentations were exact, detailed, and pedantic. Aleksandrov was a splendid teacher. Unlike Aleksandrov, Kolmogorov was very much interested in celestial mechanics and, more generally, in cosmogony and cosmology. This being so, it is difficult to say whose student Sitnikov was when it came to his proof of Schmidt's cosmogonic result.

There were also other outstanding mathematicians not connected with the Moscow school. They were, for the most part, algebraists and geometers. For example, V. F. Kagan was a fine geometer. It should be noted that he gave a complete system of axioms of geometry independently of Hilbert and his loss of priority was due to the slowness of the Russian process of publication.

N. N. Luzin was exclusively interested in the theory of functions of a real variable. All of his students began their work in this area but most of them later left it. Nevertheless, the initial set-theoretic training of the leading Moscow mathematicians was readily recognizable during the period I am describing; the set-theoretic approach was dominant and the analytic approach was relegated to the background.

I will now describe the mathematical concerns of the leading Moscow mathematicians and of some of their students during the period under consideration.

The group of mathematicians who continued to work in the area of functions of a real variable consisted of Luzin, P. S. Novikov, Men′shov, Bari, and A. A. Lyapunov.

There were three definite branches of the theory of functions of a real variable, namely, metric theory of (set) functions, descriptive set theory, and constructive theory of functions.

The key concerns of the metric theory of functions were measure and integral. The focus of descriptive set theory was the investigation of properties of sets and functions depending on their construction. The themes of these two areas were tied to the following three fundamental problems: (1) the connection between the integral and the derivative, (2) convergence of the Fourier series of a function to that function, and (3) the continuum hypothesis.

The object of the first of these problems was to extend the Newton-Leibniz theorem to a maximal class of functions. During the period under consideration this problem was solved in great generality. We are by now accustomed to regard as natural the connection between absolutely continuous functions and the Lebesgue integral, and to view the Lebesgue integral as the natural limit of generalization of the Cauchy integral. But at the time that viewpoint was not yet established. The Denjoy integral, the Denjoy-Khinchin integral, the Boks integral, all more general than the Lebesgue integral and all now basically and justly forgotten, were then still subjects of research. For example, if a function on an interval has a derivative at each of its points, then (by Lagrange's theorem) this derivative determines it uniquely [up to a constant]. But the algorithm for the construction of such an antiderivative is a rather complicated and artificial procedure (the restricted Denjoy integral), and, when it comes to applications to other areas of mathematics, a function that is everywhere differentiable, without estimates for its derivatives, turns out to be an object of little interest.

No new fundamental works were published in this area. Old results continued to be presented in special courses for students and post-graduates and to be published in monographs. For example, a translation of S. Saks' *Theory of the integral* (of 1937) appeared in Moscow in 1948. In addition to other issues, this work dealt in considerable detail with the issues just described. Though, it should be noted that, as a rule, at the time, Moscow mathematicians who worked in the area of the theory of functions of a real variable stayed within the confines of the Lebesgue integral and knew little of the subtle matters noted above.

Things were different when it comes to the two other fundamental problems. From the viewpoint of a contemporary mathematician—not a specialist in the theory of functions of a real variable—Fourier (trigonometric) series occupy a key position in the theory of generalized Fourier series. This is because nowadays the most natural norm usually employed in the study of the convergence of such series is the L^2-norm. But during the period under consideration, at least in the area of the theory of functions of a real variable, the importance of pointwise convergence was regarded as equal to that of convergence in the mean. A long time ago du Bois-Reymond had posed the following problem: Is it true that the Fourier series of a continuous function converges to it pointwise almost everywhere (a.e.)? [The term "almost everywhere" is due to Lebesgue.] This problem was later given the following modified form by Luzin: Is it true that the Fourier series of a function in $L^2[0, 2\pi]$ converges to it pointwise a.e.? (As noted above, Kolmogorov had shown that the answer is "no" for functions in $L^1[0, 2\pi]$.) This turned out to be a very difficult question. It was answered in the affirmative much later (in 1966) by Carleson. Attempts to answer this problem transformed the theory of trigonometric series into an independent subject that continues to flourish to this day. During the period under consideration these questions were studied in Moscow by Men′shov, Bari, and their students. It was then that Bari wrote her major monograph *Trigonometric series.*

I come now to descriptive set theory. Let me begin with my view of some of the history of the subject. The sets used in analysis are the Borel sets. We can describe them as the elements of the smallest σ-algebra which contains all of the open sets (we are speaking of sets on a line). These sets can be built up constructively using (now largely forgotten) transfinite induction. (In this connection it is of interest to mention L. Young's caustic remark in his *Lectures on the calculus of variations and optimal control theory* that, having learned Zorn's lemma, Mr. Bourbaki proclaimed that there is no alternative to the road passing through it, and that as a result of this proclamation all post-graduate students are ignorant of far deeper facts.) It is, however, possible to construct sets more general than Borel sets. Such a construction was first given by P. S. Aleksandrov. A. A. Suslin showed that the resulting so-called A-sets are, in general, not Borel sets.

Luzin embarked on a detailed investigation of sets using his "sieve operation". He summarized his findings in his (previously mentioned) monograph *Lectures on the theory of analytic functions.* All this took place long before the period I am describing. The reason I bring up these matters is that they were the bone of contention between Luzin and his students. The discovery of constructively generated sets other than Borel sets was sensational. The fact that the discovery was made by his students rather than by himself pained

Luzin enormously. He would rather ascribe the discovery to Lebesgue than to his students, which is what he did in the introduction to his monograph on A-sets.

Luzin had great mathematical intuition, the ability to anticipate the right orientation, and organizing talent. All these enabled him to create a school of great distinction. What he lacked, and what some of his students had, was breakthrough power. When he posed problems and guessed the nature of the relevant results he tended to think that he was also the author of the proofs actually given by some of his students. This led to massive misunderstandings. His most talented students rightly felt hurt. I did not witness all of these incidents, but, as editor of his works, I had to listen to proponents of both sides. The pro- and anti-Luzin camps formed long before I came to Mekh-Mat. The anti-Luzin camp was headed by Aleksandrov and Kolmogorov. The pro-Luzin camp included P. S. Novikov, A. A. Lyapunov, L. V. Keldysh, and Bari. The others felt that he was often wrong but tended to absolve him.

This is how I first realized that "something is rotten in the kingdom of Denmark". I am sorry to say that, in addition to this priority argument I have witnessed—then and later—quite a few others. I think that [many of them] share a common cause: sometimes a mathematician proves a theorem and presents its proof at a seminar or in a conversation. Such first proofs tend to be difficult and fuzzy. A listener may not at first understand the proof, but thinks about it intensely (he may have thought about the problem earlier) and manages to fix and simplify it. He may then sincerely believe that the proof is his, and a priority argument is likely to erupt. What surprises me about priority arguments is not their arising but their acrimony that sometimes leads to a complete break in social relations. Later I discovered that this, unfortunately, happens all over the world. More's the pity.

After this lengthy detour I return to descriptive function theory. This was its last period in Moscow (it is still of peripheral interest in the rest of the mathematical world). Many leading mathematicians (in our country, this was primarily P. S. Novikov) realized that the difficulties connected with the solution of the main open problems of this theory (such as the continuum hypothesis, and the question of measurability of projective sets) are connected with mathematical logic, and so turned to logic. What follows are my relevant recollections.

I attended P. S. Novkov's special course on descriptive function theory. This was a brilliant and substantial course. (As a curiosum I mention Novikov's extreme economy in the use of symbols. He managed with just "E" and "$\mathscr{E}$". On the other hand, Men′shov, say, used all letters of all known alphabets. Hence the amusing description (by A. L. Brudno) of the beginning of each of Men′shov's statements of theorems: "For every Latin letter there is a Greek letter such that . . .") At the end of the course we made arrangements for an examination. Novikov gave me an assignment and invited me to his house. By chance, we arrived at the house at the same time. When we got out of the elevator, Novikov opened the door, led me into the kitchen, and sat me down at the kitchen table. "Well, now," he said, but when I was about to talk of my assignment, he interrupted me and said that he would rather tell me of his new ideas on the continuum hypothesis and began to explain to me why he thought that the continuum hypothesis cannot be deduced from the axioms of set theory (its consistency had been proved earlier by Gödel). At the time, he was unable to turn his ideas into a proof. The proof was obtained much later (in 1963) by Cohen. Nevertheless, Novikov obtained a number of related results. For example, he proved the consistency of the assertion of the existence of a Lebesgue nonmeasurable set in the class of projective sets of the second class (a projective set of the second class is the projection of the complement of a plane A-set on a line), and the existence of the uncountable complement of an A-set without perfect core. Somewhat later he proved the unsolvability of the word problem for groups.

Constructive function theory was (and is) part of the theory of functions of a real variable. It has been cultivated in Russia—primarily in Leningrad (St. Petersburg)—from Chebyshev's time. When S. N. Bernshteĭn moved from Leningrad to Moscow he brought constructive function theory with him. As I mentioned earlier, Bernshteĭn was a brilliant figure from the past. In my time he was not seen at the university and the young regarded him as a museum rarity. Of course, there were other Moscow and MSU mathematicians who studied constructive function theory but they were not first-class mathematicians. I will not discuss the subject of constructive function theory in this essay.

A. A. Lyapunov was the last mathematician who actively studied descriptive set theory. He introduced the concepts of an R-set and of descriptively measurable sets. The latter are far more general than A-sets, but their Lebesgue measurability is inherent in their construction. The construction of these sets is extremely tedious. When Lyapunov presented this work at a seminar he used up the two hours of seminar time just defining the R-sets. During the week between two seminar sessions the listeners managed to forget the relevant definitions and another two hours would be needed to refresh their memories. This was the origin of the term "Lyapunov divergence" that was very popular for some time.

Lyapunov's works on descriptive set theory were among the last works in Moscow devoted to this topic. At the end of the period I am describing, he took up mathematical computing and programming.

Lyapunov was not only a mathematician but also a "questioner of nature", in the sense that he was interested in its overall structure. He contributed to biology and to geophysics.

Lyapunov was a man of the highest integrity. I'll illustrate this with just one example. In the late 1940s, when Lysenko made his revolution in Soviet biology, an attempt was made to start similar revolutions in other disciplines, whether or not they had anything to do with biology. This attempt was soon stopped by the physicists, who, building an atom bomb, felt their power. But before this, during a broadened assembly of the Mekh-Mat Research Council, one mathematician after another came forward and "confessed" his "sins". In particular, Kolmogorov renounced his genetics-related works on probability, and Khinchin published a paper in a slanderous anti-Einstein collection of papers. Of the leading Moscow mathematicians in some sense connected with genetics, only Lyapunov stood firm and remained steadfast. This did great harm to his career.

During the same period there developed a new branch of the theory of functions of a real variable. Its rise is linked with the name of A. S. Kronrod, Luzin's last student. Obviously, functions of many—mainly two—variables were studied before Kronrod, but they were closely linked to a given (orthogonal) coordinate system. The number of relevant concepts (the Tonelli variation, the exact and approximate differentials, contingents) was small and so was the number of established results. The most substantial of these results was probably V. V. Stepanov's theorem on exact and approximate differentials. During the period under consideration, and before the appearance of Kronrod's works, the greatest step forward in this area was I. Ya. Verchenko's theorem on the existence of an almost everywhere approximate tangent plane to the graph of a function of two variables with bounded area. Kronrod's papers completely revolutionized this area of mathematics.

After writing brilliant fundamental papers in the theory of functions of two variables, Kronrod abruptly discontinued this work and went over completely to mathematical computing and programming.

Kronrod was not only an eminent mathematician but also a charismatic figure. He attracted many young mathematicians. One of his distinctive characteristics was his absolute rejection of all authorities. This offended some "gurus", in particular, P. S. Aleksandrov. One indication of his great annoyance was that he referred to Kronrod's mathematical interests as "trifling". I have already stated my view that people change rather quickly. This was the case with Aleksandrov, who, within a few years' time, became one of the noblest figures in Moscow mathematical circles. He publicly acknowledged the importance of the approach developed by Kronrod. This took place at the meeting of the Research Council on the occasion of the defense of the doctoral dissertation of Kronrod's student A. G. Vitushkin. Aleksandrov wanted to shake hands with Kronrod but Kronrod rejected this attempt at reconciliation. And in vain.

I noted earlier that P. S. Aleksandrov belonged to the older generation of Luzin's students and that he discovered A-sets. In the early 1920s, Aleksandrov abandoned the theory of functions of a real variable in favor of general topology which was to become his permanent field of interest.

In the 1920s, a number of young Soviet mathematicians were sent abroad.

In addition to Aleksandrov, this group included P. S. Uryson, who perished tragically in 1924. (While swimming in the Atlantic in stormy weather, Uryson was tossed against a rock.) Like Aleksandrov, Uryson was primarily interested in general topology. In the few years before his untimely death Uryson succeeded in producing a number of first-rate works. Here it suffices to mention his research in dimension theory. Uryson's range of interests was very broad and he left his mark in a number of areas of mathematics. For example, he obtained the following sufficient condition of regularity of a boundary point in the case of the Dirichlet problem for the Laplace equation: A point in a space of dimension $n \geqslant 4$ is regular if it can be reached from the exterior of the domain by a "funnel" given in local coordinates by $\{x \in \mathbf{R}^n : 0 < x_n < h, (\sum_{i=1}^{n-1}) < x_n/|\ln x_n|^{1/(n-3)}\}$. Uryson obtained this convenient criterion—very close to the exact result—before the appearance of Wiener's criterion.

When I came to Mekh-Mat, P. S. Aleksandrov was already the long-acknowledged head of the topological school. His seminar—"the topological *kruzhok*"—included all Moscow topologists. Though Aleksandrov was a general topologist, this *kruzhok* was attended by various kinds of topologists, including combinatorial ones. The finest combinatorial topologist was L. S. Pontryagin. A word about Pontryagin. Pontryagin was blind but his phenomenal memory was a substitute for sight. He was a very gifted person. His achievements in topology are well known and truly remarkable. I had no close contact with him. I took his special course and attended his lectures at meetings at the MMS (the Moscow Mathematical Society). The clarity and precision of his presentation, delivered in a beautiful, sonorous baritone, were captivating. I know his works—in particular, his homotopy classification of the mappings of a polyhedron into a sphere—only from reports of my friends. But certain events had a disastrous effect on Pontryagin's mathematical fortunes (this is my own opinion and the opinion of some of my friends). Specifically, French mathematicians created a new theory of so-called fiber spaces that made possible simple solutions of problems that Pontryagin struggled with. His reaction was painful. He completely dropped topology and shifted to optimal control theory. There he discovered the much-touted "maximum principle". Of course, this could not even be compared with his achievements in topology. In his old age he became a fierce and vicious anti-Semite and stated his bias quite frankly in a lengthy article in an issue of the journal *Uspekhi Mat. Nauk* published on the occasion of his seventieth birthday.

Kolmogorov exerted a strong influence on the topological works of Aleksandrov's students as well as on all topology. Kolmogorov and Aleksandrov were close friends during most of the years they lived in the same dacha near Moscow.

Kolmogorov was extremely gifted. He was one of those scientists who grasp mathematics as a whole. For, in principle, the number of different ideas in mathematics is rather limited, and the impression of great variety is due to the different "languages" used to present the same idea in different branches of mathematics. I know only a few people who can see the inner connections in the whole of mathematics.

Kolmogorov's official position was that of the head of the department of probability theory. Probability and statistics were his narrow specialities. He made major contributions to mathematical logic, topology, mechanics (both classical and celestial), and the theory of functions of a real variable. He was also interested in linguistics (including applied linguistics), meteorology, and genetics. Kolmogorov is known for having created, among other things, the axiomatic theory of probability. His axioms are the basis of this theory to this day. "Law-abiding" mathematicians have no doubts about its correctness. Kolmogorov did. He was fully aware that the theory was an abstraction and was very much interested in how close it was to nature. His reasoning on the relation between the infinite and the large is of great interest. One of his lectures at the MMS dealt with this issue. (I have not come across his publications dealing with this topic.) He divided all numbers into "small", "medium", and "large". Small numbers are numbers we actually count (the first few thousand). Medium numbers are 2 to a small power. We cannot "count" them even by means of a computer but can imagine them as sets of distinct objects—say, the set of atoms in the visible part of the universe. A large number is 2 to a medium power. This is a pure abstraction hardly different from the actual infinite.

When I was a third-year undergraduate student I attended Kolmogorov's lectures in the course "Analysis III". One can get an approximate idea of its contents from the book by Kolmogorov and Fomin, titled *Elements of the theory of functions and of functional analysis*, based on this course.

There were some ten students in the course. This is because Kolmogorov talked fast, indistinctly, and "to the blackboard". By that time I had acquired a considerable knowledge of the theory of functions, and so found Kolmogorov's lectures very interesting. From time to time he would practically run along the board, bend to one side, and comment on some theorem, the comments directed to himself rather than to his listeners. This was the most interesting part of his lectures. For instance, he made some striking remarks connected with the construction of spaces of fractional dimension of grater and great homogeneity (an issue he was interested in in connection with problems of mechanics).

Kolmogorov grasped ideas remarkably fast. Here is an example. I was required to take the "minimal candidate examination" [preliminary exams] in the area of functions of a real variable and Kolmogorov was to be the examiner. At that time, the program of the examination was prepared by the student (the examiner was free to approve the program or not). I included in my program Whitney's theorem on the C^r extension of a function from a closed set to the whole space. The partition of unity was used for the first time in the proof of this theorem. The theorem appeared in the *Transactions of the AMS* in 1934 and Kolmogorov did not know it. (I noted earlier that American mathematical literature was not very popular in the USSR.) I knew the theorem because I used it in my work. To avoid superfluous explanations I included it in my program. Kolmogorov, as a curious man, asked me to talk about this very theorem. The examination went on for a few minutes. The proof of the Whitney theorem is rather difficult but he grasped it fully literally from just a few words. He was enthusiastic about the partition of unity.

It is said that M. V. Keldysh reasoned even faster than Kolmogorov. A. A. Lyapunov was another fast reasoner. In contrast, Men′shov and Petrovskiĭ reasoned slowly. What would often happen in the seminar of Men′shov, Bari, and P. S. Novikov was this: The lecturer at the board comes to a difficult point. There is a short pause. The first to speak up is Lyapunov—he has grasped it and begins to explain it to the others. He is gradually joined by all listeners with the single exception of Men′shov who sits motionless, cranes his neck, sticks out his chin, and blinks. Finally he exclaims in a thunderous bass "Aah, I've got it!" The lecture continues. Similar scenes occurred in Petrovskiĭ's seminar. True, sometimes the ending would be different: Petrovskiĭ fails to understand some point in the lecture. The lecturer and all the listeners explain the difficulty to him. Gradually the explanations peter out. After some time it turns out that the theorem being proved is false.

The most important scientific work at Mekh-Mat took place in seminars. The most popular seminar was that of I. M. Gel′fand.

Gel′fand's narrow specialty is considered to be functional analysis. However, like Kolmogorov, he takes a holistic view of mathematics and is well aware of its connections with physics (in the broad sense of the term). At a relatively recent meeting of the MMS in honour of Gel′fand's seventieth birthday, V. I. Arnol′d used the following image to compare the mathematical temperaments of Kolmogorov and Gel′fand: When Kolmogorov ends up in an unfamiliar mountainous locality he attempts to climb the highest peaks. Gel′fand begins with road construction. Having built the road he comfortably scales the peaks.

This is indeed so. Gel′fand's role as organizer of the work in mathematics and in its applications is very important. His immediate mathematical interests are closely related to questions of theoretical physics. In the period under consideration Gel′fand was also interested in computational mathematics in connection with problems in applied physics. He (and other mathematicians, such as Kronrod) invented the sweep method of solution of boundary value problems in differential equations. Later, Gel′fand closely studied the medical applications of mathematics, and, in particular, its applications to cardiology.

We now turn to Gel′fand's seminar. It was one of the "sights" of Mekh-

Mat. It always took place in a large (lecture) hall and the hall was sometimes full to overflowing.

In his seminar Gel′fand ignored niceties. He would sometimes take the chalk from the lecturer and present his own ideas. I remember that D. A. Raĭkov, a respected professor, brought with him to his lecture a chess clock and started it whenever Gel′fand interrupted him. It turned out that Raĭkov took up 10 minutes of his two-hour lecture and Gel′fand took up the rest of the time. Sometimes Gel′fand's "answer" to a question was "don't ask idiotic questions!"

I was used to the democratic seminars of Men′shov, Petrovskiĭ, and Kronrod in which all participants had equal say. Small wonder that what went on in Gel′fand's seminar struck me as strange. Nevertheless, this seminar was undoubtedly very interesting.

Gel′fand ardently followed all that was new in mathematics anywhere in the word and could unfailingly distinguish between the important and the unimportant. He was the first to grasp L. Schwartz's ideas on generalized functions. He organized a large team of workers and soon the well-known series of monographs was in print. He had a great many students and found work for all of them. He either supplied them with ideas or worked with them. The bibliography of his works contains hundreds of items co-authored with others.

Another seminar that determined the "face" of Mekh-Mat was undoubtedly the Kronrod seminar ("*kruzhok*"). Officially, this *kruzhok* was devoted to the theory of functions of a real variable, but, in fact, it dealt with a great variety of problems in mathematics and its applications. The participants were exclusively young people. Many mathematicians who later became mathematical leaders began their scientific work in this *kruzhok*.

My immediate teachers were Kronrod and Petrovskiĭ. They were very different. They belonged to different generations in the sense that Kronrod's career began when Petrovskiĭ's ended. Their characters differed sharply. Petrovskiĭ was gentle and tactful. Kronrod was far from tactful. When it came to mathematics they agreed that when embarking on the solution of a problem one should not be burdened beforehand by [too much] knowledge. True to his effervescent temperament, Kronrod loudly proclaimed "the evil of excess knowledge". Petrovskiĭ talked cautiously: "You see, what very often helped me solve problems was lack of knowledge. Had I known that people had unsuccessfully tried to solve a certain problem in the past, I certainly would not have attempted it." There are many detailed official biographies of Petrovskiĭ. Here I'll describe briefly my impressions from the time I got to know him and mention some of the things he told me about himself.

Petrovskiĭ was born in 1901 in the provincial city of Sevsk. His grandfather was a rich merchant ("merchant of the first guild"). Petrovskiĭ was his grandfather's favorite. The grandfather expected him to carry on his business. But already as a student in a science gymnasium, Petrovskiĭ, who was fortunate to have very good science teachers, got interested in the sciences. Upon graduation in 1917, just before the October revolution, Petrovskiĭ entered Moscow University where he intended to study biology and chemistry. When the revolution broke out, Petrovskiĭ was the first in his family to understand the danger the revolution held for his rich family. He returned to Sevsk and insisted that his family abandon its property and escape to the south. This is how they were saved. Then the family moved to Elizavetburg. For a few months Petrovskiĭ worked at odd jobs and then entered the local technical school for machine construction. There, for the first time, he came across a serious mathematics book and found it interesting.

Later he returned to Moscow and worked as a janitor. In 1922, as a proletarian, he entered Moscow University. After some wavering he settled on mathematics.

I already described Petrovskiĭ as intellectually deliberate. He preferred reading books to attending lectures. At the university he became a student of D. F. Egorov. Unlike many of Luzin's students, he was not brilliant. Nevertheless, Egorov prepared for him an intensive program and made him read many books in different areas of mathematics. As can be inferred from his bibliographies, the list of books he studied was rather impressive.

In spite of the fact that Egorov's name is associated with a theorem in the theory of functions of a real variable that is always paired off with Luzin's theorem, Egorov was not close to Luzin's school and gravitated toward analytic methods. The fact that Petrovskiĭ's initial training was analytic rather than set-theoretic manifested itself later as a kind of slight "accent". Nevertheless, he never told me that he regarded his work with Egorov as good analytic training. But he did say that when attending the seminars of Luzin's students he would sit in the back of the room and listen.

His first scientific assignment was his diploma work, done in Egorov's seminar. It dealt with a regularity condition of a boundary point for the Dirichlet problem in the plane, was closely related to the work of Luzin's student L. A. Lyusternik, and made up of set-theoretic and barrier techniques. Petrovskiĭ's result was not particularly impressive, all the more so because it could be embedded in Wiener's criterion (then—as I mentioned earlier—unknown in Moscow). Nevertheless, one could already discern in this work Petrovskiĭ's research potential. He eventually succeeded in independently formulating and proving the following result: The solution of Dirichlet's problem is completely determined by the values at the regular boundary points.

Petrovskiĭ's next paper (published in 1929) dealt with pure theory of functions of a real variable and was written in connection with his preparation for the post-graduate examination (he was Egorov's research student). In it Petrovskiĭ solved a problem posed by Lebesgue in his book *Integration and determination of primitive functions*. The paper was good but not distinguished. V. I. Glivenko, a student of Luzin, helped Petrovskiĭ prepare this paper for publication.

These were Petrovskiĭ's only student papers. All his later publications were mature works on a very high level. Petrovskiĭ did not live in an intellectual vacuum. His contacts with other mathematicians, in particular Kolmogorov, Bernshteĭn, and, especially, A. Ya. Khinchin, can be inferred from his choice of topics. This is especially true of some of his works in the initial period. Nevertheless, his methods and choice of material were entirely his own. The first in this series of papers, dating back to 1933, was a paper on algebraic geometry. It was followed by papers in probability theory. There is no doubt, however, that Petrovskiĭ's most significant papers deal with partial differential equations. He created a general theory of systems of partial differential equations, introduced the classes of hyperbolic, parabolic, and elliptic systems of such equations, posed the fundamental problems for these classes, and investigated their solvability. In spite of this, he gravitated toward the solution of concrete problems for concrete equations. Many years later, speaking of these systems in a session of his seminar, he exclaimed vexedly (quoting Gogol's Taras Bulba): "I gave them life and I'll . . . ." He stopped with bitterness, struck by the realization that they were no longer under his control.

During his most creative period, Petrovskiĭ celebrated each year with a new, first-rate work. This gave rise to the legend, then current at Mekh-Mat, that Petrovskiĭ chooses a problem in the spring, solves it in the summer, lectures on it in the fall, and publishes it in the winter.

I got to know Petrovskiĭ well when I became his student in 1947. He was a very interesting and versatile person marked by completely disinterested scientific curiosity. I should add that such scientific disinterestedness was, at the time, characteristic of many mathematicians. This attitude is illustrated by the following typical statement of Men′shov addressed to his students: "Don't expect from mathematical work any gain other than the satisfaction you get from its study."

To come back to Petrovskiĭ. He was curious not only about mathematics but also about other disciplines, both in the area of the sciences and of the humanities. During the last years of his life he conducted a joint seminar with the physicist I. M. Lifshits. He took part in archeological digs in Novgorod. Petrovskiĭ was an avid bibliophile. (After his death, his widow donated his splendid library to the university. It turned out that it contained 30,000 volumes.) He knew a great deal about painting. This was all the more remarkable because he was color blind.

Before meeting Petrovskiĭ I studied mainly the theory of functions of several real variables and some general topology. Petrovskiĭ was very tactful. He did not limit my freedom of choice but slipped me problems from the area of partial differential equations. As a result of his gentle direction I began to study the qualitative theory of partial differential equations. Later, my candidate and doctoral dissertations were both based on this theory.

To anticipate: I wish to note that subsequently Petrovskiĭ and I were to experience a major disappointment in connection with our joint work on the number of limit cycles of an ordinary differential equation with rational right-hand side. It was known that Dulac had proved the finiteness of this number, and we tried to obtain a good estimate for it. We thought that we had found such an estimate and published our result. Unfortunately, it contained an error. It turned out that Dulac's work also contained an error. The error in Dulac's work was found by my former student Yu. S. Il′yashenko, who also managed to show—quite recently—that Dulac's result is nevertheless true. (Il′yashenko presented this paper at the ICM in Kyoto in 1990.) It follows from Il′yashenko's proof that the tools used are clearly insufficient to attack this problem. The question of the number of cycles remains open to this day.

Notwithstanding the bitterness engendered by the mistake, I have fondest memories of my cooperation with Petrovskiĭ during those years. His enthusiasm was contagious. Every morning at eight the telephone would ring. Petrovskiĭ was calling to find out about possible new inventions in the last 24 hours and to share his ideas. It must be borne in mind that at the time he was the rector of the university and, as such, had to shoulder a tremendous administrative load. Now let me return to the period under consideration.

In spite of signs of discord in personal relations, the Moscow mathematicians of that period represented *en masse* an integrated and powerful group. They were unified by enthusiasm and scientific honesty. The MMS played a very important role in unifying their efforts. At that time the Society had still retained the old traditions that went back to its founder Brashman, and to Zhukovskiĭ, its president for many years. Even its meetings took place in the same auditorium. Current scientific work was done in seminars and the most significant results were presented during meetings of the Society. To be a member of the Society was a high honor indeed. A necessary requirement was the presentation by the candidate of two lectures dealing with his work at meetings of the Society. When it came to the choice of lectures, the Society's governing board was extremely captious. A candidate accepted for membership in the Society would get a personal letter from its President.

The unifying power of science was visibly demonstrated during the celebration of Men′shov's fifty-fifth birthday in 1947. It took place in A. A. Lyapunov's home. (Lyapunov had a large house in what was then the outskirts of Moscow.) Those present included all of the leading mathematics professors of Mekh-Mat and three young people, namely Kronrod, Lunts, and me. There was great merriment and tomfoolery. After all, most of the guests were only slightly over forty. Men′shov seemed a patriarch. But even he gave in to the general request and sang in his deep voice a humorous song abut a "little jar of venom". In those postwar years one tried to save electricity, and so the electric meters were fitted with control devices. Since all the lights were on, the device clicked and the lights went out. The Lyapunovs had a little dachshund. Someone—not one of us three young ones, and therefore one of the "important professors"—put the dachshund on the table. When the lights went on, everybody was delighted to see the little dog on the table contentedly licking the plates.

In the meantime a new generation was growing up. In the late 1940s and early 1950s a number of talented young mathematicians defended their doctoral dissertations. They included A. S. Kronrod, E. B. Dynkin, N. Ya. Vilenkin, M. I. Vishik, and M. I. Graev. Kronrod skipped the candidate degree. Dynkin's dissertation dealt with Lie groups and Lie algebras. In this work he introduced for the first time the "Dynkin diagrams" so popular today. In addition to these topics, Dynkin also studied probability and later made it the central topic of his investigations. He was an excellent teacher and even at that time had many students.

The university was soon to move from its old quarters on Mokhova Street, in the center of Moscow, to a new location in Lenin Hills. The move took place in 1953.

A new era began in the life of Mekh-Mat. It was to experience a new advance which materialized in the 1960s. There appeared a new generation of leading mathematicians such as V. I. Arnol′d, S. P. Novikov, Yu. I. Manin, and A. A. Kirillov. Now their average age is 50. They are older than most of those who sat at the table with the little dog. Unfortunately, so far there is no equivalent replacement for them in sight.

Translated by ABE SHENITZER
with the help of H. GRANT

Reminiscences of Soviet Mathematicians

B. A. ROSENFELD

J. N. Spielrein. Apart from my school teacher Tat′yana Yul′evna Eichenwald, who had earlier been a student of N. N. Luzin and member of his "Luzitania", I first encountered mathematicians when I was a student at the Moscow Energy Institute (MÈI), which I entered in 1935. The mathematics department at MÈI was headed by Jan [Yan] Nikolaevich Spielrein, a corresponding member of the Academy of Sciences and one of the founders of the institute. He taught higher mathematics in our "assembly line". MÈI arose from the merger of the Electrical Engineering Faculty of Moscow Higher Technical College (MVTU—now MGTU, Moscow State Technical University) and one of the faculties of the Institute for National Economy at the initiative of the electrical engineer Karl Adol′fovich Krug and Spielrein, and was located in the buildings earlier belonging to these institutes. Spielrein also founded the Electrophysical Faculty of MÈI, the faculty with the highest degree of mathematical training, and subsequently separated into the Electrovacuum and Radio Faculties. Spielrein had studied mathematics at the Sorbonne and electrical engineering at the Technische Hochschule in Darmstadt, worked for a while in Germany and published a thick textbook on the vector calculus while he was there. This book contained a detailed presentation of the theory of vector and tensor fields, large parts of which had been worked out by the author himself, and many applications of this theory in physics and engineering. In the USSR Jan Nikolaevich began to publish a Russian version of this book, but only succeeded in finishing the first part of "Vector calculus for physicists and engineers", containing vector algebra. In the 1930s the vector presentation was just beginning to force out the coordinate presentation in teaching analytic geometry, theoretical (rational) mechanics and electrical engineering. Jan Nikolaevich was a passionate defender of vector methods and ruthlessly ridiculed the solution of geometric, physical, and engineering problems by coordinate methods, and at the same time he also propagandized linear operators, which he called "affinors". The students loved Jan Nikolaevich and those in his circle called him "Spielvector". At his lectures I learned a lot that did not figure in the usual program for technical universities, such as the differential geometry of lines and surfaces, differential equations and the application of vector diagrams and Heaviside's method to solving them, and different ways of integrating. Besides the usual two forms of products of vectors, the scalar (inner) product ab and the vector (cross) product $[ab]$, Jan Nikolaevich also defined the tensor product $a \cdot b$, which is a linear operator, and the result of the action of this operator on a vector x is a vector $a \cdot bx$. Instead of the form $[ab]$ he preferred to write the vector product in the form ab, where a is considered as a skew-symmetric linear operator. I was completely charmed by this man. In 1938 Jan Nikolaevich was suddenly arrested and the internal newspaper of the institute printed an article in which it was stated that the "fascinatingly popular Spielrein" had been "unmasked" as a "well concealed enemy of the people". Apparently, his foreign studies and work played a role. I never heard anything more about him. In my student years Spielrein was already not a young man and could hardly have lived very long in prison.

Others who taught me at MÈI were the algebraists Vladimir Konstantinovich Turkin and Petr Evgen′evich Dubuque.

A. P. Kotel′nikov. The mathematics department was located in a single room with a "brother department", as Spielrein termed it, the department of theoretical mechanics. The most brilliant figure in this "brother department" was not its head, Moiseĭ Izrailevich Besprozvannyĭ, but the unpretentious professor Aleksandr Petrovich Kotel′nikov (1865–1944). Kotel′nikov was the son of the Kazan University professor Petr Ivanovich Kotel′nikov (1809–1879), a collaborator of Lobachevskiĭ, who was the first mathematician to publicly defend the non-Euclidean geometry of Lobachevskiĭ in 1842. A. P. Kotel′nikov's son Vladimir Aleksandrovich is an important electrical engineer and Academician, and one of the founders of information theory. My friends who were senior students told me about Kotel′nikov's books *Screw calculus* and *Projective theory of vectors.* I borrowed these books from the library. It turned out that they were his Master's and Doctoral dissertations defended in Kazan in 1895 and 1899. In the first of them he considered vectors whose coordinates are "parabolic complex numbers" $a + b\varepsilon$, $\varepsilon^2 = 0$ (such numbers are now called "dual numbers"), and in the second he considered vectors whose coordinates are ordinary complex numbers and "elliptic complex numbers" $a + be$, $e^2 = +1$ (such numbers are now called "split or complex numbers" or "double numbers"). Kotel′nikov called such vectors "screws" of Euclidean, hyperbolic (Lobachevskiĭ) and elliptic (Riemann) spaces, respectively, and he called the vector products of these complex vectors "screw products". If the usual vectors together with their vector product form the Lie algebra of the group of rotations of Euclidean space, then the screws with their screw products form the Lie algebras of the groups of motions, respectively, of Euclidean, hyperbolic and elliptic space. The "Kotel′nikov interpretations" of the manifolds of lines in these three spaces as spheres in Euclidean space whose coordinates are complex numbers of the above-mentioned three types are based on this. Using screws, in his doctoral dissertation Kotel′nikov constructed a statics in hyperbolic and elliptic spaces. At that time I had not yet studied non-Euclidean geometries, and a very strong impression was made on me by the "parabolic complex numbers", possessing the property that for an analytic real function $f(x)$ the same function of a dual argument has the form $f(x+y\varepsilon) = f(x) + yf'(x)$. I was particularly pleased that the derivative $f'(x)$ of the function $f(x)$ appeared here, from which I deduced that using these numbers one could found analysis without a theory of limits (at that time I did not know about the theory of functions of a real variable). I went to A. P. Kotel′nikov with some question about *Screw calculus* and Aleksandr Petrovich was extremely happy that this work of his, written in the previous century, had found a contemporary reader. Subsequently I worked together with Kotel′nikov at MVTU. Later I wrote first a paper and then, with three co-authors, a book about his life and work. My co-authors T. B. Putyata, B. L. Laptev and B. N. Fradlin complemented the information that I had with information about the Kazan and Kiev periods in Kotel′nikov's life and about his works in mechanics. In addition to MVTU and MÈI, Kotel′nikov worked at the Central Aerohydrodynamics Institute, where he edited the collected works of the institute's founder, N. E. Zhukovskiĭ (Joukowsky), and also participated actively in the editing and writing of commentaries on the works of Lobachevskiĭ. Kotel′nikov's interpretations were applied by the opponent of both his dissertations, D. N. Zeĭliger (Seeliger), and other geometers to the differential geometry of lines of Euclidean and non-Euclidean spaces, and dual numbers have been applied in geometry and homological algebra to define spaces, groups and algebras obtained by passage to a limit.

G. M. Shapiro. When I was a first-year student at MÈI, at the beginning of the spring semester, I was at a concert in the Moscow State University (MGU) building, in which the Mechanics-Mathematics Faculty ("Mekh-Mat") was located. During the intermission I decided to go to the third floor of this building to see what Mekh-Mat looked like. There I found out that on February 25 docent G. M. Shapiro was going to start a course on tensor analysis. The acquaintance with vectors and affinors in the previous semester had brought me close to tensors; without tensors it was impossible to look into the theory of relativity, in which I was then very interested, but I could not understand from the books what it was. Therefore I joined this course and began to study it together with several upper-class students from Mekh-Mat. Genrikh (Heinrich) Mikhaĭlovich Shapiro (1903–1942) was a student of V. F. Kagan. He worked as a docent at Mekh-Mat and simultaneously was a professor at Moscow State Pedagogical Institute (now the Pedagogical University), where he taught algebra. In his textbook *Course in higher algebra* I read the very astonishing news that Omar Khayyam, known to me up to then only as a poet, was the author of a treatise on cubic equations. Others who took this course together with me were A. L. Zel′manov, later a well-known specialist in relativistic cosmology, and I. M. Gul′, the author of a book about Lobachevskian geometry. In June we had exams; the university students received their grades in record-books, and Genrikh Mikhaĭlovich gave me a "certificate" that "the MÈI first-year student B. A. Rosenfeld passed the course on tensor analysis with a grade of 'distinguished' ".

At the beginning of the war Shapiro volunteered for the National Guard and participated in the heavy combats west of Moscow. He was demobilized in January 1942 because of his health, found his family in evaculation in Kuĭbyshev (Samara), and began pedagogical activity there, but in August of that year, completely exhausted, he suddenly died.

V. V. Stepanov. After my successful experiment with tensor analysis I decided to expand this experiment to other courses at Mekh-Mat. At that time Mekh-Mat worked in two sessions. The first- and second-year students took their classes during the first session, when I was busy at the MÈI, and, during the second session, when I was free, the more advanced students had their classes. I asked Spielrein what books I needed to study in order to attend third-year lectures at the University. He advised me to study analytic geometry from G. Salmon's books and mathematical analysis from the *Cours* of É. Goursat. Salmon's textbooks had been translated into Russian in the 1890s. The role of vectors was played in those textbooks by the now almost forgotten "method of truncated notations"; Goursat's *Cours* had been translated at the beginning of the 1930s under the editorship of V. V. Stepanov, but by that time it had been supplanted in the universities by the *Cours* of de la Vallée-Poussin. Of course, I knew nothing of this and, having conscientiously studied these old-fashioned textbooks, in September I began to attend the second part of the course on the integration of differential equations (IntDE) given by that very same Vyacheslav Vasil′evich Stepanov (1880–1950) who had edited the Goursat translation, I. I. Privalov's course on the function theory of one complex variable, and other third-year lectures. At that time Privalov's textbook had just appeared; he gave his lectures by following this book exactly, so that the students did not have to take notes during the lectures, but only marked the readings in the book. Stepanov's textbook had appeared at that time only in the form of lithographed notes from the Correspondence Division of the university. Stepanov did not repeat what was in the book, but led lively discussions with the students, alternating his jokes and proverbs "as we say in Smolensk". When I told him that I wanted to study simultaneoudly at the MÈI and at Moscow State University, he was pleased and promised me that if I successfully passed "IntDE" he would also accept the course in analysis for the first two years from me, but, of course, using de la Vallée-Poussin, not Goursat. I passed these courses at the beginning of 1937.

V. V. Stepanov played a fundamental role in my becoming a mathematician. In 1939, when I had taken all the mathematics courses at Mekh-Mat (I took the humanities courses at MÈI) and the State exams and had handed in my documents to become a graduate student, Vyacheslav Vasil′evich, who at that time was Director of the Scientific Study Institute in Mekh-Mat, insisted upon my acceptance as a graduate student and looked after me like a father during the whole course of my graduate studies.

Stepanov, who was later a Corresponding Member of the Academy of Sciences, was a widely educated mathematician. He attended all the lectures at sessions of the Moscow Mathematical Society and was an active participant in discussions there, including things that were then, in the 1930s, little known. As a consequence I got to know his wife Yulia Antonovna Rozhanskaya, who was a student of N. N. Luzin and a friend of my school teacher T. Yu. Eichenwald.

V. F. Kagan. At this time at Mekh-Mat the "Seminar on vector and tensor analysis" was in regular operation, founded and supervised by Veniamin (Benjamin) Fedorovich Kagan (1869–1953). Since I loved vectors and tensors, I naturally started to attend all of the advanced courses and seminars of Kagan and his students, and then also the "big seminar". My earliest scientific activity was associated with Kagan's school. For many years Kagan worked at Odessa University, where in 1900 his textbook on Lobachevskian geometry appeared, and in 1905–1907 his two-volume monograph *Foundations of geometry*, which he defended as his Master's thesis. In this book he presented his famous axioms of Euclidean geometry, based on the concept of distance. The scientific publisher "Mathesis" that he directed, published many Russian translations of recent mathematical literature in the pre-Revolutionary period. Therefore, after the organization of the USSR State Press in 1923, Kagan was invited to Moscow to direct the mathematics editorial section. From that time on he worked at MGU, where he organized two new sections ("*kafedra*", German "Lehrstuhl"), the section of differential geometry at Mekh-Mat and the section of mathematics at Fiz-Fak (the Physics Department at MGU); until the war he was chairman of both sections, and of the first of them until his death (now S. P. Novikov's section is organized on the basis of the differential geometry section). Already in Odessa, Kagan had written a book on the theory of relativity and began to be engaged with tensor differential geometry, while in Moscow he organized a new scientific school, aimed at developing this domain of geometry. In 1927 he organized the "Seminar on vector and tensor analysis and its applications to geometry, mechanics and physics". Kagan was a deputy of the Moscow Soviet; I was present at one of his reports to voters. After 1933 under his editorship, the Proceedings (Trudy) of his Seminar began to be published (until the war these Proceedings were published in foreign languages). In 1934 the Moscow International Conference on Tensor Differential Geometry took place, which he had organized. During the 1930s Kagan created the theory of subprojective spaces. These are Riemannian spaces which are the closest generalizations of spaces of constant curvature, i.e., non-Euclidean spaces. In later years he wrote two monographs, *Foundations of the theory of surfaces in tensorial presentation and foundations of geometry*, substantially different from the book of the same title that he had written half a century before. I often spoke with Veniamin Fedorovich, visited him at home, and also knew his grandchildren when they were young; now they are well-known teachers, the physicist Grigoriĭ Isaakovich Barenblat and the mathematician Yakov Grigor′evich Sinaĭ. Veniamin Fedorovich was the opponent at the defense of my Candidate's (Ph.D.) dissertation.

P. S. Aleksandrov. The President of the Moscow Mathematical Society at that time was Pavel Sergeevich Aleksandrov [Alexandroff] (1896–1982), considered the leader of Soviet mathematicians. A student of N. N. Luzin, Aleksandrov, together with his friend Pavel Semënovich Uryson (1898–1924), founded the important Soviet school of topology. I took some of Pavel Sergeevich's courses in topology. In his lectures I felt like I was at the "leading edge" of the world of mathematics, an impression that was strengthened by the distinctive timbre of his voice. Perhaps this was because, it was said, that at the time of the Civil War Aleksandrov was an artist and even a major theatrical producer. Aleksandrov was a Corresponding Member of the Academy of Sciences after 1929 and a Full Member after 1953. In 1969 he was given the title of a Hero of Socialist Labor. In his autobiography, published in "Uspekhi Mat. Nauk", Pavel Sergeevich told about his failed marriage to Ekaterina Romanovna Èiges, the sister of his high-school (gymnasium) mathematics teacher, and that she got to know him with his friend, the well-known lyric poet Sergeĭ Esenin. I have learned from "Eseninophiles" that Esenin's famous poem "Letter to a Woman" was dedicated to Ekaterina Èiges and that the phrase "You live with a serious, clever man" refers to Pavel Sergeevich. The cordial relationship that Aleksandrov had with the memory of Esenin is shown by the fact that he protected the son that Esenin had with the poetess Nadezhda Davydovna Vol′pin, Aleksandr Vol′pin, who had studied topology under Pavel Sergeevich's direction, and in order for him to become a graduate student, named him "Esenin-Vol′pin" in all the documents. As a result A. S. Vol′pin became a well-known specialist in mathematical logic (which he learned while he was in exile in Kazakhstan). At the present time he is a professor in Buffalo, New York.

I also encountered Pavel Sergeevich after I graduated from university and graduate school. He presented some of my papers to the "Doklady" of the USSR Academy of Sciences. Aleksandrov was a great lover of aquatic sports. I last met him on the Courland Spit (Kursu Neringa) in Lithuania, where he was recovering from injuries suffered when a motorboat passed over him while he was swimming under water.

A. N. Kolmogorov. Another great mathematician whose lectures I attended when I was a student was Andreĭ Nikolaevich Kolmogorov (1903–1987). Kolmogorov presented a course on probability theory. He had transformed this area of mathematics, which until that time had been a "side-branch" of mathematics, by showing that probability theory could essentially be considered as a constituent part of the theory of functions of a real variable, but the student course that he gave each year was no longer of any interest to him. In a completely different vein he gave advanced courses in which he presented his recently obtained results. Kolmogorov worked in almost all disciplines of mathematics; along the way he obtained two significant results in the foundations of geometry. He was made a Full Member of the Academy of Sciences in 1939 and in 1963 was given the title of a Hero of Socialist Labor.

In his youth Kolmogorov, like his friend P. S. Aleksandrov, was a great lover of sports. Their friendship began in 1929, when they made a joint trip along the Volga and in the Caucasus.

Andreĭ Nikolaevich presented many of my papers to the "Doklady" and the "Izvestiya" of the Academy of Sciences, the first before the war and the last in 1983.

In his last years Andreĭ Nikolaevich focussed his attention on mathematics teaching in schools. He organized a remarkable boarding school in which he himself taught geometry and issued several new school textbooks, which significantly raised the level of mathematics teaching in the entire country. Kolmogorov's school textbooks are different in an essential way from earlier ones, for which Kolmogorov and his coworkers faced persecution on the part of functionaries from the Ministry of Education. This led Kolmogorov to Parkinson's disease and significantly shortened his life.

N. N. Luzin. Aleksandrov's and Kolmogorov's teacher, the founder of the Moscow school of set theory, Nikolaĭ Nikolaevich Luzin (1883–1950), was no longer at Mekh-Mat when I was a student, and I saw him only once. In 1936, when I attended the lectures of G. M. Shapiro, there was a vociferous campaign at Mekh-Mat for the banishment of Luzin from the University and the Academy, analogous to the "stuggle with cosmopolitanism" in 1948. Luzin was accused of all the deadly sins. My oldest friends told me about Luzin, that he gave lectures like an artist, that at the time of the "Luzitania" he was called "the great god, Professor Luzin". They told me that he was elected to be an Academician in 1929, not just as a mathematician, but also as a philosopher, and only then joined as a member of the mathematics division.

I saw Luzin in 1945 at the festivities in honor of the 220th anniversary of the Academy of Sciences. I was late to the session he addressed, since it took me a long time to get there from the war department, where I served at that time. I entered the room while a man whom I had never before seen was speaking. He spoke in French, but with a horrible Russian accent. It turned out that this was N. N. Luzin. I know that he had lived in France for many years and could not believe that he spoke French with such a terrible accent, but nevertheless it was he.

Luzin did not have good relations with his students Aleksandrov and Kolmogorov. While Luzin was alive, Aleksandrov could not be elected to the Academy, despite all of Kolmogorov's efforts. At the time of one of the votes for Aleksandrov's admission to the Academy, Luzin promised Kolmogorov that he would vote "For", but Kolmogorov found out that he had voted "Against", for which Kolmogorov publicly slapped him in the face.

L. S. Pontryagin. When I was a student and a graduate student, I took the important course "Continuous Groups" and other courses given by Lev Semënovich Pontryagin (1908–1988). Blinded at the age of 14, Pontryagin under Aleksandrov's guidance was able to become one of the most famous mathematicians in the world. At his lectures Pontryagin asked one of the students to write on the blackboard while he kept the entire lecture in his mind (he remembered what letter had been used to denote some mathematical concept in the previous lecture; the only thing he did not remember was whether that letter had been a capital letter or a small letter). In the years before the war Pontryagin was a very pleasant man, for the antisemitism that he became famous for in the last years of his life had not completely surfaced. I often chatted with him at that time. He told me that he had been named Lev in honor of Lev Tolstoĭ, who had celebrated his eightieth birthday within a few days of his own birth, and that his ancestors' family name was originally "Portnyagin" (*portnyaga*=taylor in Russian), but they decided to change it to make it more euphonious. He told me: "You can be proud of the fact that you finished Mekh-Mat while simultaneously attending a Technical University, but in fact you have lost very much. If you had not wasted time in technical subjects, but had studied mathematics during that time, you would have been a much better mathematician." When, after I had already passed my Candidate's (Ph.D.) degree, I attempted to join the group of students working for a Doctor's degree (doctoranture) at the Mathematics Institute of the Academy, Pontryagin said that I should not be accepted, since the work that I presented was not adequate. And when I argued that if I had had stronger work, I would have defended it as a Doctoral dissertation and it would not be necessary to admit me to the doctoranture, Pontryagin said: "You don't understand at all why doctoranture is needed. It is not at all needed in order to prepare a Doctoral dissertation. It is needed in order to prepare mathematicians of the caliber of Khristianovich, but not those of the caliber of Piskunov." (Piskunov, the author of several textbooks for technical universities, was a professor in one of the military academies after finishing his doctoranture. Khristianovich became an Academician.) At that time I felt very hurt by Pontryagin, but afterwards understood that in both these cases he was right.

A. G. Kurosh. I took the first-year higher algebra course of Aleksandr Gennadievich Kurosh (1908–1971) shortly after I took Stepanov's mathematical analysis. However, I attended his advanced courses and participated in his seminar on lattice theory (which he, like Garrett Birkhoff, called "structures"). I also visited him in his pre-war apartment, in a little wooden house in the "Sokol" ("Falcon") settlement, and in a new one, in one of the buildings of the new Moscow University edifice.

I was also well acquainted with Kurosh's wife Zoya Mikhaĭlovna Kishkina (1917–1987), who studied at Mekh-Mat one year after me, and after defending her Candidate's dissertation worked, herself, as a docent in the algebra department at Mekh-Mat.

When I took the candidate's oral exam, the committee included Kurosh, and I could not answer his question: "Give an example of a noncommutative field", to which he said: "Well, I guess that it will never be necessary for you to deal with the quaternions." And, as they say, "like a reflection in water", almost all my life I have had to deal with the quaternions and their generalizations.

Kurosh was the author of many textbooks in different branches of algebra, translated into many languages and often cited. Therefore he was of course interested in the "index of citability" as an estimate of a mathematician's significance, and gave several lectures on this theme in various auditoria, from which it appeared that the highest rating was his.

Kurosh died of a heart attack brought on by the unpleasantness that followed his signing of the well-known letter of 99 Soviet mathematicians in defense of A. S. Esenin-Vol′pin.

B. N. Delone (Delaunay). At that time analytic geometry was taught at Mekh-Mat by Sergeĭ Sergeevich Byushgens (1862–1963) and Boris Nikolaevich Delone (1890–1980). Both were authors of textbooks. Byushgens, a differential geometer, wrote a textbook in the classical style. Delone, a specialist in number theory, was led to geometry through the geometry of numbers and crystallography, in which he made important discoveries, and wrote a very original textbook using vectors and affine transformations. The students were very amused when he called the formula for the double vector product $[a[bc]] = b \cdot ac - c \cdot ab$ "*bats* ('band') minus *tsab* (scratch)",[1] and affine transformations were demonstrated using the example of transformations of a picture of a kitten.

I passed Byushgen's exam on analytic geometry, which was on material very close to that which I had studied from Salmon's books. But I often spoke with Delone on various occasions (usually he invited me to his home on the Bol′shaya Polyanka and I returned home to Arbat from there on foot late at night). Boris Nikolaevich told me a lot about his life. His ancestor Delaunay was a brother of the last commander of the Bastille, who was killed at the time of the French Revolution. The son of this Delaunay was a surgeon in Napoleon's army at the time of the latter's Russian invasion, was taken prisoner in Russia and married into the Tukhachevskiĭ family of the Smolensk nobility (so that B. N. Delone was related to the famous field-marshal). A

[1] *Editor's note*: Russian pronounciation of *bac*, *cab*.

B. N. DELONE

portrait of one of his ancestors hung in Boris Nikolaevich's house. When I asked him, "Who is this old man?", he answered me, "Well, this 'old man' is some marquis."

B. N. Delone's father Nikolaĭ Borisovich was a professor of mechanics at the Kiev Commercial Institute (where my father was a student). Boris Nikolaevich told me that at the time of the Civil War, when Kiev was in the hands of the Whites, his father was imprisoned because he gave lectures to workers when Kiev was under Soviet rule. Boris Nikolaevich's mother liberated him from prison: she found out that the wife of the commander of Kiev, General Dragomirov, was a friend of hers from the Institute of Noble Girls and told her about his misfortune. As a result Dragomirov sent to the prison a soldier with an invitation for Professor Delone to dinner.

Boris Nikolaevich told me a lot about his friend Yakov Viktorovich Uspenskiĭ (1883–1947), a Leningrad (St. Petersburg) mathematician and Academician, who married on one of his trips abroad, brought his wife back to Leningrad, and since life in Leningrad did not please her, went abroad with her again. Afterwards I learned that, after leaving Leningrad, Uspenskiĭ lived in the USA and was a very esteemed professor at Stanford University.

Boris Nikolaevich was one of the opponents at the defense of my Doctoral dissertation. When I brought him the dissertation, he took a Doctor of Science diploma from a drawer of the writing table and said: "You will get this after the defense"; then he took a professor's certificate and said: "You will get this soon after"; then he took the document of a Corresponding Member of the Academy of Sciences and said: "You will hardly get this", and then he showed me a small book and said: "And this I prize above all!" It was his book *Guidebook to the mountains of the western caucasus*. Like his student, the head of the Soviet differential-geometry school of "geometry in the large", Aleksandr Danilovich Aleksandrov (born 1912), Boris Nikolaevich was a well-known alpinist.

A. P. Norden. My first research supervisor was V. F. Kagan's student Aleksandr Petrovich Norden (born 1904). At that time Norden was a docent in Kagan's department at Fiz-Fak. At Mekh-Mat he led a seminar and gave an advanced course "Geometry of Spaces of Lines", in which he studied the geometry of manifolds of straight lines in three-dimensional non-Euclidean spaces, elliptic and hyperbolic. Here I again encountered the Kotel′nikov interpretations and other interpretations of these manifolds, but by this time, having taken Kagan's course on the foundations of geometry and Norden's advanced course, I now had a good understanding of non-Euclidean geometries. In fact, all the themes of both my dissertations (Ph.D. and D.Sc.) arose from this course. Norden was a splendid research supervisor and throughout his long life he has inspired a great many geometers.

Norden had an extraordinarily interesting genealogy. One of his ancestors was the famous "Moor of Peter the Great", Abram Petrovich Gannibal, one of whose daughters was married off to the Swedish Baron August Nordén.

Aleksandr Petrovich's closest ancestors were landowners in Saratov Province. For this reason Norden as a "social outcast" was not wanted in a university, but after he had worked in one of Saratov's factories and had obtained "proletarian training", he was given the chance to learn.

In 1945, after Petr Alekseevich Shirokov (1895–1944), the head of the geometry department at Kazan University and one of the founders of tensor differential geometry in the USSR, died, Norden was invited to head this department and he moved to Kazan, where he still lives. He headed the department for many years, but now it is chaired by Aleksandr Petrovich Shirokov (born 1926), the son of P. A. Shirokov. Under Norden's leadership the Kazan geometry school, created by P. A. Shirokov, became one of the most prominent Soviet scientific schools. Norden succeeded in applying tensorial methods to projective differential geometry and conformal differential geometry. He is the author of several university-level textbooks and monographs, the most important of which is *Spaces with an affine connection*.

Since the theme of my works, as said, came from Nordan, I also include myself in this school.

P. K. Rashevskiĭ. While I was a graduate student my supervisor was Petr Konstantinovich Rashevskiĭ (1907–1983), a very strong student of Kagan's. Rashevskiĭ directed Kagan's department after the death of S. P. Finikov, managing it from 1953 to 1964. The son of the author of well-known high-school textbooks on mathematics Konstantin Nikolaevich Rashevskiĭ (1874–1956), Petr Konstantinovich became a professor at Moscow University as early as 1934. Like Norden, he was the author of several university textbooks, for courses in differential geometry, in Riemannian geometry and tensor analysis, and in the geometric theory of partial differential equations. Both Rashevskiĭ's lectures and his books are distinguished by their exceptional clarity. His Doctoral dissertation was devoted to "polymetric geometry". Rashevskiĭ was the first to consider reductive spaces, i.e., spaces with an affine connection, for which both the curvature tensor and the torsion tensor are covariant constant. Rashevskiĭ called these spaces "symmetric spaces with torsion" (the term "reductive spaces" was introduced by K. Nomizu, who discovered them independently of Rashevskiĭ).

However Rashevskiĭ's main object was to use tensor analysis as a foundation for the creation of a new mathematical apparatus for quantum physics. This problem apparently arose as a consequence of the successful application of tensor analysis to the general theory of relativity. Rashevskiĭ worked on the creation of this apparatus for many years and spent the best years of his life on it. But when the work was complete and published as a large paper in "Uspekhi Mat. Nauk", physicists said that they did not need it, and that the apparatus they already had was completely established. This was such a shock that Rashevskiĭ was never able to recover.

I did not live far from Petr Konstantinovich, and often visited him. Many times we talked while walking from his Malyĭ Levshinskiĭ Lane to my

Krivoarbatskiĭ Lane, going all the way to my door, but not once did I succeed in persuading him to come in. Rashevskiĭ never wanted to enter as a guest. The only exception that I know of occurred in 1949, when Kagan's eightieth birthday was celebrated, and Rashevskiĭ went to Veniamin Fedorovich's house. There one found two tables and Petr Konstantinovich, very merry, going around both tables and clinking glasses with all the guests, each of whom assumed that Rashevskiĭ was seated at the other table, when in fact he was seated at neither of the tables.

Rashevskiĭ was married to Natal′ya Mikhaĭlovna Arnol′d, a first cousin of the well-known mathematician Igor Vladimirovich Arnol′d (1900–1948), the father of Academician V. I. Arnol′d. The house where the Rashevskiĭ lived had a façade on Kropotkin Street (Prechistenka), dating from when it was the manor house of a country estate. The Rashevskiĭs' appartment was in the courtyard and consisted of two rooms which had belonged to one of the servants and a corridor joining these rooms with the house. Rashevskiĭ's study was in this corridor, with drafts from all sides. When the new Moscow University building and several apartment buildings for professors were built in 1953, Rashevskiĭ was offered an apartment in one of these buildings, but he refused it. And only later, when both his daughers were married, did he move to a new apartment in Volgin Street.

While I was a graduate student, Rashevskiĭ told me several times that the papers of mine that were to be published and which extended my previous topic, inspired by A. P. Norden, were not a Candidate's dissertation; but when I asked him what topic would be a dissertation, he answered: "Read my Doctoral dissertation." I did not obey him, and, continuing my earlier work, wrote both a Candidate's and a Doctoral dissertation on that topic, as well as several books. I did not read Rashevskiĭ's Doctoral dissertation until I was already a professor.

When I embarked on my third year of graduate study, the war began. Moscow University was evacuated to Ashkhabad (Turkmenistan), where I defended my Candidate's dissertation. At that time Rashevskiĭ had been evaculated to Tomsk, and I was left without a supervisor. The only geometers in Ashkhabad were Kagan and his student Ya. S. Dubnov, who became my opponents.

In Tomsk Petr Konstantinovich found his very talented student Nikolaĭ Nikolaevich Yanenko (1921–1984). Having returned from the army to Moscow, Yanenko entered graduate study with Petr Konstantinovich as a supervisor, and wrote Candidate's and Doctoral dissertations in differential geometry. He worked first at the Mathematics Institute of the Academy of Sciences, where he got interested in numerical mathematics; then he moved to Novosibirsk, where he became the director of one of the institutes of the Siberian Division of the Academy of Sciences, an Academician, and a Hero of Socialist Labor. When Yanenko was asked why he lost interest in geometry, he answered: "But could I have become an Academician and a Hero if I had continued to be interested in geometry?"

Rashevskiĭ was the first opponent at the defense of my Doctoral dissertation. When I had completed work on it and many geometers stated that the results that I had obtained were sufficient for a defense, Rashevskiĭ did not agree with them and required applications of my general results in the geometry of manifolds of "subspaces" ("figures of symmetry") to the differential geometry of families of these figures. And only when these applications had been obtained would he agree to be my opponent—now I was able not to fear the outcome of the defense.

Ya. S. Dubnov. The second opponent at the defense of my Candidate's dissertation was Yakov Semkënovich Dubnov (1887–1957). He was the son of the well-known historian Semën Markovich Dubnov (1860–1941), the author of *The history of the Jewish nation*, who lived before the war in Riga and was killed there by the Germans. Dubnov had already been a student of Kagan's in Odessa. Dubnov's lectures, his *Foundations of the vector calculus*, were models of methodology. I very much liked his theory of "tensors with vector components" and his tensor theory of line congruences (2-parameter families of lines in a 3-dimensional space). At the beginning of my study at Mekh-Mat I took his course on differential geometry (which I consequently read many times).

Dubnov died in Saratov while visiting V. V. Wagner, where he had gone the day before his seventieth birthday, in order to avoid the celebration in Moscow.

V. V. Wagner. Viktor Vladimirovich Wagner (1908–1981) was appointed as the third opponent at the defense of my Doctoral dissertation, but knowing of his weak health, I was afraid that he might not travel from Saratov; therefore I asked that another Moscow opponent be appointed for me, and this turned out to be B. N. Delone.

Wagner's father was a Germain from along the Volga River, a railroad machinist by trade, while his mother was a Czech of the Vondráček family, not Catholic like most Czechs, but Orthodox. In this way Wagner avoided deportation at the beginning of the war with all the Germans: when it came for the family to be evicted, his mother showed his birth certificate in the Orthodox Church, from which it followed that he was not German but Russian, and they were left in peace.

In 1934, when Wagner defended his Candidate's dissertation on nonholonomic geometry, his supervisor V. F. Kagan asked the Dutch geometer Jan Schouten, then visiting the USSR, to be Wagner's opponent. The dissertation impressed Schouten so much that he recommended that Wagner be awarded a Doctor's degree, not a Candidate's. This recommendation was accepted by the Academic Council, but he was confirmed in this degree only in 1938. In 1937 Wagner had been awarded the Lobachevskiĭ prize for young researchers in Kazan; at the time the main prize was awarded to Élie Cartan (1869–1951). After receiving the title of professor, Wagner headed the geometry department at Saratov University for many years.

Wagner's work turned out to have a very great influence on me, particularly his work on congruences of planes in higher-dimensional spaces (a congruence of m-dimensional planes in an n-dimensional space is an $(n-m)$-parameter family of such planes). I talked with Wagner when he came to Moscow (usually he stayed with Dubnov) and I also visited him in Saratov. First I was an opponent for his student Iraida Frolova. Wagner told me that in her application for graduate study he required that at the time of her training in graduate school "Ira was not involved with looking for a husband; the department took this on itself". And he got her together with his student Georiĭ Zhotikov, a talented mathematician, but, as it later turned out, a heavy drinker, which brought Frolova a lot of pain. Afterwards Wagner did not accept unmarried women as graduate students.

S. P. Finikov. The second opponent at the defense of my doctoral dissertation was Sergeĭ Pavlovich Finikov (1883–1964). Sergeĭ Pavlovich was the sole disciple of the old Moscow school of geometry in Kagan's departments; the remaining geometers of Mekh-Mat who belonged to this school were in the department of higher geometry, which was headed at that time by P. S. Aleksandrov.

The old Moscow school of geometry, which was called the school of "classical differential geometry", was founded in the second half of the nineteenth century by Karl Mikhaĭlovich Peterson (1828–1881), a native of Riga, who had graduated from the German University in Dorpat (now Tartu) in Estonia and taught mathematics in the German secondary school in Moscow. Peterson was the teacher of the Moscow University professor Boleslav Kornelievich Mlodzeevskiĭ (1858–1923) and one of the founders of the Moscow Mathematical Society and the journal "Matematicheskiĭ Sbornik". Mlodzeevskiĭ's students were Dmitriĭ Fedorovich Egorov (1869–1931), N. N. Luzin, and S. P. Finikov. Egorov was an honorary member of the Academy of Sciences and President of the Moscow Mathematical Society for many years. In 1931 he was exiled to Kazan and died there shortly afterwards. Luzin and Finikov were great friends.

In 1925–1926 Finikov took a long scientific journey in France and Italy, during which he got to know and became friends with many geometers of Western Europe. He was particularly close friends with Élie Cartan. Finikov mastered Cartan's method of exterior forms and created an enormous scientific school of differential geometers who worked with this method. At the present time Finikov's students and their students work in many cities of the former USSR and in many countries. I sometimes attended the "classical differential geometry" seminar, which Finikov directed in the University, and got to know many of the participants.

In 1945 when Cartan was in the USSR for the 220th anniversary jubilee of the Academy, Finikov introduced me to Cartan.

While I was a graduate student in Kagan's department and regularly attended the meetings of this department, Finikov argued with Kagan and his students at every meeting about almost every question and his remarks were sometimes very caustic. Therefore, when I learned that the Academic Council of Mekh-Mat had appointed Finikov as my opponent, I fell into a panic. But Sergeĭ Pavlovich liked my dissertation very much, in particular the proof that the pairs of line congruences in 3-dimensional projective space, which he had defined and called " T pairs", are expressed in the Plücker interpretation by focal pseudocongruences of lines in a 5-dimensional space, and the Calapso transformations are expressed in this interpretation by Laplace transformations. Finikov praised me a lot at the defense and, since the opinions of Rashevskiĭ and Finikov, who belonged to two opposing schools of geometry, and also the opinion of Delone, who was neutral, and the written comments of Wagner were all positive, the dissertation was confirmed by VAK (Higher Attestation Committee) in three months. And after the defense Finikov always had a very good relationship with me. I gave numerous lectures in his seminar and visited him at his house on Sobach′aya Ploshchadka (Place).

D. I. Perepelkin. Among the geometers who participated in Finikov's seminar, I was best acquainted with Dmitriĭ Ivanovich Perepelkin (1990–1954). When we first got to know each other, he gave me a set of reprints of this papers, including his beautiful work on the geometry of m-dimensional surfaces in n-dimensional Euclidean space. His wife Anastasiya Nikolaevna, also a geometer, was ten years older than he, and therefore, apparently for respectability, Dmitriĭ Ivanovich acquired a very beautiful beard. I visited him at the Moscow State Pedagogical Institute, where he then worked. I met him especially frequently when he was the referee of the manuscript of my book *Non-Euclidean geometries*, in the course of which he gave me a large number of very useful comments. After his death I supervised the work of his graduate student I. I. Zhelezina, devoted to the geometry of quasielliptic and quasihyperbolic spaces, as a result of which I mastered the area of the geometry of quasisimple Lie groups, which was new to me and which I later investigated further.

A. I. Mal′tsev. The well-known algebraist and mathematical logician Anatoliĭ Ivanovich Mal′tsev (1909–1967) had a significant influence on me. I met him during the war, when I served in the army and visited Moscow from time to time. At that time he was a professor at Ivanovo Pedagogical Institute and a research professor at the Mathematical Institute of the Academy of Sciences. He was the referee of my papers on Lie groups and symmetric spaces in the "Izvestiya of the Academy of Sciences". I often spoke with him at that time, not only about concrete questions of the theory of Lie groups, but also about more general mathematical questions; he gave me a great deal of useful advice. I am particularly grateful to Anatoliĭ Ivanovich for suggesting that I be introduced to and speak with Élie Cartan. When Mal′tsev became an Academician and one of the leading mathematicians of the Siberian Division of the Academy of Sciences, I no longer saw him.

S. A. Yanovskaya. When I was a student at Mekh-Mat, I attended lectures on the history of mathematics and mathematical logic, which were given by Sof'ya Aleksandrovna Yanovskaya (1896–1966). At that time the lectures on the history of mathematics seemed boring to me (the course was required, and at these lectures students played quiet games like "battleship"). Of course, I had no idea that in the future I myself would become a historian of mathematics.

Sof'ya Aleksandrovna was an active participant in the Civil War and was even shot once by the Whites. She was saved by the fact that she was of very small stature and wore a very tall hat. The shots occurred on a bridge, the bullets hit the hat, Yanovskaya herself fell into the river and stayed hidden in the rushes until it was dark. In one of the stories of I. I. Babel′, in which the action occurs at Odessa at the time of the Civil War, a certain Sonya Yanovskaya is mentioned. When I asked Sof'ya Aleksandrovna whether this story was about her, she answered: "Isaak Babel′was my friend."

During the war Sof'ya Aleksandrovna was evaculated to Perm′. She transferred several of the most gifted students to Moscow University. Some of these students became outstanding mathematicians, such as Eugene B. Dynkin, Academician of the National Academy of the USA and professor at Cornell University, and Ol′ga Aleksandrovna Oleĭnik and Mikkhaĭl Mikhaĭlovich Postnikov, both professors at Moscow University.

At MGU Sof'ya Aleksandrovna directed the department of the history of mathematics and, after this department was abolished and the Section for the History of Mathematics and Mechanics was created, she transferred to the logic department. In the department there was a seminar that Yanovskaya directed, first jointly with M. Ya. Vygodskiĭ, and then with A. P. Yushkevich. I often talked with Sof'ya Aleksandrovna, mostly about philosophical and historical topics.

M. Ya. Vygodskiĭ. I became acquainted with Mark Yakovlevich Vygodskiĭ (1898–1965), a geometer and historian of mathematics, when I was a graduate student. Like Yanovskaya, he was an active participant in the Civil War. He was an intelligence officer, delegated by the Reds to an institution of the independent (Musavatist) Azerbaijan Republic (until the Revolution he had lived in Baku, where he completed high school (gymnasium)). The Revolution found him a first-year student at Warsaw University, which at the time of World War I had been evacuated to Rostov-on-Don. In Rostov Vygodskiĭ became a student of the geometer and historian of mathematics, Dmitriĭ Dmitrievich Mordukhaĭ-Boltovskoĭ (1876–1952), who turned out to have exceptional influence on his entire scientific career. After the Civil War was over Vygodskiĭ moved to Moscow and completed the Mathematics Division of Fiz-Mat at Moscow University, while simultaneously teaching stenography at the Sverdlov Communist University, and his first book was *A course in parliamentary stenography*. After completing university, he taught mathematics at the same Sverdlov University and at the Institute of Red Professorships, and then became a graduate student of O. Yu. Schmidt in the history of mathematics. Later Vygodskiĭ taught mathematics in different technical universities in Moscow and organized a seminar in the history of mathematics at Mekh-Mat jointly with Yanovskaya. He also worked as principal editor at the State Publishing House for Technical-Theoretical Literature (GITTL), where his initiative led to the publication of translations of many classics of science with commentaries, some of them his own translations.

Vygodskiĭ's textbook *Foundations of a calculus of infinitesimals* was very popular at the beginning of the 1930s, and was based on the application of actual infinitesimals used in the higher levels in contemporary "nonstandard analysis". His book *Galileo and the inquisition* had great response. Shortly after this book appeared he was expelled from the Party, accused of having participated in the institutions of the White government in Baku, and also accused of having errors in his book on Galileo. With great difficulty he rejoined the Party after the war, by proving that he worked in the Musavatist institutions, completing a job for the Red Army, and then he quietly "mechanically left" the Party. I often talked with Mark Yakovlevich and he explained to me his divergences from the Party view and why he did not wish to be a member. After the war he published a very original textbook *Differential geometry* and several first-class mathematics reference books.

In the last years of his life he lost his position in Moscow and Leningrad and moved to a job at the Pedagogical and Polytechnical Institutes in Tula. His best known students are the geometer Èduard Genrikhovich Poznyak and the historian of mathematics Feodosiĭ Dement′evich Kramar (both were his students in Alma Ata during the war).

A. P. Yushkevich (Youschkévitch). In the realm of the history of mathematics the greatest influence on me was that of Adolf Pavlovich Yushkevich (born 1906). The son of the well-known social-democratic philosopher Pavel Solomonovich Yushkevich (1873–1945) who was a mathematician by training (he was the author of a paper on mechanics in hyperbolic space), Adolf Pavlovich was born in Odessa, and lived for many years of his childhood in France, to which his father had emigrated. After completing Moscow University A. P. Yushkevich taught mathematics at MVTU and in 1943, when I first got to know him, he was the head of the mathematics department at MVTU. When I returned to Moscow from Ashkhabad, the place to which Moscow University had been evacucated, as a Candidate of science, Yushkevich hired me as an assistant, and I worked in his department for several years. This was a model of a well-managed department in a large technical university.

Adolf Pavlovich wanted to work as a graduate student with S. A. Yanovskaya, but she would not accept him. Yushkevich began to study the history of mathematics independently and quickly became one of the leading historians of mathematics. In 1955–1958 he was President of the International Academy of the History of Science. His Doctoral dissertation on the history of mathematics in Russia was defended in 1940 and afterwards appeared as a separate monograph. His research on the history of mathematical analysis is well known, and his book *History of mathematics in the middle ages* has been translated into many languages.

In 1945 the Institute for the History of Natural Science of the USSR Academy of Sciences (IIE) was organized. In this form the Institute for the History of Science and Technology, which was previously in Leningrad, was reconstituted, only partically for the time being. That earlier institute was closed in 1938 in connection with the declaration that its director, Nikolaĭ Ivanovich Bukharin (1888–1938), was declared an "enemy of the people". The new institute was organized in Moscow and the old institute became the Leningrad Branch of the Moscow Institute. Yushkevich worked at the new institute from the day it was founded, at first jointly with his job at MVTU, but in 1953, when the IIE was transformed into the Institute for the History of Natural Science and Technology (HET), he finally left MVTU and moved to IIET on full salary.

I was not a very diligent assistant, since I was writing my Doctoral dissertation at that time. For a couple of years after my defense, convinced that I would not succeed in obtaining a professor's job in Moscow (these were Stalin's last years, when the question of nationality played a very important role in finding work), I received an invitation from Rashevskiĭ's student Maksud Alievich Javadov (1902–1972), who was pro-rector of Azerbaijan University at that time, and moved to Baku. I gave courses on geometry and a course in the history of mathematics, and quickly obtained the title of professor. While preparing the history course, I discovered that the prominent mathematician and astronomer Nasīr ad-Dīn at-Tūsī had worked in the thirteenth century in southern Azerbaijan. I asked whether there was anyone in Baku engaged in studying his works, and I was told that there was no one so occupied. But then a mathematician came to me who was studying at-Tūsī's proof of Euclid's fifth postulate, on which the theory of parallel lines is based. Then I obtained the proposition to participate in the preparation for publication of Russian translations of at-Tūsī's fundamental mathematical works. I went to Moscow and met with Yushkevich at Mekh-Mat; I told him about this work. He invited me to give a lecture on this topic at the seminar at MGU, and then suggested that I submit a paper "On the Mathematical Works of Nasīr ad-Dīn at-Tūsī" for the next volume of "Istoriko-Matematicheskie Issledovaniya" (IMI) (Studies in the History of Mathematics), which he had begun to publish shortly before, together with G. F. Rybkin. My paper appeared in the fourth volume (1951). Yushkevich also advised me to consult D. E. Smith's paper "Euclid, Omar Khayyam, and Saccheri", in which at-Tusī's work on parallel lines is mentioned, and it is shown that the famous "quadrilateral of Sacceri" (a quadrilateral with two right angles and equal lateral sides) and three of Saccheri's hypotheses concerning the remaining two angles of this quadrilateral ("the hypotheses concerning acute, obtuse and right angles") occurred in Khayyam's geometric treatise, quoted by at-Tūsī. In Baku I went to the History Institute of the Academy of Sciences, in order to correct the transcription of the Arabic title of this treatise. The old Arabist there, although he did not speak Russian, upon reading the title of Khayyam's treatise, stood up, took me by the hand and led me to the Manuscript Division of the Academy of Sciences, which was in the next block. There he spoke to the director in Azerbaijani, and in five minutes Khayyam's treatise lay in front of me, printed before the war in Iran (the director had found it in Tabriz, where he spent the war). I decided to translate this treatise and found an Arabist in Baku, an emigrant from Iran who knew Arabic and French, but no Russian. I spoke French with him. He made me a French translation of the treatise, from which I made a Russian translation. But the Azerbaijanis who were engaged with at-Tūsī required that my publication of this treatise be joint with them. Then I understood that it was necessary for me to study Arabic myself, and arranged with my translator to teach me Arabic. Soon with his help I was able to make direct translations from Arabic into Russian of three treatises of Khayyam: commentaries on Euclid, where in addition to the theory of parallel lines there was a presentation of an original theory of relations and a proposal for generalizing the concept of number up to real numbers; the treatise on algebra, discussed above in connection with G. M. Shapiro (in Baku I found an edition of the Arabic text of this treatise with a French translation), and Khayyam's treatise on the composition of an alloy by weighing it in air and water. These translations were published, with commentaries by Yushkevich and me, in the sixth volume of IMI (1953). After this, a whole series of my translations from Arabic appeared.

In 1955–1964 I worked as a professor at Kolomna Pedagogical Institute and continued to work on translations of Arbic mathematical treatises, and also wrote papers on the history of interpretations of hyperbolic geometry, on the life and works of A. P. Kotel′nikov, and on geometry in the works of Leonhard Euler. In 1964 I was invited to take a vacant job as research professor at IIET. I worked at this institute from 1964 to 1990 in the history of mathematics section, which was organized by Adolf Pavlovich and first directed by him and later by his student Sergeĭ Sergeevich Demidov. In this section I participated in collective works of the section and wrote the books *History of non-Euclidean geometry* (1976; English translation 1989) and *Mathematicians and astronomers of the medieval Islam and their works* (1983). For my work on the history of mathematics I was made a Corresponding Member of the International Academy of the History of Science in 1971 and a Full Member in 1978.

S. V. Fomin. One very strong student in the group of mathematicians with whom I studied at Mekh-Mat was Sergeĭ Vasil′evich Fomin (1917–1975). Serezha was the son of a history professor. First he was in the group of astronomers, but after he obtained interesting results in algebra he transferred to the mathematicians' group. At the competition of student works at Mekh-Mat in the fifth year, he and I received second prizes.

Four members of our group went on to graduate study: Serezha and I (he to Kolmogorov, I to Rashevskiĭ), Oleg Golovin and Mosya Pesin. Oleg Nikolaevich Golovin (1916–1988), the brother of the famous atomic bomb maker Igor Nikolaevich Golovin, was a graduate student of Kurosh; after returning from the front he worked in the algebra department of MGU and after Kurosh's death was chairman of this department. Moiseĭ Il′ich Pesin (1913–1941) was a graduate student of Norden; at the beginning of the war he entered the National Guard and perished at the front.

During the war Fomin worked at the Military Scientific-Study Institute. After the war he worked at Mekh-Mat, was a member of the editorial board of the journal "Uspekhi Matematicheskikh Nauk" (I usually submitted my papers to this journal through him), and also in several other places. This overwork served as the reason for his first heart attack. Because he signed the letter of 99 mathematicians in Esenin-Volpin's defense Fomin was expelled from the Party, which led to his second heart attack (after this the Central Committee of the Party reinstated him in the Party). After he was expelled from Mekh-Mat he worked in one of the biological institutes of the Academy of Sciences and became interested in the application of mathematical methods to biology. He was not allowed to go abroad for a long time, and when this ban was lifted he successively made two foreign trips, and then traveled to Vladivostok (Russian Far East), where he died of his third heart attack while playing chess.

G. E. Shilov. A very strong student in the class one year before ours was Georgiĭ Evgen′evich Shilov (1917–1975). I became acquainted with him already when I was in school. He completed seven grades in the same school I attended from eighth to tenth grade. He completed Rabfak ("Workers' Faculty") and entered Mekh-Mat in the year I went into ninth grade. When he visited our school I asked him what subjects were studied at the university, and he answered: set theory. In school and in university he was called Yura Bosse, but when he had finished university he explained that Bosse was his stepfather's name, and he took his father's last name (his first scientific paper was signed Yu. Bosse).

Yura loved music and literature very much and introduced me to his circle, which met in the University Club, and its members played, sang and read poems. Besides Yura this circle included Yura's classmates Boris Gurevich and Zorya Shapiro and the young docent I. M. Gelfand. Before university Boris Lazarevich Gurevich taught literature and athletics in a secondary school, and after the war, when he was a military meteorologist, he defended a dissertation on a topic given by Shilov and worked in Odessa University as a meteorologist in the Geographical Faculty. Izrail Moiseevich Gelfand (born 1913) is now a world-famous Academician. Zorya Yakovlevna Shapiro, a well-known mathematician, was Gelfand's wife for a long time. The friendship between Shilov and Gelfand lasted until Shilov's death and the topics of Shilov's scientific work were closely connected with the topics of Gelfand's work. Shilov's second wife, who directed the children's study "Orlenok"[2] at Moscow University, was also a great lover of music, and after Shilov's death she organized several soirées in his memory at the State Musical Museum. Shilov was also interested in the connections between music and mathematics, and he wrote the little book *Simple gamma* on this topic.

In 1950–1955, when I worked in Baku, Shilov worked in Kiev University. Shilov was the author of many books, some of which have been translated into English. Many of his former students are currently professors.

V. A. Rokhlin. Vladimir Abramovich Rokhlin (1919–1984) was a student at Mekh-Mat in the class after mine, and later worked together with me at Kolomna Pedagogical Institute. He was born in Baku and finished secondary school there. His father was a leader of the Baku Mensheviks and an implacable enemy of Shahumyan and other "Baku Commissars"; after the establishment of the Soviet regime he was the Chairman of the Baku Soviet. "Do you know the Armenikend district in Baku?" Volodya asked me (of course, I knew this district of new construction from the 1920s very well). "This district was constructed on my father's initiative," he told me. Volodya was two years younger than his classmates in school and university. This happened because, when his older sister Ida went to school, Volodya went with her and quickly turned out to be the very best student in the class. He recalled for me a song from his childhood: "Ida and Volya Levinson sat together on the phaeton." In old Baku cab-wagons where called phaetons. Levinson was his mother's family name. Volodya told me that the father of the famous Russian writer Korneĭ Chukovskiĭ was his mother's uncle (with the same family name); Chukovskiĭ himself in his memoirs writes only that his father was a student from St. Petersburg and his mother was a young servant in this student's family; her family name was Korneĭchuk, and the future writer "extracted" his first and last names from her family name. I was always amazed by how much Volodya resembled Chukovskiĭ, so Volodya's story about this relationship explained the reason for the resemblance to me. Volodya and Ida finished school together and moved to Moscow, where Volodya entered Mekh-Mat and Ida entered the Oil Institute.

Volodya was one of the strongest students in his class and wrote his first paper on topology while he was a student. He entirely deserved the Stalin Grant, which was given to the best students, but did not receive it, since in 1937 his father was arrested as a former Menshevik. But this could not prevent him from pursuing graduate studies.

When the war began, Rokhlin went into his second year of graduate study. Although he was given a postponement of levy until the end of his graduate study, signed by the Deputy Minister of Defense Shchadenko, by 1941 he was

[2]The Eaglet

in the National Guard and participated in the combat near Smolensk. He was wounded there and hid in the barn of a compassionate peasant woman. He told me that for several months he was in a crooked state and during this time he thought over his Candidate's dissertation and obtained the main results. At this time the Smolenski region was occupied by the Germans and immediately after he had recovered from the wound and attempted to cross to the front line, he fell into a German prisoner-of-war camp. He told me that he passed as an Azerbaijani and twice the Germans wanted to shoot him as a Jew; once he was betrayed by a prisoner of war who had known him in the Soviet Army, and the other time they suspected he was a Jew because of his appearance. In both cases his "racial purity" was verified by a doctor, but, as he related, "there are decent people among the German doctors". He escaped from German camps three times, and a fourth time, when parts of the Soviet army approached their camp and the prisoners had the idea that the camp was about to be liquidated, and they were all to be shot, they offered to guard the camp with wooden and iron sticks. After this Rokhlin was in the active Soviet army and participated in the capture of Berlin. But after the end of the war he ended in a Soviet camp for "special check-up". When he wrote to Moscow about this, Kolmogorov and Pontryagin went to the Minister of State Security with a request that Rokhlin's check-up be speeded up. The check-up was accelerated, but Kolmogorov and Pontryagin had to go again to the same minister, as a result of which an order appeared for Rokhlin to be "sent out on business under the auspices of the Academy of Sciences of the USSR".

Upon his return to Moscow, Rokhlin became a graduate student of Pontryagin and, after the defense of his Candidate's dissertation in 1947, worked in the Mathematics Institute of the Academy of Sciences. But he was quickly expelled from there because he was a Jew, and he taught at the Timber Technology Institute near Moscow in the Mathematics department headed by the geometer Nikolaĭ Vladimirovich Efimov (1910–1972). Several years later, having defended his Doctoral dissertation, Rokhlin became a professor at the Arkhangel′sk Timber Technology Institute, and then, after working for several years at the Ivanovo and Kolomna Pedagogical Institutes, he was invited to Leningrad University by A. D. Aleksandrov, who was the Rector at that time. Rokhlin was one of the outstanding topologists. His life in German and Soviet camps had significantly weakened his health and he died at the age of 65.

I. M. Yaglom. One of my closest friends was Isaak Moiseevich Yaglom (1921–1986). I became acquainted with him and his twin brother Akiva when I was in my first year of graduate study, and the brothers were second-year undergraduates, Isya at Mekh-Mat and Kika at Fiz-Fak. Isya was one of the leaders of high-school mathematical circles[3] and also involved me in

[3]See footnote 12 to Arnol′d's article concerning these mathematical circles, or *kruzhoks*.

this; one year he and I directed a high-school circle in geometry. While I was in evacuation in Ashkhabad, the brothers were in Sverdlovsk (Ekaterinburg) and worked in the Leningrad Central Geophysical Observatory evacuated in this city. In 1942 MGU moved from Ashkhabad to Sverdlovsk and the brothers were able to continue their university studies, and both studied the program of both faculties. After finishing, Isya became a graduate student of V. F. Kagan and Kika became a student of Kolmogorov. The brothers were as alike "as two drops of water" and they were often confused for each other. When Isya defended his Candidate's dissertation they had one suit coat between them; at the time of his inaugural speech Isya wore this coat, which he then gave to Kika. After the Academic Council decided to award the degree and everyone went to congratulate Isya, Kagan embraced and kissed Kika, thinking he was Isya.

After completing graduate school Isya worked in the mathematics editorial section of the Publishing House for Foreign Literature (now called Mir), at Mekh-Mat, in Orekhovo-Zuyevo and Moscow State Pedagogical Institutes, in the External Metallurgical Institute, in Yaroslavl′ University and at the Academy of Pedagogical Sciences. When Isya worked at the Publishing House, he and I jointly translated the second volume of Schouten and Struik's *Introduction to the new methods of differential geometry* (the first volume has been translated by Z. Kainer, and we also jointly translated a collection of Élie Cartan's works, but only my translations were published in the book *Geometry of Lie groups and symmetric spaces*. Isya was expelled from MGU because he was a Jew, and from MGPI in connection with the letter in defense of Esenin-Vol′pin; in the technical institute he was "not in harmony" with the authorities, and he resigned from Yaroslavl′ University, since it became difficult for him to travel there from Moscow.

I. M. Yaglom was not only the leader of high-school mathematical circles for many years; he also established an important series of books, the "Mathematical Circle Library" ("Biblioteka Matematicheskogo Kruzhka"), a large part of which he wrote, sometimes with his brother and other co-authors. These books, and also other books written by him for a general audience, have played an important role in raising the mathematical culture in both the USSR and other countries, since many of them have been translated into English and other languages. It is also necessary to point out his great editorial work. After becoming the director of the editorial board of the Central Editorial Office of Physics and Mathematics Literature of the publisher "Nauka" (formerly Fizmatgiz), Pontryagin with his anti-Semitic tendencies stopped "Nauka" from publishing the *Mathematical circle library* and other of Yaglom's books, and it was only with difficulty that he was able to have his new books printed by other publishers.

I. M. Yaglom obtained important mathematical results in non-Euclidean and symplectic, and in differential and integral geometry, and also wrote interesting historical books on Hamilton and Grassmann, and on Klein and

Lie. The most important of these results I consider to be his research in the area of degenerate projective metrics and the application of algebras with degenerate norms to geometry. More than once he turned geometers' attention to the importance of these geometries. In particular, the topic of I. I. Zhelezina's dissertation, which I discussed above, was suggested to D. I. Perepelkin by Yaglom, with whom he often spoke while he worked at Orekhovo-Zuyevo Pedagogical Insitute, where A. N. Perepelkin directed the mathematics department. These geometries were the topic of many dissertations of his and my students. They were described in our joint paper "Projective Metrics" in "Uspekhi Mat. Nauk", and then in some of his and my books. The paper "Projective Metrics" and the book *Projective transformations* served as Yaglom's Doctoral dissertation. At the present time the transformation groups of these geometries and the corresponding algebras are called quasisimple and r-quasisimple groups and algebras.

Yaglom was a close friend with many of those who fought for human rights in the USSR, particularly with Andrei Sakharov, who gave the eulogy at the memorial service. Yakiva M. Yaglom now lives near Boston, Massachusetts.

R. I. Pimenov. One of the most interesting Soviet mathematicians I have known is Revol′t Ivanovich Pimenov (1931–1990). When I learned of his existence, he was in a camp serving a ten-year sentence for his dissident activities. During one of my visits to Leningrad Viktor Abramovich Zalgaller, a student of A. D. Aleksandrov, told me that a talented geometer had sent from a camp a paper that no one in Leningrad was able to evaluate, and asked me to get to know him. The next day one of Pimenov's friends, Èrnst Orlovskiĭ, brought me this paper. The author had independently been led to quasi-non-Euclidean and r-quasi-non-Euclidean spaces, which I was investigating at the time along with I. Yaglom, although his approach was of course not the same as ours. I sent my opinion to A. D. Aleksandrov, Aleksandrov vouched for Pimenov to the President of the Academy of Sciences, M. V. Keldysh, and Keldysh vouched for him before the higher authorities, and Pimenov was freed with four years to go before the end of his sentence. Upon his return to Leningrad, Pimenov started to work in the Leningrad Division of the Mathematics Institute of the Academy of Sciences and soon afterwards defended his Candidate's dissertation and then his Doctoral dissertation. He visted Moscow several times, talked with me and Yaglom, and gave lectures at our seminars. In Moscow, Pimenov defended his Doctoral dissertation on further generalizations of quasi-non-Euclidean and r-quasi-non-Euclidean spaces, namely quasi-Riemannian and r-quasi-Riemannian spaces and their applications to physics. But soon after receiving a positive answer from VAK (the Higher Attestation Committee) he was again arrested and sent to Syvtyvkar. The trial took place in Kaluga, and many fighters for human rights, including Sakharov, traveled to it. Pimenov told me that it was here that Sakharov met his future wife, the physician Elena Bonner, a school friend of Pimenov's wife, treating the Pimenov's son.

In Syvtyvkar Pimenov worked in the Komi Branch of the Academy of Sciences and did a lot of scientific work. At the beginning he did not have the right to leave the city, but then he obtained this right, and his work was published. When I traveled to Syvtyvkar to give lectures at the Pedagogical Institute, he attended my lectures, and invited me to his home, where he showed me his publications in "Kontinent".

The last time I saw him was in Kishinev at a geometry conference. He had a thick beard. He told me that he was fighting with VAK to confirm the positive opinions of his dissertation. Not long before my trip to the USA he telephoned me that VAK had finally confirmed his Doctoral degree and that he was at a session of the Supreme Soviet of the Russian Republic in Moscow, to which he had recently been elected, and worked there in the commission to reconsider the constitution of the Russian republic. Pimenov's student, Nikolaĭ Alekseevich Gromov, defended his doctoral dissertation in 1990 in Minsk (Belarus). The results of his thesis were presented in Gromov's book *Contractions and analytic continuations of classical groups* published in Syktyvkar in 1990. On 19 December 1990 Revolt Ivanovich Pimenov died from stomach cancer. His book *Foundations of the theory of a temporal universum* was published in Syktyvkar in 1991.

Translated by J. S. JOEL

A. N. Kolmogorov

V. M. TIKHOMIROV

In his article "The architecture of mathematics" describing his program, Nicolas Bourbaki notes with regret that there is no mathematician, even among those with the broadest erudition, who would not feel himself a stranger in certain fields of the enormous world of mathematics. The exceptions to this rule are the geniuses like Poincaré and Hilbert who left their mark in every field of mathematics. To even so short a list we may with complete justification add the name of Andreĭ Nikolaevich Kolmogorov.

The breadth of Kolmogorov's scientific range of interests is unique and reminiscent of the universality of the geniuses of the Renaissance. Trigonometrical and orthogonal series, set theory, mathematical logic, the theory of measure and integral, topology, the theory of approximations, celestial mechanics, homology theory, turbulence, ergodic theory, superposition of functions, functional analysis, ... And in addition to all this the vast field of probability, which Kolmogorov transformed completely, starting from its foundations; information theory; the theory of automata; and still more—numerous works on applications to biology, geophysics, production control, the theory of marksmanship, the theory of versification, etc. And in almost every one of these fields, we owe to Kolmogorov fundamental ideas and outstanding results.

Kolmogorov was born in 1903. His mother died in childbirth. His father had no part in his upbringing, but from his childhood he was surrounded with love, kindness and attention. The responsibility for his upbringing was taken on by his mother's sister, who adopted him. So in his early childhood he lived on the estate of her father, his grandfather. In this home, the goal was to develop in the child curiosity and an interest in books and nature. Later he said, "I learned early on the joy of mathematical 'discovery', noticing at the age of five or six the regularity in the identities $1 = 1^2$, $1 + 3 = 2^2$, $1 + 3 + 5 = 3^2$, $1 + 3 + 5 + 7 = 4^2$, and so on." In the home where the little boy grew up, his relatives organised a small school incorporating the latest educational theory. A journal was published "Vesennie lastochki" (The

This is a translation of the revised and extended version of the article by V. M. Tikhomerov, which originally appeared in *Russian Mathematical Surveys*, **43**, no. 6, 1988, pp. 1–39; it is reprinted here with permission from the London Mathematical Society and BLDSC.

A. N. Kolmogorov

swallows of spring), and the five-year-old pupil "edited" the mathematical section. There he published his "discoveries".

When he reached the age of six he moved with his aunt to Moscow. There he was entered in one of the most progressive grammar schools of the time, which was run by a group of radically inclined intelligentsia. Unlike the majority of grammar schools of that time, it was coeducational and there were many interesting experimental teaching methods.

The boy's range of interests was extremely broad. He took a serious interest in biology and physics, and at the age of 14 he was familiar with higher mathematics. He was interested in chess, he was fascinated by social problems (and, in particular, he wrote a constitution for an utopian island community—a commune, where the principles of higher justice were to be put into practice); he took part in the elections for the Constituent Assembly of 1917. His interest in history was very profound. "The first scientific report which I made at the age of 17 at the University of Moscow", he recalled later, "was a report to the seminar of Professor S. V. Bakhrushin about the history of Novgorod in the fifteenth and sixteenth centuries."

Kolmogorov then made a small discovery, which was recognized by S. V. Bakhrushin, one of the greatest Russian historians of that time. The young man asked the professor whether he should publish his result. The reply he received was: "Certainly not, young man! You have found only one proof. That is very little for a historian. You need at least five proofs." It is possible that the disappointment he felt at that moment influenced the whole of his future destiny, and he began to study mathematics, where one proof suffices.

Nevertheless he did not immediately decide to devote his life to mathematics. At the same time that he was studying at the University of Moscow, he enrolled (in 1920) in the Faculty of Metallurgy at the Institute for Chemical Technology, because he had the intention to study practical engineering. But very soon his interest in mathematics outweighed "his doubts about the relevance of the profession of mathematician". He continued his studies only at the University and from then on his whole life was tied to the University of Moscow.

He immediately succumbed to an atmosphere of enormous creative enthusiasm. He attended the lectures of the outstanding scholar, his future supervisor, Nikolaĭ Nikolaevich Luzin. He came into lively scientific contact with Luzin's students, who were later to play an important part in the development of topology—Pavel Sergeevich Aleksandrov and Pavel Samuilovich Uryson.

Once in 1920 at Luzin's lecture, the 18-year-old freshman disproved one of the lecturer's hypothetical assertions. He presented his argument with confidence to the circle of mathematicians, and there for the first time he attracted some attention. P. S. Uryson invited Kolmogorov to become his student. But it was the problems put forward by Luzin that drew the young man's attention. He began to construct a general theory of operations on sets, wishing to advance it beyond the boundaries at which P. S. Aleksandrov and M. Ya. Suslin had stopped, and he wrote a long paper on this subject, dated February 1922. Kolmogorov also attended V. V. Stepanov's seminar on trigonometrical series. There he solved one of the problems posed by Luzin, and after that Luzin "with some solemnity" (as Kolmogorov wrote) suggested that he become his student.

In the summer of 1922 (the work is dated 2 June 1922), Kolmogorov found an outstanding result—he constructed a Fourier series divergent almost everywhere, and this immediately brought him international repute. It was at this moment that his incomparable creative career began.

In 1925 Kolmogorov graduated from the University of Moscow and became Luzin's research student. The same year saw the beginning of his work on probability theory. In 1929 he set out on a long boat journey along the Volga with Pavel Sergeevich Aleksandrov, and this marked the beginning of a friendship which lasted until the last days of Aleksandrov's life (P. S. Aleksandrov died on 16 November 1982). In 1935 Kolmogorov and Aleksandrov acquired a house not far from Moscow on the bank of the river Klyaz′ma, and it was in that house that most of their creative life was lived. For more than 50 years this house was a center of mathematical activity for several generations of Soviet mathematicians and for many guests from abroad.

From 1931 Kolmogorov was Professor at the University of Moscow. In 1939 he was elected a full Member of the Academy of Sciences of the USSR. At the University, Kolmogorov founded several departments. From 1954 to 1958 he was Dean of the Faculty of Mechanics and Mathematics at the University of Moscow. From 1964 until 1966 and from 1973 to 1985 he was President of the Moscow Mathematical Society.

Between June 1930 and March 1931 Kolmogorov went on academic journeys abroad (Göttingen, Munich, Paris). There he met Hilbert, Carathéodory, E. Landau, P. Lévy, Fréchet, Lebesgue, Borel, Weyl, and other outstanding mathematicians. Kolmogorov often spoke of the exceptional significance for his research work of that scholarly atmosphere and that he felt privileged to have spent those years establishing personal contacts with scholars from different countries.

In 1954 he spent two months as Professor at the Humboldt University in Berlin. In 1958 he was Professor at the University of Paris, for the Spring term.

Kolmogorov attended the international Mathematical Congresses in Amsterdam (1954), Stockholm (1962), Moscow (1966) and Nice (1970). At the Amsterdam Congress Kolmogorov was invited to conclude the scientific program of the Congress with his lecture on dynamical systems (the lecture by J. von Neumann had opened the scientific program).

More than twenty scientific organisations have elected Kolmogorov as an Honoray Member. He was elected a Member of the Royal Netherlands Academy of Sciences (1953), of the Royal Society of London (1964), of the

National Academy of Sciences of the USA (1967), of the Académie des Sciences of Paris (1968), an Honorary Member of the Roumanian Academy of Sciences (1965; he had been a Corresponding Member since 1957), a Foreign Member of the Polish Academy of Sciences (1956), an Honorary Member of the Hungarian Academy of Sciences (1965). He was a Doctor Honoris Causa of the Universities of Paris, Stockholm, Warsaw, Budapest, of the Calcutta Mathematical Society, of the Royal Statistical Society of London, etc.

Kolmogorov was awarded the State Prize of the USSR (jointly with A. Ya. Khinchin, 1941), the P. L. Chebyshev Prize of the Academy of Sciences of the USSR (jointly with B. V. Gnedenko, 1949), the Lenin Prize (jointly with V. I. Arnol′d, 1965), the N. I. Lobachevskiĭ Prize (1986), and the following international prizes: the Balzan Prize (1963, the Prize was awarded simultaneously but in other sections to Pope John XXIII, the historian S. Morrison, the biologist K. Frisch, the composer P. Hindemith), the Wolf prize, the Helmholtz Gold Medal and many others.

Kolmogorov devoted the last two and a half decades of his life mainly to the problem of instruction and teaching in secondary schools.

—— • ——

The creative power of a scholar can be seen in his achievements in the arts, whether in his originality as a writer, as a painter, or as a poet. The legacy of a great scholar, as any cultural phenomenon, has its own history, its own architecture, and possibly its own undeciphered plan and is linked to profound conceptions, with consequences in culture in general.

At a superficial glance Kolmogorov's genius may seen to be a pile of isolated and unrelated parts. Kolmogorov himself, it would seem, sometimes confirmed this verdict. "For me, prone to lack singleness of purpose ... " he wrote once to his teacher N. N. Luzin, and he repeated something similar many times. But if we consider his creativity more intently, we see that his efforts were all subject to one general plan and the diversity of what was planned and achieved was the result of the infinite number of goals he set for himself.

When the Academy of Sciences of the USSR took the decision to publish his collected papers, it was at first assumed that this publication would consist of two volumes. The question of how to divide the material into two parts was, naturally, passed to the author himself to solve. Kolmogorov suggested putting into one of the volumes papers on mathematics and mechanics, and, in the other, papers on probability theory and information theory. This division reflected the fact that our "enormous world of mathematics" is in fact itself divided into two parts, as it were into two kingdoms.

In one, deterministic phenomena are studied, and in the other, random ones. A century ago the Kingdom of Chance occupied a very modest territory in our world of mathematics. Until the end of the century there was still no fundamental role for random phenomena in the natural sciences (although the foundations of thermodynamics were already laid). Then the

Kingdom of Chance began to colonise greater and greater areas and now the world of chance is comparable to the rest of the mathematical world. (And the Congresses of the Bernoulli Society, as Kolmogorov once remarked, are comparable to the International Mathematical Congresses.)

Kolmogorov was a leader of progress in both kingdoms, and he was the discoverer of many previously unknown regions. And at the end of his life he put forward a grandiose program of simultaneous and parallel study of the complexity of deterministic phenomena and the statistical determination of random phenomena. In this program there was concentrated the experience of practically all his creative work.

The basic idea is that the Kingdom of Order and the Kingdom of Chance have no real boundaries, that our mathematical world is one and in principle indivisible.

The attempt to reveal the essence of the concepts of "order" and "chaos" crowns Kolmogorov's original work. In this conception, he encompassed the ideas of the theory of probability, the foundations of mathematical logic and the theory of algorithms, the ideas and methods of information theory, the results of ergodic theory and of the theory of dynamical systems, and intense speculations about the natural sciences. But many of Kolmogorov's first works (on the theory of functions, on set theory, geometry, etc.) can be interpreted as a prologue to the realization of all this immense program.

Observing the creativity of great men brings us closer to discovering the nature of man. There are two opinions about the nature of genius. Thomas Mann—Kolmogorov's favourite author—dedicated his last work "*Doktor Faustus*" to one of them. For Mann, a genius is a person, who in order to achieve his goal, which requires surmounting impossible difficulties, is compelled, as was Faust, to make a pact with dark, demonic forces (in the novel the hero consciously infects himself with a fatal disease, so that his inflamed brain can reflect the transcendental images and ideas).

But there is another point of view. In Russian culture it is most clearly expressed in Pushkin's creative work (indeed in his own personality). In Pushkin's short tragedy *Mozart and Salieri*, Mozart is the shining chosen one who is given the gift of hearing the divine voice. Most probably Pushkin did not know the words written by Mozart in one of his letters, but he had a profound understanding of the essence of the very phenomenon of Mozart's genius. Mozart wrote, "Ideas come to me in quantities and with unusual facility. Where do they come from? I know nothing at all of this ... The work grows, I hear it more and more distinctly ... Then I comprehend it in one glance."

To a large extent the great discoveries made by Kolmogorov saw the light of day in a similar way. Kolmogorov was undoubtedly one of the "shining" geniuses.

Over and over again one and the same motif crops up in his publications and discussions. Once the remarkable Soviet geometer Boris Nikolaevich Delone, appearing before his pupils, expressed the idea that the only difference between a great scientific discovery and a good Olympiad problem is that solving the Olympiad problem requires 5 hours but finding a powerful scientific result requires an investment of 5,000 hours. (Apparently this was the way Delone himself worked, and this was the manner in which many others such as Hilbert worked—we shall compare further the creativities of Hilbert and Kolmogorov.) Kolmogorov often recalled those words of Delone. But each time the talk turned to these famous 5,000 hours, he said, with a feeling of some awkwardness and even of irritation, that he himself was unable to concentrate his thoughts for such a long time on one and the same problem. In one interview Kolmogorov answered the question "How do you work?" with the words: "You read some books, you prepare your own lectures with some new variations or other, and suddenly, from the soil of this everyday work, some unexpected idea emerges, and, vaguely as yet, some completely different approach can be seen. Then having worked it out, almost everything else is neglected—and one thinks, thinks endlessly along the lines which have just appeared. Fortunately, I usually had the opportunity to do this, but in the whole history of my scientific discoveries such complete oblivion, cut off from everything else, might last for a week, sometimes possibly for two—not more."

And then the sudden inspiration would come. Thus it was with the discovery of the trigonometrical series divergent almost everywhere, which was preceded by three days of uninterrupted reflexion and complete concentration; thus it was the final result, connected with Hilbert's thirteenth problem, and with the entropy invariant for dynamical systems and many others. After concentrating enormous energy there suddenly came the moment when the whole picture of the phenomenon appeared in its complete form. And then what?

Wilhelm Ostwald's classification of scholars is well known. He divides them into classicists and romantics. "The first," Ostwald wrote, "can be compared to a mill, carefully grinding with logical millstones the initial ideas, in order to extract from them the most far-reaching possible consequences and to develop the theory to the limits of possible completion and perfection. The latter one must rather equate with generators of new ideas. Having stated the idea, they quickly lose interest in it, however brilliant it may have been, and take no part in the further development of it."

This opinion was often quoted. I am not completely satisfied with Ostwald's classification, but the label "romantic" suits Kolmogorov very well.

This accumulation of energy over a short span of time, of which we spoke earlier, led to a powerful explosion, causing gaping holes in what one would think of as impregnable bastions. Then suddenly there rushed in tens and even hundreds of followers. Usually Kolmogorov did not himself pursue the matter further; it was as though a creative tiredness would set in and he already had in his mind another goal in view.

Something of that kind happened to him almost all the time. That was the case with the solution of Hilbert's thirteenth problem: when he had almost reached his goal, he declared that he would leave the completion of the work to his successors; and it was the same with problems in classical mechanics, and with ergodic theory and with many other subjects.

Kolmogorov was very much interested in the essence of creative talent. He distinguished algorithmic, purely geometric and logical faculties in man. Now, a division into two groups is more common (analytic and descriptive geometric), corresponding to the different functions of the two hemispheres of the brain. I find it difficult to say which of the qualities mentioned above were the most prominent in Kolmogorov's case. Analytic papers with an abundance of calculations and transformations exist side-by-side with papers of his where "the art of a consistent, correctly partitioned logical argument" plays a fundamental role. And besides, in his research in function theory and topology, for example, there are geometric constructions which stand out by their beauty.

By the way, Kolmogorov was left-handed from birth, and as a child he taught himself to use his right hand, which he did well. We cannot rule out that he may be one of that rare breed of men each of whose hemispheres fulfills, as it were, both functions, so that asymmetry is practically nonexistent.

An essential part of an intellect is its quickness of thought. P. S. Aleksandrov once spoke jokingly of "quick" and "slow" geniuses, and he included Hilbert among the latter. Kolmogorov was undoubtedly one of the "quick" geniuses, but what stuck me personally most of all about him was not so much the quickness of his thinking as the quickness of his grasp and his perception. I was often a witness of the way scarcely perceptible emanations or very vague rough drafts, which came to his attention, were immediately incorporated into an ordered and complete system. (Sometimes one had the impression that Kolmogorov was one of those people who have as it were "a priori" knowledge—he knew, it seems, everything, although it was not quite clear when he had learned all that.)

And it was unusually interesting to observe how suddenly, at the right moment, there surfaced in his consciousness and were fused together Menger's universal tree and Kronrod's constructions, or the ideas of Poincaré-Bogolyubov-Kryolv and the Newton-Kantorovich method, the Pontryagin-Shnirel′man designs and Shannon's theory.

It is natural to raise the questions: who had the greatest influence on Kolmogorov, on "whose shoulders" did he stand, on what is his creativity mainly based? This subject is still awaiting a more careful study, and we shall only touch upon it. In the first volume of Kolmogorov's selected works there are 60 articles. In none of them does he express any thanks to his teachers or colleagues (for help, for useful advice, for suggesting the problems, or anything else). In 15 of these articles there is no bibliography at all, and in the others the work of 93 mathematicians is mentioned. Of scholars of former ages only Aristotle and Leibniz are cited, of modern times (and in a not very important context) only Grothendieck. All others are famous scientists from a preceding generation and from his own generation. Of his predecessors the most significant references are to Hadamard, Birkhoff, Borel, Brouwer, Hilbert, Carathéodory, Lebesgue, Luzin, Taylor, Von Kármán, Hardy, and Hausdorff; in other volumes to Chebyshev, Bernstein, von Mises, and Fisher. I think it is precisely these whom we must include among those of his predecessors to whom he was most indebted. Somewhat surprising is the absence of references to Poincaré. This is largely because Kolmogorov learned of Poincaré's ideas by reading the works of Chazy and Charlier. The other mathematicians to whom Kolmogorov refers where part of the current scientific scene. Here we must mention the great influence that the works of Krylov-Bogolyubov and de Rham had on him.

After these general remarks let us turn to the concrete results Kolmogorov obtained. It is impossible to describe in one article all of his contributions to science with any completeness, but, whilst inevitably restricting myself, I should like to present Kolmogorov's original work as a whole, picking out the basic components and linking them together. The clarity and wide importance of the goals which Kolmogorov set himself are so striking that it is difficult to deny oneself the pleasure of trying to explain the significance of the results he obtained to the average reader irrespective of his or her speciality.

There are different ways of speaking of his scientific achievements. We can give a chronological account, or we can consider separately each scientific direction, or we can classify the papers according to their importance, and so on. I prefer to follow a different path. In the original mathematical work of an outstanding scientist it is possible to differentiate several components: *results* (solutions to difficult problems, finding new formulae), *ideas* (introduction of new concepts and interpretation of old ones, enunciation of problems, working out the beginnings of new scientific directions), *theories* (where the aim is to explain a group of phenomena), *methods and conceptions.* All these are present in Kolmogorov's work. And we shall investigate what makes up each of these components.

—— • ——

We have already spoken of the *conceptions.* Now we shall speak of *theories.* At the very outset of his creative work Kolmogorov was obsessed by "a vague desire to study mathematics, which has links with physics and the natural sciences". This led him to perceive the deep and mysterious secrets of nature.

Here we must first of all speak of his papers on classical mechanics. Can the solar system last forever? This is probably the central problem of astronomy. Is perpetual motion possible in simpler planetary systems, consisting, say, of only three bodies? Or, will perhaps the evolution of a planetary system always (on a set of full measure) end in catastrophe? It was natural to try to consider first the motion of systems close to integrable ones. Problems of this kind date back to Newton and Laplace, and Poincaré called the problem of describing the evolution of Hamiltonian systems for small perturbations "the fundamental problem of dynamics".

In his papers Kolmogorov solved the fundamental problem of dynamics for the three-body problem for the majority of initial conditions in the case of general position. As a direct consequence of Kolmogorov's theorems it was found that a satellite of Jupiter moving in the plane of the circular orbit of Jupiter along an elliptical orbit, perturbed by Jupiter but not perturbing it, always remains in a elliptical orbit. Kolmogorov's theory proved to be applicable to a large number of problems in mechanics and physics (the problem of the stability of rapid rotation of an asymmetric rigid around a fixed point; the problem of magnetic surfaces in systems of tokamak type, and so on). Kolmogorov gave a survey lecture, to which we have referred, on his results at the International Congress of Mathematicians in Amsterdam. The ideas of Kolmogorov, developed in the papers of V. I. Arnol′d and J. Moser were called the KAM theory (the theory of Kolmogorov, Arnol′d and Moser), the name now used by almost every mathematician.

Kolmogorov made a significant contribution to the creation of the theory of stochastic processes. In 1931 he published his paper "Analytic methods in probability theory". The initial methodological position is remarkable for its simplicity and depth. Deterministic processes, in which initial position determines their further development, are replaced by processes where "the state x of the system at some moment of time t_0 stipulates only a known probability for the occurrence of a possible state y at some later moment $t > t_0$". This consideration led Kolmogorov to a definition of a Markov process. He wrote it out in the general form of an integral equation, which had been found for special cases by Smolukhovskiĭ. From this integral equation one can deduce direct equations (already seen in the work of such powerful physicists as Planck, Einstein, Fokker, Smolukhovskiĭ) and inverse equations, which where not known to the physicists. In his work Kolmogorov combined the theory of heat, due to Fourier, the theory of Brownian motion, due to Einstein and Smolukhovskiĭ, and a description of probabilistic random walks, on which Markov and his followers had worked, and also the ideas of Bachelier and Wiener, who constructed the first examples of stochastic processes. In their article on the occasion of Kolmogorov's fiftieth birthday, P. S. Aleksandrov and A. Ya. Khinchin wrote: "In the whole of probability theory in the twentieth century it is difficult to name any other piece of research that has turned out to be so fundamental for the further development of science and its applications than this paper by Kolmogorov."

In the course of his work on the theory of Markov processes, Kolmogorov began a fruitful collaboration with physicists (S. I. Vavilov, M. A. Leontovich and others). Kolmogorov loved to assert that in his joint paper with Leontovich the "physical" part was due to him and the mathematical part to Leontovich.

Physical intuition lead Kolmogorov even in his research along purely abstract lines. According to him homology theory occurred to him from visualizing descriptions of a flow of a liquid over a manifold.

A clear understanding of the physical picture of undamped oscillations with a continuous spectrum served as a lode-star for the creation of the theory of stationary processes. Wiener was very proud of his construction of the theory of interpolation and filtration of stationary stochastic processes and it was with a feeling of some bitterness that he acknowledged Kolmogorov's priority in this area. Stationary processes, that is processes whose probability characteristics remain constant in time, are the idealization of a great number of stochastic natural phenomena (in the atmosphere, in the ocean, and so on); they occur constantly in technical applications, for example in radio engineering. One of the most important problems is the predication of the future by observing the process over a period of time. It was this theory that Kolmogorov and Wiener worked out. It found innumerable applications in different fields of science and technology.

The theory of stationary processes lead Kolmogorov on to turbulence problems. The motion of fluids and gases is subject to deterministic laws; however the very character of the motion turns out to be so complex that it calls to mind a stochastic process. Kolmogorov's papers on turbulence developed the research by the great engineers of this century—Taylor and von Kármán. They completely transformed it and had an enormous influence on the whole further development of this important field of the natural sciences. Among Kolmogorov's concrete results one can point to the widely known "two-thirds law", which has the character of a law of nature: in a turbulent flow (under certain conditions) the mean square of the difference of the velocities at two points, at a distance r (of mean sizes), is proportional to $r^{2/3}$.

Kolmogorov made a very significant contribution to the natural sciences in his research inspired by biological problems. His work in biology led him to the solution of a series of problems in mathematical biology and to remarkable results in pure mathematics. In a joint paper with I. G. Petrovskiĭ and N. S. Piskunov, the stimulus for the writing of which was their work on biological problems, they were the first to construct a mathematical theory of the stability of solutions of travelling wave type for the diffusion equation with a nonlinear right-hand side. (The list of papers on this subject extends at present to some thousand items.) Here the physical side, the formulisation and description of the qualitative picture of the phenomenon is due to Kolmogorov. ("I had noticed how a Bickford fuse burns," he said of this, although the work arose from attempts to describe the propagation of a gene.)

Biological problems led Kolmogorov to formulate a theory of branching processes.

Had Kolmogorov done nothing more than create the theory of stochastic processes and the theory of turbulence and laid the foundations of KAM theory, he would still have been one of the great stars in the world of science.

In his monograph "Limit distributions for sums of independent random variables" Kolmogorov presented the results of the initial stage of the summation of random variables, which had been begun in the work of Jacques Bernoulli, de Moivre and Laplace, and had been continued by Poisson, Chebyshev, Markov, Bernstein and others.

During the war Kolmogorov made an important contribution to the theory of ballistics, and, according to the experts, his results completely transformed this field. Reduced to tables, graphs and nomograms, they helped to achieve victory in the Great Fatherland War.

And finally we must mention his contribution to information theory, where (with Khinchin and others) he laid the mathematical foundations of the theory and obtained some interesting formulae and theorems.

—— • ——

What are the fundamental *results* obtained by Kolmogorov?

Let us begin with the metric theory of functions. The best known work from the first period of his scientific activity was his example of a Fourier series divergent everywhere.

If $x(\cdot)$ is integrable on $[-\pi, \pi]$ $(\Leftrightarrow x(\cdot) \in L_1([-\pi, \pi]))$, then from it one can construct a Fourier series:

$$\sum_{h\in\mathbb{Z}} x_k e^{ikt}, \quad \text{where } x_k = (2\pi)^{-1}\int_{-\pi}^{\pi} x(t)e^{-ikt}\,dt.$$

Fourier series date back to the eighteenth century, but the theory of their summation began to develop in the middle of the nineteenth century.

The following result discovered by the 19-year old Kolmogorov caused a sensation.

THEOREM 1. *There is a function integrable on* $[-\pi, \pi]$ *whose Fourier series diverges almost everywhere.*

In 1926 Kolmogorov constructed an example of a series that diverges everywhere.

The question of whether for every square integrable function $(x(\cdot) \in L_2([-\pi, \pi]))$ the Fourier series converges almost everywhere was raised by N. N. Luzin in 1915, was solved by Carleson in 1966, and was completed by Hunt (for any function $x(\cdot)$ in $L_p([-\pi, \pi])$ for $p > 1$, its Fourier series converges to $x(\cdot)$ almost everywhere). In particular, it was not known until Carleson's theory whether it is possible to replace $L_1([-\pi, \pi])$ in Theorem 1 by $C([-\pi, \pi])$, the set of continuous functions on $[-\pi, \pi]$.

In the field of trigonometric series, in which H. Weyl, Hardy, Luzin, Men′shov, Bari and many other outstanding mathematicians worked, two results stand out by reason of their power and completeness: Kolmogorov's example and the Carleson-Hunt theorem.

In connection with Luzin's problem it is appropriate to mention another remarkable result which Kolmogorov found jointly with D. E. Men'shov.

THEOREM 2. *There is an orthogonal system on the unit interval consisting of functions of unit modulus, and a square integrable function on the same interval such that the Fourier series of this function with respect to the given system diverges everywhere.*

In the same note Kolmogorov stated the following result: there is a function in $L_2([-\pi, \pi])$ such that the Fourier series with respect to a rearranged trigonometric system diverges almost everywhere (so that Carleson's theorem reflects a very deep fact concerning *harmonics in their natural order*).

Kolmogorov did not publish his proof of this theorem, and this result was proved in 1960 by the Polish mathematicians Zahorski.

The next theorem, like the two previous ones, is very simple in its formulation, but its significance is not so immediately obvious to anyone who is not connected with harmonic analysis.

One of the most actively developing branches of harmonic analysis studies problems on the convergence of Fourier sums and of other important integral operators in various function spaces. One of the most important operators is the Hilbert transform H, associating to a function x the function $\tilde{x}$ that is conjugate to x in the sense of complex analysis. We have the formula

$$Hx(t) = \tilde{x}(t) = -(2\pi)^{-1} \int_{-\pi}^{\pi} x(t+\tau) \cot(\tau/2)\, dt.$$

Various generalisations of the Hilbert transform (singular integral operators) play a very important part in classical analysis. The Hilbert transform maps $L_p([-\pi, \pi])$ into $L_p([-\pi, \pi])$ for $p > 1$ (M. Riesz), or, as we say, this is a transformation of strong type (p, p) for $p > 1$. For $p = 1$, it is not a transformation of strong type $(1, 1)$. However, we have the following result.

THEOREM 3. *Let $x(\cdot)$ be a function integrable on $[-\pi, \pi]$. Then the measure of those points where the Hilbert transform $Hx(\cdot)$ is greater than α in modulus satisfies the inequality*

$$\operatorname{meas}\{t\colon |Hx(t)| > \alpha\} \le C\alpha^{-1} \|x\|_{L_1[-\pi, \pi]}.$$

In modern language this inequality means that the operator H is of weak type $(1, 1)$. Theorem 3 proved to be one of the starting points for a very broad field of mathematics, with which are linked the names of M. Riesz, Marcinkiewicz, Zygmund, Calderon, Hardy, Littlewood, Paley and others, whereas the very concept of weak type was very important in the theory of singular operators.

It is very difficult to know which result to choose when talking of Kolmogorov's work on probability. Kolmogorov deemed this field to be his "narrow speciality", and in it he was unquestionably a world leader. His input here was enormous.

At the beginning of his work in this field he found some fundamental theorems crowning the research of Chebyshev, Markov, Bernstein and others. In classic probability theory there were two fundamental themes—the law of large numbers and limit theorems. In this century, a third one has been added—the law of the iterated logarithm discovered in the simplest situations by Khinchin. In particular, Kolmogorov found a necessary and sufficient condition for the law of large numbers, and very significantly extended the limits of applicability of the law of the iterated logarithm. He made other fundamental discoveries. We shall mention only his definitive results, relating to the strong law of large numbers. We owe the concept of the strong law to Borel. A sequence of independent random variables satisfies the strong law of large numbers if its sequence of arithmetic means converges almost certainly to some number (unlike the law of large numbers, where it converges in a weaker sense—in measure).

THEOREM 4. *Let* $\{\xi_n\}_{n\in\mathbb{N}}$ *be a sequence of independent equally distributed variables. Then the condition of finiteness of the mathematical expectation of* $|\xi_1|$ *(and hence of all* $|\xi_n|$*) is necessary and sufficient for it to satisfy the strong law of large numbers.*

One of the basic and natural questions is: Which distributions are limits of sums of mutually independent normalized terms? It turned out that under natural assumptions the answer is: Only the infinitely divisible distributions, introduced by Bruno de Finetti. A random variable subordinate to such a law can be represented as a sum of n independent equally distributed terms, for any n. The problem then was how to describe infinitely divisible laws.

THEOREM 5. *Let* $F(*)$ *be an infinitely divisible distribution with finite dispersion, and* $\varphi(*)$ *its characteristic function,* $\varphi(*) := \exp|\varphi(*)|$. *Then we have the formula*

$$\phi(t) = i\beta t + \int_{\mathbb{R}} (e^{itx} - 1 - itx)\, d\lambda(x)/x^2,$$

where $\beta \in \mathbb{R}$ *and* $\lambda(*)$ *is the Borel measure on* $\mathbb{R}$, $\lambda(0) = 0$.

Later, using a completely different method, P. Lévy removed the requirement of finite dispersion,[1] and then Khinchin showed that Lévy's general result could be easily obtained by Kolmogorov's method. "In this way," Khinchin wrote, "this method gives the simplest and clearest to-date substantiation of the canonical form of infinitely divisible laws, making their totality easy to visualize."

[1] Here the last term is replaced by

$$\int_{\mathbb{R}} \left(e^{i\lambda x} - 1 - \frac{i\lambda x}{1+x^2}\right) \frac{1+x^2}{x^2}\, d\lambda(x).$$

We spoke earlier of the theory of stochastic processes, and here we shall mention only two of Kolmogorov's remarkable results, concerning a stationary random sequence: *a criterion for the regularity of a sequence* (that is, a criterion that observation of a sequence implies the influx of new information) and *a formula for the error of prediction.*

The question of regularity is primary in the theory of stationary sequences and processes. M. G. Krein solved it for processes.

In mathematical statistics Kolmogorov proved one of the most fundamental results in this discipline. Given a random variable ξ, whose distribution function is F, let $\{x_i\}_{i=1}^n$ be the results of n independent observations of this variable, arranged in increasing order. The function $F_n(x) = 0$ if $x < x_i$, $F_n(x) = k/n$ if $x_k \leq x < x_{k+1}$, $F_n(x) = 1$ if $x \geq x_n$, is called the empirical distribution function.

One of the fundamental questions in statistics then arises: To what extent does the empirical distribution function $F_n(*)$ "resemble" the function $F(*)$ itself? We then ask: Is it true that $P[\sup|F(x)-F_n(x)| < \varepsilon) \to 1$ as $n \to \infty$? Glivenko gave the answer to this question, and in the same issue of the journal Kolmogorov not only answered this question, but he also gave the limit distribution of the variable $D_n = \sqrt{n}\sup|F_n(x)-F(x)|$. More precisely, the following theorem holds:

THEOREM 6. *Let $F(*)$ be a continuous distribution function. Then the probability $\Phi(\lambda)$ that $D_n < \lambda$ as $n \to \infty$ uniformly in λ tends to*

$$\Phi(\lambda) = \sum_{h\in\mathbb{Z}}(-1)^h \exp(-2k^2\lambda^2)$$

(*Kolmogorov's distribution*).

The question of how to measure the degree of approximation of $F_n(*)$ to $F(*)$ occurred to such great scholars as von Mises and Cramer, but they were far from answering it. It is only natural that Kolmogorov's test of goodness of fit is found in all textbooks on mathematical statistics.

There is one result in topology of which Kolmogorov spoke with satisfaction and of which he was, I think, proud. A map $f: X \to Y$ of a topological space X onto a topological space Y is called *open* if the image of any open set under this map is open. "The problem of whether the dimension can increase under an open map was of very great interest to P. S. Aleksandrov. For some time we worked together on proving the nonexistence of an open dimension-raising map. During these attempts the reasons for our failure were gradually revealed. This analysis of our failures led in the end to a counterexample." Thus they proved the following result.

THEOREM 7. *There exists an open map of a one-dimensional compactum onto a two-dimensional one.*

This work acted as a stimulus in other areas of research, among which we must first mention the work of L. V. Keldysh. In particular, by a very complicated procedure, she constructed an open map of a one-dimensional compactum onto a square. Many years later, I. M. Kozlovskiĭ constructed an open map from a one-dimensional compactum onto a square by a very transparent and simple method, using Kolmogorov's construction as a fundamental element.

In 1969 Kolmogorov obtained a result on summation of random variables, of which he had been dreaming since the thirties. He was clearly satisfied when he reported on this work at a seminar. We have already spoken of infinitely divisible distributions. We denote the totality of such distributions by D. Let $F(*)$ be the distribution function of the random variable ξ, and $\Phi_n(*, F)$ the distribution function of the sum of n independent random variables that have distribution $F(*)$. We recall that only infinitely divisible distributions can be limits of sums of independent equally distributed random variables. Is it true that the distributions of sums of equal independent terms are uniformly close to D? The answer is given by the following theorem.

THEOREM 8. *There is a constant C such that for any distribution $F(*)$ and for any $n \in \mathbb{N}$ there is a $\Psi(\cdot) \in D$ such that*

$$|\Phi_n(x, F) - \Psi(x)| \leq Cn^{-1/5}.$$

At once the question arose about the exponent $1/5$: Can it be improved? Among those working on this were Yu. V. Prokhorov, L. D. Meshalkin and Kolmogorov himself. The definitive solution was found by T. V. Arak (1981–1982); the optimal exponent turned out to be $2/3$.

Do there exist functions of several variables? That would seem to be a pointless question—of course they exist. Let us phrase this question more precisely. Some functions of three variables can be expressed in terms of functions of two variables, say, $f_1(x, y, z) = \varphi_1(x, \psi_1(y, z))$ or $f_2(x, y, z) = \varphi_2(\psi_2(x, y), \chi_2(y, z))$, and so on. In these cases we say that f_1 and f_2 are superpositions of functions of two variables. Confidence that functions of three variables are not reducible to functions of two variables (in the sense that not all are representable as superpositions of functions of two variables) was so great that Hilbert (Hilbert again) indicated a concrete analytic function of three variables which (as he thought) could not be represented as a superposition of continuous functions of two variables (Hilbert's thirteenth problem). Alas! this expectation was not to be fulfilled. It turned out that *any* continuous function of any number of variables can be represented as a superposition of continuous functions by using only one function of two variables, namely $s(x, y) = x + y$, and the remaining functions are of one variable. Here is the story of this discovery.

In 1956 Kolmogorov proved in an article that any continuous function of four variables can be represented as a superposition of continuous functions of three variables. Refuting Hilbert's conjecture was reduced to a concrete problem about the representation of functions defined on universal trees in $\mathbb{R}^3$.

In Kolmogorov's opinion, this paper of 1956 was his most complicated technical achievement, requiring the longest period of uninterrupted thought in the whole of his life. As we have already said, he left the final step to his successors. In the spring of the following year a third-year undergraduate student, V. I. Arnol′d, obtained the necessary result about functions on universal trees, and thus Hilbert's thirteenth problem was completely solved (that is, Hilbert's conjecture was disproved). Soon after that Kolmogorov devised a comparatively simple, natural and very beautiful construction, which led to the following remarkable theorem, which we have already mentioned.

THEOREM 9. *For any integer* $n \geq 2$ *there are continuous functions* $\psi_{ij}(*)$ *defined on* $I = [0, 1]$ *such that any function* f *continuous on* I^n *can be represented in the form* $f(x_1, \dots, x_n) = \sum_{i=1}^{2n+1} \chi_i(\sum_{j=1}^{n} \psi_{ij}(x_j))$, *where the function* $\chi_1(*)$ *is continuous on* $\mathbb{R}$.

We conclude this section devoted to results with a theorem relating to the ergodic theory of dynamical systems, in which a solution was given to a problem raised by von Neumann and going back more than 25 years.

A dynamical system is a pair (T, A), where $T = (T, \Sigma, \mu)$ is a measure space, and $A\colon T \to T$ is a measure-preserving map. In the space $L_2(T)$ the map A generates a natural unitary operator: $Ux(t) = x(At)$. The spectrum of this operator is called the spectrum of the dynamical system. The problem of the classification of dynamical systems up to an isomorphism was studied for many years as the basic problem in ergodic theory. At the beginning of the 1930 von Neumann proved that two dynamical systems with a common point spectrum are isomorphic. It was assumed to be plausible that the spectrum always uniquely determines the dynamical system. "Bernouilli schemes" provide an example of dynamical systems with a special kind of spectrum. We suppose that independent tests have been carried out, for which 1 occurs with probability p, and 0 with probability $1 - p$, $0 < p < 1$. Such a procedure gives a measure on the set of all sequences of 0's and 1's, and a shift by one step in such sequences generates a dynamical system, known as a Bernoulli scheme.

THEOREM 10. *Bernoulli schemes for different* p *are nonisomorphic.*

This led to the solution of von Neumann's problem and signalled the beginning of a period of rapid growth in this theory.

We have cited only ten results from the rich collection of Kolmogorov's theorems. Again we must express our astonishment at the number of fundamental original problems (therefore so simple in their formulation) that he managed to solve.

—— • ——

Now it is time to turn to the *ideological aspects* (*ideas*) of Kolmogorov's creativity. In his early period Kolmogorov spent a lot of energy on the interpretation (within the framework of the axiomatic method) of the limits of the fundamental concepts of mathematics. We give his answers to questions about the most primary structures.

What is an integral? Recall the evolution of this concept: Archimedes, Newton, Leibnitz, Cauchy, Riemann; then the Lebesgue integral, then the integrals of Stieltjes, Hellinger and others; later the integration of vector-valued functions (in infinite-dimensional Fréchet space). Kolmogorov proposed the following simple definition of the integral, which covers all the above. Let T be some set and $\mathfrak{M}$ a distinguished system of subsets of T such that $A \in \mathfrak{M}$, $B \in \mathfrak{M} \Rightarrow A \cap B \in \mathfrak{M}$. Further, assume given a generally speaking many-valued function f with values in a vector space, defined on any $A \in \mathfrak{M}$. We define the integral of f on the set T for a given system $\mathfrak{M}$ as the limit (in the sense of Shatunovskiĭ-Moore-Smith) of the Riemann sums $(Rf)_{\mathscr{T}}(T) = \sum_{n\in\mathbb{N}} f(T_n)$, where $\mathscr{T} = \{T_n\}_{n\in\mathbb{N}}$ is a partition of T, i.e., $T = \bigcup_n T_n$, $T_i \cap T_j = \varnothing$, $i \neq j$.

We denote this integral by $\mathfrak{M}\int f\,dT$. This construction subsumes all the concepts of integration we spoke of earlier (Riemann, Lebesgue, Stieltjes, and so on). It implies, in particular, the following most unusual fact: the Lebesgue integral is a limit of Riemann (not Lebesgue!) sums. Kolmogorov had a high opinion of his paper "Study of the concept of the integral" and often expressed regret that its definition had not yet taken root.

Is it possible to extend substantially the concepts of derivative, integral and summation of divergent series? This was the problem considered by Kolmogorov in a paper of 1925. It turned out that the search for effective methods of differentiation, integration or summation of series came up against the same difficulty as did the effective construction of non-measurable functions. For example, there is a method of summing divergent series that satisfies the conditions

$$\text{(a)}\quad \sum_{n\in\mathbb{N}} x_n = x_1 + \sum_{n\in\mathbb{N}} x_{n+1}, \qquad \text{(b)}\quad \sum_{n\in\mathbb{N}} a x_n = a \sum_{n\in\mathbb{N}} x_n,$$

and if by using this it is possible to define the sum of the series of the form $\sum_{n\in\mathbb{N}} \sin 3^n x$, then it is possible to construct an effective example of a function that is nonmeasurable in the Lebesgue sense.

What is the k-dimensional measure of a set situated in n-dimensional Euclidean space? Here too, in his paper of 1933, Kolmogorov approached the problem from an axiomatic point of view. The following result holds.

Definition. Let μ be a set function defined on all Suslin sets $E \in \mathbb{R}^n$ ($\Leftrightarrow E \in \mathrm{Sus}(\mathbb{R}^n)$) and satisfying the following axioms:

(1) $E \subset \bigcup E_n$, $(E, E_n \in \mathrm{Sus}(\mathbb{R}^n)) \Rightarrow \mu(E) \leq \sum_{n\in\mathbb{N}} \mu(E_n)$.

(2) $E_i \in \mathrm{Sus}(\mathbb{R}^n)$, $E_i \cap E_j = \varnothing$, $\bigcup_{i=1}^{s} E_i \subset E \Rightarrow \sum_{i=1}^{s} \mu(E_i) \leq \mu(E)$,

(3) $\mu(I^n) = 1$, where $I^k = [0, 1]^k$ denotes the k-dimensional Euclidean cube,

(4) $E' := \varphi(E)$, $|\varphi(x) - \varphi(x')| \leq |x - x'| \Rightarrow \mu(E') \leq \mu(E)$ (i.e., if E' is a "nonexpanded" image of E, then its measure is not greater than that of E). Then we say that μ is a k-dimensional measure. At the time when the paper referred to appeared, many different k-dimensional measures had been constructed (lengths of curves, Hausdorff measures, Carathéodory measures, and so on), satisfying the axioms (1)–(4). In this paper Kolmogorov obtained the following fundamental result: for any natural number $k < n$ there are two special k-dimensional measures, the maximal measure $\overline{\mu}^k$ and the minimal measure $\underline{\mu}^k$, such that any k-dimensional measure μ of any Suslin set E lies between $\underline{\mu}^k(E)$ and $\overline{\mu}^k(E)$: $\underline{\mu}^k(E) \leq \mu(E) \leq \overline{\mu}^k(E)$. Both measures $\underline{\mu}^k$ and $\overline{\mu}^k$ are defined in a completely natural way.

Let us turn to descriptive set theory. We shall not talk about it in any detail; this field used to be at the epicentre of the interests of the Moscow Mathematical School, but now relatively few are interested in it. In 1916 P. S. Aleksandrov in Russia and F. Hausdorff in Germany independently solved the continuum problem for Borel sets, proving that any Borel set is either countable or has the power of the continuum. For this Aleksandrov and Hausdorff used the property that any Borel set can be obtained by means of some constructive procedure. A clear description of this procedure was given by Suslin: he called it the A-operation. Given a sequence of sets $\mathscr{E} = \{E_{s_1 \cdots s_n}\}$, numbered "by processions" $\{s_1, \dots, s_n\}$, $s_i \in \mathbb{N}$, we construct the set $A(\mathscr{E})$ according to the following rule:

$$A(\mathscr{E}) = \bigcup (E_{s_1} \cap E_{s_1 s_2} \cap E_{s_1 s_2 s_3} \cap \cdots \cap E_{s_1 \cdots s_n}),$$

where the union is carried out over all the sequences $\{s_n\}_{n \in \mathbb{N}}$. Then we say that $A(\mathscr{E})$ is obtained from $\mathscr{E}$ by an A-operation.

In the same year, 1916, Suslin obtained a result that made a very great impression on his contemporaries. He proved that there are sets obtained from closed subsets of an interval by means of an A-operator that are not Borel sets. Sets obtained from closed ones by means of an A-operator were called Suslin sets. Thus the class of Suslin sets (or A-sets) proved to be broader than that of Borel sets (B-sets) and, what is more, it was soon discovered that many natural procedures of analysis lead to A-sets and A-measurable functions that are not B-measurable.

In the autum of 1921 Aleksandrov gave a course of lectures at the University of Moscow on descriptive set theory. Apparently he spoke in this course of the Γ-operation he had introduced—an operation "complementary" to the A-operation. Kolmogorov attended this course. Later he wrote: "In 1921–1922 I drew up a broad plan of research into descriptive set theory." He immediately tried to work out a maximal general scheme of set-theoretic operations (δ_s-operations).

Every map $f \colon \mathbb{N} \to \mathbb{N}$ generates a sequence $\{f(n)\}_{n \in \mathbb{N}}$. Let $\mathscr{F} = \{f\}$ be some family of such maps and $\mathscr{E} = \{E_n\}_{n \in \mathbb{N}}$ some sequence of sets (for definiteness—subsets of the interval $I = [0, 1])$.

Definition. If the set $\Phi_{\mathcal{F}}(\mathcal{E})$ is formed by the rule

$$\Phi_{\mathcal{F}}(\mathcal{E}) = \bigcup_{f \in \mathcal{F}} \left(\bigcap_{n \in \mathbb{N}} E_{f(n)} \right)$$

then we say that it is *obtained from* $\mathcal{E}$ *by means of the* δ_s*-operation generated by* $\mathcal{F}$. An A-operation is obviously a δ_s-operation. Kolmogorov defined in a general form a complementary δ_s-operation, of which Aleksandrov's Γ-operation is a special case.

Kolmogorov's first mathematical research was devoted to these questions. The work was finished in January 1922, while he was a second-year undergraduate student. This paper contains the germ of the general theory of operations on sets, which was later developed by Hausdorff, Kantorovich, A. A. Lyapunov and others.

In it he proved the following fundamental result.

There is a set of the form $\Phi_{\mathcal{F}}(\mathcal{E})$ *the complement of which is not a set of the form* $\Phi_{\mathcal{F}}(\mathcal{E}')$ *for any sequence* $\mathcal{E}' = \{E'_n\}_{n \in \mathbb{N}}$ *of (closed) sets* E'_n.

Two of Kolmogorov's papers, from 1925 and 1932, are on mathematical logic. The 1925 paper is in fact the first paper on mathematical logic by a Soviet mathematician. It was of very great significance; witness its inclusion in Heijenoort's famous book on mathematical logic *From Frege to Gödel*.

From the beginning of the century there was no end to arguments about the essence of mathematics, about the applicability of logical laws, about paradoxes in set theory. All this led to a new line, "intuitionism", which recognized the invalidity of the application of "tertium non datur" (the law of the excluded middle) in the field of transfinite induction. In his paper "Tertium non datur" Kolmogorov proposed an interpretation of classical mathematics in which all its propositions are converted into intuitionistic propositions. Thus Kolmogorov proved that all finite conclusions obtained by means of transfinite application of the "tertium non datur" principle are true and can be proved without its help.

In the 1932 paper, Kolmogorov proposed investigating "the logic of the solution of problems" (say, problems on construction) side-by-side with "the logic of proof". To the principle of syllogism: "if b follows from a, and c follows from b, then c follows from a" there corresponds "if the solution of problem a reduces to a solution of problem b, and the solution of problem b reduces to a solution of problem c, then the solution of problem a reduces to a solution of c". Kolmogorov introduced the corresponding symbolism and constructed a calculus of problems. The form of this calculus coincided with the system, formulated just a bit earlier by Heyting, of the axioms of intuitionistic logic. This gave another interpretation of intuitionistic logic.

Kolmogorov published two papers on geometry. One of them ("On the foundations of projective geometry", 1932) reflects his youthful love of projective geometry "which was old-fashioned, but on which Aleksei Konstantinovich Vlasov lectured in a truly talented way" (Kolmogorov always listed Vlasov among his teachers). In the early 1930s "synthetic structures" were being developed intensively, when, we might say, algebra united with topology (recall the representation theory of compact groups). Kolmogorov was in at the beginning of "topological geometry". How beautiful and unexpected the links between geometry and topology can be is shown by the following result of Kolmogorov.

Let three systems of elements (called points, straight lines and planes) satisfy the projective axioms of combination, where each of the systems is a connected compactum and the relations of combination are continuous.[2] Then a geometry that has the three given systems as objects is isomorphic to either $\mathbf{RP}^3$ or $\mathbf{CP}^3$ or $\mathbf{HP}^3$, i.e., either real, or complex, or quaternion projective geometry.

Here the most important thing is the idea of combining geometry and topology, along with the aesthetic element included in the formulation. The main point of the proof is the reference to Pontryagin's theorem (proved on Kolmogorov's suggestion) giving a description of all the connected locally compact topological skew-fields with countable base (there are only three: $\mathbf{R}$, $\mathbf{C}$ and $\mathbf{H}$). These facts undoubtedly belong to the golden treasure-house of topological algebra. This fully justifies the thought repeatedly expressed by Kolmogorov: aesthetic motives not infrequently have interesting consequences.

Four papers by Kolmogorov are concerned with functional analysis and in each of them there are important ideological elements.

In a paper of 1931 Kolmogorov gave the first criterion for compactness in L_p. Later M. Riesz gave a criterion similar to the Arzela criterion—more usual and natural, but Kolmogorov's result remains significant, since it can easily be extended to any measure space, whereas Riesz's cannot.

The best-known of these papers on functional analysis is undoubtedly a paper from 1934, which contains the definitions of a *linear topological space* and a *bounded set in it* (in his historical survey, Bourbaki singles out the importance of these concepts for the further development of functional analysis) and in which the following classical criterion of normability is proved: *for a linear topological space to admit a norm it is necessary and sufficient that it have a bounded convex neighborhood of zero.*

At the sources of the theory of normed rings created later by I. M. Gel′fand is their joint paper of 1935, which contains the following beautiful result:

Let T_1 and T_2 be two topological spaces (satisfying the first axiom of countability). For T_1 and T_2 to be homeomorphic it is necessary and sufficient that the rings $C(T_1)$ and $C(T_2)$ (of continuous functions on T_1 and T_2,

---

[2] We say that a straight line passing through two points depends continuously on these two points, and so on.

respectively) or the rings $C^b(T_1)$ *and* $C^b(T_2)$ *(of continuous bounded functions) be algebraically isomorphic; that is, an algebraic isomorphism of the algebras of continuous functions determines a topological homeomorphism of the compacta themselves.*

In Kolmogorov's papers on approximation theory the formulation of the problem of revealing the connection between the concepts of approximation and smoothness is central; it is to Kolmogorov that we owe the very important concepts of *n-diameter* and *ε-entropy*.

Kolmogorov made a fundamental contribution to algebraic topology. To him we owe the introduction of what is perhaps the central concept in the whole of this field—that of cohomology, one of the concepts fundamental to the whole of modern mathematics. Kolmogorov also defined the most significant operation in cohomology groups—*the product operation*, giving a ring structure.

The American topologist Alexander had the same idea at the same time. Kolmogorov completed his own theory with the remarkable discovery of the duality for closed sets in compact spaces.

In the mid-1950s Kolmogorov turned to the theory of algorithms and the theory of automata and set himself a task close in spirit to the one we discussed in connection with the theory of operations on sets. It is very well expressed in the notes to his 1958 joint paper with V. A. Uspenskiĭ. "Authors striving to interpret the concepts of 'a computable function' and 'an algorithm' have tried to survey the existing variants of the definitions in the literature and finally convinced themselves that there is no hidden possibility of extending the range of the concept of 'computable function'. The result of this work is the broadest (apparently, of all possible) definitions of the basic objects of the theory—algorithms and automata."

In ergodic theory Kolmogorov introduced two fundamental concepts—*metric entropy* (Kolmogorov's idea was definitively formulated by Ya. G. Sinaĭ) and a *quasiregular dynamical system*.

In almost every field to which he turned his attention, Kolmogorov initiated one or several principal scientific lines, on which tens or hundreds of research workers began to work, survey articles and monographs appeared and scientific conferences were arranged. For example, in the 1930s alone and in the field of probability alone he laid the foundations of the theory of probability and of stochastic processes; he began to develop Markov processes, the theory of Markov chains with a countable number of states; in statistics, the theory of statistical criteria and (jointly with Khinchin) the theory of stationary processes. These directions appear as separate chapters in modern courses on probability theory, mathematical statistics and stochastic processes, and if we look at the content of these courses, we get the impression that the sum total of these new directions created by one man in the course of less than ten years can be compared with the sum total of the new

V. M. TIKHOMIROV, A. N. KOLMOGOROV,
AND S. SADIKOVA-PROKHOROVA

directions that have emerged in these sciences over the last fifty years.

And Kolmogorov's ideas themselves always had a remarkable vitality. They were quickly taken up, where overtaken by successors, were interwoven with others, forming scientific directions and schools.

—— • ——

There is a multitude of *methods*, constructions and lemmas that Kolmogorov devised to solve individual problems, and they are constantly being used in further research. He developed particularly many methods in the theory of probability and stochastic processes (starting, say, with the well-known Kolmogorov inequality); he introduced many constructive ideas into the theory of functions, topology and other fields. For example, the number of uses of the comparison theory—an auxiliary result, which he obtained when proving the theorem on the intermediate derivative—is exceptionally large. The same applies to the principle of introducing a new invariant of entropy type to prove the nonisomorphism of various aggregates. But what had the greatest effect was his method for solving functional equations with small denominators—a special kind of implicit function theorem—which is the basis of KAM-theory.

—— • ——

More than once in this article there has appeared the name of one of the greatest mathematicians of all time—the name of David Hilbert. Here

I should like to draw some comparisons between the genius of Hilbert and that of Kolmogorov.

Intellectual achievements cannot be completely classified (they make up "a partially ordered set"), geniuses cannot be compared and my comparison of them arises not from a wish to contrast one with the other or to elevate the one at the expense of the other, but from an attempt to demonstrate how far creative styles can differ, by what different paths they can achieve their great aims.

Hilbert had a purposefulness not matched I dare say by anyone else's. His scientific biography is sharply divided into periods, dedicated to work in one field.

Hilbert's first creative period was devoted to algebra and arithmetic (the theory of invariants 1885–1893, algebraic number theory 1893–1898); in the first decade of the century in the course of two summers (1908–1909) he solved Waring's problem. Kolmogorov did not write any papers on algebra or arithmetic, just as Hilbert did not work in the field of random phenomena.

The next stage of Hilbert's interests was geometry (1989–1902). It ended with his classic book *Grundlagen der Geometrie.*

Kolmogorov did work in the field of geometry and in two special areas of it— topology and algebraic topology, as we recall, in the 1930s, while working at the same time in the field of analysis, probability theory, function theory and many others.

Then came Hilbert's "analytic" period (1900–1910), when he developed the calculus of variations and infinite-dimensional analysis (in particular, the theory of integral equations).

Kolmogorov turned to problems of mathematical analysis and the theory of functions in two main periods of his life (trigonometrical and orthogonal series, the theory of approximation, functional analysis in the 1930s, dynamical systems, superpositions, ergodic theory in the 1950s).

Then came for Hilbert the period when he solved mathematical problems in the natural sciences, mainly in mathematical physics (1910–1922).

Kolmogorov studied many problems in the natural sciences—biology and physics in the 1930s, turbulence, geology and meteorology in the 1940s, celestial mechanics in the 1950s, and oceangraphy in the 1970s.

Hilbert's last period was devoted to the logical foundations of mathematics. Kolmogorov was interested in logic both in the 1930s and 1950s, and in the very last period of his creativity. He, like Hilbert, concluded his creative life with work on logic.

As we said, Hilbert did no work on random phenomena. The theory of probability, on the contrary, was a constant concern of Kolmogorov's, he was involved in it in the 1920s, the 1930s, and in the 1940s and 1950s, and moreover, he was involved not just in classical problems, but also in random processes, in statistics, in information theory and in numerous applications.

Hilbert could only work on one subject at a time; Kolmogorov, at the heights of his creativity, could work down to the foundations, to the roots, to the very essence of the subject; Kolmogorov was striving to go forward, new goals beckoned him, he tried hard to conquer unattainable heights, leaving it to others to incorporate the new territories. But both achieved most brilliant results, they introduced most fundamental ideas, they worked out most deep theories, they created most fruitful methods and basic concepts.

And they were both great teachers. The number of mathematicians influenced by them cannot be counted. But the number of their actual students can be compared. The list of Hilbert's disciples is given in his collected works. The specialities of the students correspond exactly to the various interests of their teacher at the given time. And now we should like to present a list of Kolmogorov's pupils:

A. M. Abramov (education),

V. M. Alekseev (classical mechanics),

A. M. Arato (probability theory),

V. I. Arnol′d (superpositions, classical mechanics), Corresponding Member of the Academy of Sciences of the USSR,

E. A. Asarin (complexity),

G. M. Bavli (probability theory),

G. I. Barenblatt (hydrodynamics),

L. A. Bassalygo (information theory),

Yu. K. Belyaev (stochastic processes),

V. I. Bityutskov (probability theory),

E. S. Bozhich (mathematical logic),

L. N. Bol′shev (mathematical statistics), Corresponding Member of the Academy of Sciences of the USSR,

A. A. Borovkov (probability theory), Corresponding Member of the Academy of Sciences of the USSR,

A. V. Bulinskiĭ (stochastic processes),

N. A. Dmitriev (stochastic processes),

R. L. Dobrushin (probability theory),

A. N. Dvoĭchenkov (theory of functions),

E. B. Dynkin (stochastic processes),

V. D. Erokhin (approximation theory),

M. K. Fage (functional analysis),

S. V. Fomin (ergodic theory),

G. A. Gal′perin (dynamical systems),

I. M. Gel′fand (functional analysis), Member of the Academy of Sciences of the USSR,

B. V. Gnedenko (probability theory), Member of the Academy of Sciences of the USSR,

O. S. Ivashev-Musatov (theory of functions),

A. T. Kondurar′(theory of functions),

M. V. Kozlov (stochastic processes),

V. V. Kozlov (probability theory),
V. P. Leonov (probability theory),
L. A. Levin (complexity),
A. I. Mal′tsev (mathematical logic), Member of the Academy of Sciences of the USSR,
P. Martin-Lef (complexity),
A. V. Martynov (probability theory),
R. F. Matveev (stochastic processes),
Yu. T. Medvedev (mathematical logic),
L. D. Meshalkin (probability theory, ergodic theory),
V. S. Mikhalevich (probability theory), Member of the Academy of Sciences of the USSR,
M. D. Millionshchikov (turbulence), Member of the Academy of Sciences of the USSR,
A. S. Monin (oceanology, turbulence), Corresponding Member of the Academy of Sciences of the USSR,
S. M. Nikol′skiĭ (approximation theory), Member of the academy of Sciences of the USSR,
A. M. Obukhov (atmospheric physics, turbulence), Member of the Academy of Sciences of the USSR,
Yu. S. Ochan (set theory),
Yu. P. Ofman (complexity),
B. Penkov (probability theory),
A. A. Petrov (probability theory),
M. S. Pinsker (information theory),
A. V. Prokhorov (study of prosody),
Yu. V. Prokhorov (probability theory), Member of the Academy of Sciences of the USSR,
Yu. A. Rozanov (stochastic processes),
M. Rozenblat-Rot (stochastic processes),
B. A. Sevast′yanov (stochastic processes), Corresponding Member of the Academy of Sciences of the USSR,
A. N. Shiryaev (stochastic processes),
F. I. Shmidov (theory of functions),
Ya. G. Sinaĭ (ergodic theory),
S. Kh. Sirazhdinov (probability theory), Member of the Academy of Sciences of the Uzbekistan Soviet Socialist Republic,
V. M. Tikhomirov (approximation theory),
A. N. Tulaikov (theory of functions),
V. A. Uspenskiĭ (mathematical logic),
I. Ya. Verchenko (theory of functions),
V. G. Vinokurov (probability theory),
V. G. Vovk (complexity),
A. M. Yaglom (turbulence),

B. M. Yunovich (theory of functions),
V. N. Zasukhin (stochastic processes),
I. G. Zhurbenko (probability theory),
V. M. Zolotarev (probability theory).

Kolmogorov did most of his research work in his house in Komarovka, outside Moscow. He loved this house very much and there he played the role of gardener. As a result of his fifty years of work in the Komarovka garden, it had huge birches, larches, maples, crab-apples, the whole garden was full of flowers—lilacs, roses, peonies. Whatever Kolmogorov planted flourished. Once, many years ago, he gave a colleague a bush of sweet briar. This colleague is no longer and now little remains of the once rich and well-tended garden. And only the sweet briar, given by Kolmogorov, to this day does not cease to flower and to bear fruit. This luck of a gardener illuminated both his original research and his work as a teacher; everything he touched flowered and bore fruit.

Let us go back to where we started.

The heritage of a great man is riches beyond price, the comprehension of which confirms our faith in mankind, and the personality of a genius sets a moral example and lets us discover directions in life. What is conducive to a genius? What sets him in motion? How splendidly and happily were combined in the fate of Andreĭ Nikolaevich Kolmogorov a cheerful childhood, a remarkable secondary school, an unusual time resulting in an atmosphere of creative élan during his youth, and the outstanding School of Mathematics at the University of Moscow, early recognition, and an unusual gift of originality ...

But let us not forget that without which even the combination of all these qualities would not lead to such a beneficent outcome: the presence of inner stimuli, resting, in the final analysis, on the fundamental moral perseverance of the individual. Kolmogorov himself once said on this subject: "It is absolutely essential in science (as in poetry, music, etc.) that the man who has the right moral qualities should accept his work as a crucial duty." And here, it seems to me, the great commandment: the greatest stimulus in life is the voluntary acceptance of one's duty.

In answer to the question: "By whom were you guided in life?" Kolmogorov replied thus: "I always considered that the most important thing is truth." Striving for truth is one of the basic aims in the life of a genius.

Kolmogorov is one of those incomparable people who embellish life by the simple fact of their existence. The mere awareness of the fact that somewhere on earth beats the heart of a man who is endowed with such perfect wisdom and such a pure soul gave wings, gave joy, gave the strength to live, protecting us from evil and inspiring us to good deeds.

Translated by A. LOFTHOUSE

On A. N. Kolmogorov

V. I. ARNOL'D

I have always wanted to understand how Andrei Nikolaevich Kolmogorov went over from one theme to another: he capriciously changed his studies to different topics in an apparently unpredictable way. For example, papers on small denominators in classical mechanics were not prepared for by any preceding ones and appeared quite unexpectedly in 1953–1954. In the same way, in 1935 there unexpectedly appeared the topological papers of Andrei Nikolaevich.

For myself I constructed a theory of the origin of papers on invariant tori: it began with the work of Andrei Nikolaevich on turbulence. In a well-known article by Landau in 1943 the rise of turbulence was "explained" by means of invariant tori—attractors in the phase space of the Navier-Stokes equation. To a laminar flow, observed as a small Reynolds number, there corresponds a stable equilibrium position (a point attractor). Landau's scenario of the transition to the turbulence is that it is a sequence of bifurcations under an increase in the Reynolds number. First there arises the limit cycle, then the attractor becomes a two-dimensional torus, and under further increase of the Reynolds number the dimension of the invariant torus increases. It may happen, remarked Andrei Nikolaevich in discussion of the Landau scenario, that already for a finite Reynolds number transition to an infinite-dimensional torus and even to a continuous spectrum occurs. On the other hand, even if the dimension of the invariant torus remains finite for a fixed Reynolds number, the spectrum of conditionally periodic motion on a sufficiently high-dimensional torus contains so many frequencies that it hardly differs from a continuous spectrum. The question of which of these two cases actually occurs was posed by Andrei Nikolaevich more than once. At the end of the 1950s on the notice board of the Faculty of Mathematics and Mechanics (Mekh-Mat) at Moscow State University there was a program[1]

[1]Here is the complete text of the program:

Themes of the seminar:

1. Boundary-value problems for hyperbolic equations whose solutions everywhere depend discontinuously on a parameter (see, for example, S. L. Sobolev, Dokl. Akad Nauk SSSR **109** (1956), 707).

2. Problems on classical mechanics in which the eigenfunctions depend everywhere

of his seminar on the theory of dynamical systems and hydrodynamics (the program included, among other things, the problem of proving the practical impossibility of long-range dynamical weather forecasting on account of its strong dependence on the initial conditions). Andrei Nikolaevich chuckled somewhat about Landau's tori: "Apparently he (Landau) did not know of any other dynamical systems."

The transition from Landau's tori to dynamical systems on a torus was a quite natural line of thought. Finally I almost believed in the theory itself and asked Andrei Nikolaevich whether this was really so. "No," replied Andrei Nikolaevich, "I didn't think that at all. The main thing was that in 1953[2] there appeared some hope. I became somehow quite enthusiastic, I had thought about problems of celestial mechanics ever since childhood, from Flammarion, and then reading Charlier, Birkhoff, the mechanics of Whittaker, the works of Krylov and Bogolyubov, Chazy, Schmidt. Several times I tried but it didn't turn out. Now I was able to make some progress."

The matter was as follows. At that time Andrei Nikolaevich introduced practical work at Mekh-Mat and selected problems for it. In a number of problems he chose the investigation of the motion of a heavy point on a torus that is symmetrical about the vertical axis. This is a completely integrable Hamiltonian system with two degrees of freedom, and motion in it takes place, as a rule, over two-dimensional tori in the phase space. These tori are conditionally-periodically covered by spiralling trajectories: the angular coordinates on them can be chosen so that they vary uniformly under the motion of a phase point.

---

discontinuously on the parameter (a survey of these problems is contained in a lecture by Kolmogorov at the Amsterdam Congress in 1954).

3. Monogenic Borel functions and quasianalytic Gonchar functions (in the hope of applications to problems of types 1 and 2).

4. The rise of high-frequency oscillations when the coefficients of the higher derivatives tend to zero (papers of Volosov and Lykova for ordinary differential equations).

5. In the mathematical theory of partial differential equations with a small parameter in front of the higher derivatives, up to now there has been a study of phenomena of the type of boundary layers and interior layers that converge to surfaces of discontinuity of limiting solutions or of their derivatives under "vanishing of viscosity". In real turbulence the solutions deteriorate on an everywhere dense set. The mathematical study of this phenomenon is assumed to be carried out at least on model equations (the Burgers model?).

6. Questions of stability of laminar flows. Asymptotically vanishing stability (at least on model equations).

7. Discussion of the possibility of application to real mechanics and physics problems of the ideology of the metrical theory of dynamical systems. Questions of stability of various types of spectrum. Structurally stable systems and structurally stable properties (in the latter direction, hardly anything is known for systems with several degrees of freedom!).

8. Consideration (at least on models) of the conjecture that, in the situation at the end of 5 above, in the limit the dynamical system turns into a random process (the conjecture of the practical impossibility of a long-term weather forecast).

[2]*Editors' note*: The year of Stalin's death.

A. N. KOLMOGOROV

At this time the theory of integrable Hamiltonian systems was not, as now, a fashionable branch of mathematics. It was thought that it was a hopelessly obsolete, outmoded and purely formal branch of analytical mechanics. To be engaged in similar "nontopical themes" was regarded as a reprehensible concession to the pressure of external circumstances for a mathematician (it was assumed that mathematicians should add together prime numbers, generalize Lebesgue integrals, investigate continuous but not differentiable groups).[3] Andrei Nikolaevich, chuckling, said that the French write "Celestial Mechanics" with capital letters, and "applied mechanics" with small letters. And he always regarded all forms of "mathematical imperialism" with some scorn, independent of its source (be it Bourbaki, Harvard, or the Steklov Mathematical Institute).

Thus Andrei Nikolaevich observed that in "integrable" problems of practical work properly defined phases on tori vary uniformly with time. He at once posed the question: Is this so if the system on the torus is not integrable, but only has an integral invariant (it preserves the measure with positive analytic density)? He solved this question in a note in 1953 about systems on a torus—the first in which small denominators appeared. Technically this note is not complicated (although it contains some lemmas necessary for the fundamental paper of 1954). Andrei Nikolaevich's conclusion was as follows: we can almost always introduce phases that vary uniformly with time, but sometimes (for a ratio of frequencies that can be abnormally well approximated by rational numbers) mixing is possible (the image of a small disc under the action of a phase flow spreads over the whole torus).

This remark about mixing, which relates to a pathological case (it hardly ever occurs) does not seem to be particularly important. But it was the source of the famous paper of Andrei Nikolaevich on small denominators, published in 1954, where he proved that invariant tori are preserved under a small variation of the Hamilton function.

Andrei Nikolaevich's arguments (recalled by him in a lecture at the International Congress of Mathematicians in Amsterdam in 1954) are as follows.

In integrable systems a motion over invariant tori is always conditionally-periodic (we can introduce phases that rotate uniformly with time). Consequently, mixing in integrable systems does not occur. In order to find out whether the phenomenon he discovered has mechanical applications, Andrei Nikolaevich decided to look for motions over tori in nonintegrable systems, where in principle mixing could be observed.

But how can we find an invariant torus in the phase space of a nonintegrable system? It is natural to start with perturbation theory, considering a system close to an integrable one. Different versions of perturbation the-

[3]Fréchet said to me 1965: "Oh, Kolmogorov—is he the young man who constructed a summable function with Fourier series that diverges almost everywhere?" All the later achievements of Andrei Nikolaevich—in probability theory, topology, functional analysis, turbulence theory, the theory of dynamical systems—were of less value in the eyes of Fréchet.

ory have been discussed repeatedly in celestial mechanics, and later in early quantum mechanics.[4]

But all these perturbation theories lead to divergent series. Andrei Nikolaevich understood that divergence can be overcome if instead of expansions in powers of the small parameter one uses Newton's method in a functional space (about which he had read not long before this in an article by L. V. Kantorovich entitled "Functional analysis and applied mathematics" in the Uspekhi Matematicheskikh Nauk).

Thus, Kolmogorov devised the "method of accelerated convergence" not for the sake of those remarkable applications to problems of classical mechanics to which it leads, but for the sake of investigating the possibility of realizing the special set-theoretic pathology in systems on a two-dimensional torus (mixing).

Andrei Nikolaevich did not solve the problem he posed to himself on the realization of mixing on weakly perturbed invariant tori since, on the tori that he found, his method automatically constructs angular coordinates that vary uniformly under the motion of a phase point. The question of the existence of invariant tori carrying the mixing flows in generic systems close to integrable ones, from which all the work of Andrei Nikolaevich developed, remains unsolved even today, as far as I know.

This technical question, in comparison with the results obtained, is insignificant and forgotten. Nobody now recalls anything about it. Physicists say (I heard this from M. A. Leontovich) that new physics most often begins with a refinement of the last decimal digit. New mathematics, as we have just seen, can also arise by refining minor technical details of preceding works. From this it is clear that the planning of fundamental research is a bureaucratic absurdity (and most often a fraud).

—— • ——

Although Andrei Nikolaevich himself regarded the hope that arose in 1953 as the main reason for the success of this work, he always spoke of Stalin with gratitude (following the old principle of not speaking ill of the dead): "Firstly, he presented each academician with a blanket in the difficult years of the war, and secondly he forgave my brawl[5] in the Academy, saying 'that's the way it goes with us also'." Moreover, Andrei Nikolaevich tried to speak well of Lysenko, when the latter fell in disgrace, claiming that he sincerely erred in ignorance (while Lysenko[6] was in power, Andrei Nikolaevich's attitude to this "fighter against chance in science" was quite different).

Repeating what Khodasevich said about Gorky, we can say about Andrei

[4]In particular detail in Born's book *Atomic mechanics*, an amusing Russian translation of which was published in the 1930s in Kharkov: for example, it includes "Трёхизмерительные разновидности" ("three-measuring races") for dreidimensionale Mannigfaltigkeiten.

[5]*Editors' note*: The Russian has "рукоприкладство" meaning the use of hands (physical force) rather than arguments. Kolmogorov had slapped Luzin's face at an election meeting.

[6]*Editors' note*: Lysenko's slogan against genetics was: "Science is the enemy of chance."

Nikolaevich that he was at the same time one of the most stubborn and one of the most unstable people.

"Sometime I will explain everything to you," Andrei Nikolaevich would say to me each time he would act in a way that clearly contradicted his principles. Apparently he was experiencing pressure from some kind of evil genius whose influence was enormous (well-known mathematicians had the role of a link transmitting the pressure). Andrei Nikolaevich did not quite live until the time when it became possible to talk about these pressures, and like almost all the people of his generation who lived through the 1930s and 1940s, he feared "them" to his last day. We must not forget that for professors of that time not to report on seditious speeches of an undergraduate or post-graduate student quite often meant being charged the next day with sympathizing with seditious ideas (following denunciation by this undergraduate or postgraduate-provocateur).

—— • ——

Andrei Nikolaevich said that he could never think intensively about a mathematical problem with absolute effort for more than two weeks. He assumed that any valid discovery could be presented on four pages of notes in the Doklady "because the human brain is not capable of creating anything more complicated at one go". Andrei Nikolaevich preserved a lively interest in a problem, in his own words, only as long as it was still unclear whether the problem would be solved positively or negatively ("as if you were on a razor's edge"). As soon as this became clear to him, Andrei Nikolaevich tried as quickly as possible to get rid of writing proofs and began to look for a way of handing over the whole field to an apprentice. At such moments one ought to avoid contact with him.

In the development of each branch of science one can distinguish three stages. The first, the pioneering one, is a breakthrough into a new field, a brilliant and usually unexpected discovery, often contradicting the accepted conjectures and dogmas. Then there follows the technical stage—protracted and labor-intensive stage. The theory is cluttered up with details, it becomes difficult of access and cumbersome, but then it involves all the more applications. Finally, at the third stage there appears a new, more general view of the problem and connections between it and other questions, apparently far from it: a breakthrough to a new field of research has been made possible.

The mathematical works Andrei Nikolaevich are characterized by the fact that he was a pioneer and discoverer in many fields, at times solving problems several hundred years old. Andrei Nikolaevich tried to avoid technical work on generalizing the theory he had constructed. He perceived his own instinctive aversion to this form of activity as a deficiency.[7] On the other hand, at the third stage, where it is necessary to interpret the results one has

[7]"Due to old age and laziness, having done something good, at best I write it immediately, but usually throw away attempts at amplification and extension" (from a letter of 8 March 1958).

obtained and sees new paths, at the stage of creating fundamental generalized theories, remarkable achievements are due to Andrei Nikolaevich.

—— • ——

An example of an unexpected breakthrough of Andrei Nikolaevich to a new field consists of his topological works, published in four notes in Comptes Rendus and reported to the Moscow Topology Conference in 1935. In these works Andrei Nikolaevich constructed cohomology theory (at the same time as J. Alexander and independently of him). After this Andrei Nikolaevich was not engaged in topology, but when the works of Milnor on differential structures on spheres appeared, they made a very great impression on Andrei Nikolaevich. After Milnor's report to the Soviet Mathematical Congress in Leningrad in 1961, Andrei Nikolaevich commissioned me, then a postgraduate student, to examine the proofs and tell him what they were about. I tried to carry out the commission, and began to study with V. A. Rokhlin, S. P. Novikov and D. B. Fuchs (and was even an external examiner of Novikov at the defense of his dissertation—on differential structures on products of spheres). But attempts to explain some of this to Andrei Nikolaevich were not crowned with success. "My papers on topology," Andrei Nikolaevich then told me, "were not understood as they should have been. You see, I proceeded from physical concepts—from hydrodynamics and electromagnetic theory, and not at all from combinatorics. Everyone has accepted the cohomology *groups*, which I then introduced, and now uses them. But in these notes something more was done—I constructed not only groups, but also a *ring*! This ring is far more important, and I think that if topologists were familiar with this, many interesting things could be obtained."

Apparently Andrei Nikolaevich picked up all the information about the development of topology after 1935 from P. S. Aleksandrov and his pupils. In any case, the evaluation of the cohomology ring by Andrei Nikolaevich, mentioned above, was remarkable: it contains a penetrating analysis of his own work, and a subsequently justified forecast of the significance of cohomology operations. (This comment on the statement of Andrei Nikolaevich that we have mentioned is due to Rokhlin, who exhibited in this case a tolerance that was unusual for him; in the 1960s, with naive irreconcilability, I militantly tried to tell my teacher what had actually been happening in topology for the last thirty years.)

—— • ——

But Andrei Nikolaevich had his own prepared points of view. For example, he told me that spectral sequences are contained in a Kazan paper of Pavel Sergeevich Aleksandrov. And that after sixty years one should not work in mathematics (this conclusion was evidently based on the experience of relations with mathematicians of preceding generations). So my attempts to explain homotopy theory to Andrei Nikolaevich finished as unsuccessfully as attempts to teach him to ride a bicycle or to put him on water skis. Andrei Nikolaevich dreamed of becoming a buoy-keeper after becoming sixty, and long before that he tried to pick up a suitable place on the Volga. But when the time came, the buoy-keepers changed from rowboats to motorboats, which Andrei Nikolaevich hated, and the project had to be abandoned. So Andrei Nikolaevich resolved to return to the profession of school teacher, which he first held.

——•——

The last mathematical work that Andrei Nikolaevich told me about (probably in 1964) had a "biological" origin. The question was about the minimal cube in which one could fit a "brain" or "computer" of N elements ("neurons") of fixed size, each of which is joined to at most k others by "wires" of fixed thickness. The number k is fixed, and N tends to infinity. Clearly, a very simple "brain" (like a "worm" of N successively joined elements) can be fitted into a cube with radius of order $\sqrt[3]{N}$. The grey matter of the brain (body of neurons) is situated on the surface, and the white (the connections) inside. This led Andrei Nikolaevich to the conjecture that the minimal radius has order $\sqrt{N}$, and that it is impossible to fit any sufficiently complicated brain into a cube of smaller radius (the words "sufficiently complicated" can be given an exact mathematical meaning).

This was the way it finally turned out (in the original estimates of Andrei Nikolaevich there were superfluous logarithms, but the final result was without logarithms, jointly with Bardzin').

Of course, Andrei Nikolaevich realized perfectly well that his theorems had little relation to the structure of a biological brain, and so he did not refer to a brain in the article. But his thoughts about grey and white matter were the source of the whole theory. It is interesting to note that this work, perhaps because of a too serious mathematical presentation, remained little known, even to specialists. When I mentioned this in Kolmogorov's obituary in Physics Today (October 1989), I received letters from American engineers, apparently working on miniaturization of computers, with a request to point out the exact reference for the work of Andrei Nikolaevich.

——•——

Recently I visited the mountains above Marseille and went again round the "calanques"—a remarkable system of fjords in a half-kilometer steep precipice of the Alpes Maritimes. Andrei Nikolaevich told me about this place in 1965; uninhabited mountains five kilometers from Marseille, a marked path, where an arrow points down, beyond the precipice. It turns out that there is a ledge for the foot there. If you put your foot on it you can see the next ledge—and little by little going down to the sea. Thus said Andrei Nikolaevich—and now in this place the University of Luminy has been constructed.

To converse with Andrei Nikolaevich was always very interesting, and I am sorry that I did not make notes of his stories. Fortunately, I kept his letters. It seems to me that some fragments, mentioned below, give quite a clear idea about their author and his attitude.

Moscow, 28 March 1965 .. I was very glad to receive your letter of 14 February on my return from the Caucasus, where I arrived on 5 March and left on the 23rd. Five of us went (Dima Gordeev, Lenya Bassalygo, Misha Kozlov and Per Martin-Lev, my 22-year-old Swedish student). In Bakuriani it snowed for the first six days, but it did not prevent us from travelling around. Per and I, in particular, overcame a great incline in Tsagveri in a ravine of the Black River. Dima Gordeev stubbornly trained for eight hours a day on a slalom hill. Later S. V. Fomin arrived and brought sunny weather. On the very first sunny day we went to the slopes of the local Tskara-Tskara range and there for three hours at a height of about 2400 meters all my boys got so burnt (walking in shorts and without them) that for the last two nights they didn't sleep as they should. On the fourth sunny day we climbed to the top of the said range (a height of about 2800 meters), which turned out to be in good time, since on the following day the mountains were covered in cloud, below which a head wind blew. In Tbilisi Misha Kozlov, Martin-Lev and I gave lectures, and we all visited all sorts of sights and had contacts with the local mathematicians. We then fit in two whole-day excursions:

a) to Betania, not far from Tbilisi, where in a forest near the remains of the snow there was a multitude of spring flowers (our usual pale blue snowdrops, small cyclamens, crocuses, early irises). Our aim, however, was a twelfth century church with frescoes.

b) to Kintsvissi, not far from Gori on the slopes on the Trialetskii range, where there is a really remarkable painting from the beginning of the thirteenth century, the impression of which, as of the work of a great and entirely original (though unnamed) artist, is comparable with the impression of the Dionysius frescoes in Ferapontov. In three cars the excursion included seven Georgians (and even one director—of the Institute of Cybernetics). But the end of the journey to the monument itself turned out to be rather difficult, so that apart from us only G. S. Chogoshvili arrived there, while the others, as far as I could tell, in the course of four hours of waiting, feasted quite well in the nearest village accessible to the automobiles. Later, however, there was a festive supper in the Intourist restauarant in Gori next to the building where Stalin was born. From Gori we at once left for Moscow (at Tbilisi our skis were loaded into the train, accompanied by a young Tbilisi mathematician who was detailed for that; probably after our departure he and our companions revisited that restaurant more than once).

You are supposed to ski over Easter (that is, in the two weeks from 18 April to 2 May). According to the guidebook in Savoie or

Dauphine you could choose a ski station there to suit your taste, but preferably from the very top (1700–2000 meters). A place in a hotel of any class (beginning with a hostel with berths in two tiers) has to be booked in advance. You would probably be completely satisfied with hired skis if you didn't intend to buy them to import back...

I don't wish to say spiteful things about [...], but also I don't wish to defend him from your assumption on account of his ability to regard as interesting only the branches of mathematics in which he is interested himself, or at least has command of. But I do wish to defend myself slightly. I am now inevitably very much occupied with how I can manage to do everything that still remains for me to do, and my plans are quite large and various. I am therefore rather stingy in my effort to study things in which I do not intend to apply my proper activity, and sometimes the even lighter efforts required, for example, to listen to survey reports with understanding (or, say, your explanations). My young friends here often have no comprehension of the inevitable age gaps, as in the desire to teach me to ride a bicycle or ride on water skis.

But I do not observe in myself the inclination to deny the *objective* interest and significance of new directions, which arises from such self-limitation. Sometimes I restrain myself from judgement, sometimes I even actively support and recommend for study by young people things that on a general impression seem to me important and promising, although they go beyond the limits of my own repertoire. If I stand up more actively and temperamentally for the importance of directions that I value from a knowledge of their structure (sometimes concealed from those passively reading finished works) and their promise, then this seems to me understandable and legitimate. Such are our "small denominators" and much else.

I send personal regards to Leray, his wife and children. With him I have formed more natural and personal relations than with other French mathematicians. However, this also happened with L. Schwartz and in another way with Favard, and of the comparatively inconspicuous people with A. Réné (probability theory and statistics with engineering and physical applications). I would like to obtain materials that would help me to write an interesting obituary of Favard. I know his mathematical work quite well, but his pedagogical and social work and his personal biography not so well. Both are quite interesting (including the active help to Spanish emigrés and, in general, very unexpected forms of activity for a mathematician)...

The concluding phrase with amendments ... "(I repent) I blame myself that (it hurts) it grieves you ... " pleases me very much. The second amendment is absolutely true, since "to hurt" me is not so easy. The change from "I repent" to "I blame myself" evidently means that to repent is not at all characteristic of you ... Uryson's grave is situated in the small town of Batz near Le Croisic, it seems that Mademoiselle Cornu is still alive—an old maid guarding this grave.

Moscow, 11 October 1965 ... Only now have I managed to answer your letter of 29 August from Chamonix, since at the beginning of September I was very busy, and then I went to Yugoslavia (Belgrade, an excursion to "old Serbia" in search of some frescoes of the thirteenth century, Zagreb, and an excursion to the shores of the Adriatic).

I have indeed made quite a detailed observation of the opinions and customs of very different circles in France and other countries, but some things you wrote interested me. In the position of a young person on an equal footing with students, I was in France only in 1930–1931, but in 1958 even though I spent a few days in a hostel for skiers with sleeping places arranged in two tiers, I was always perceived by the others as a professor (which you are also, of course, but for the present this is not written on your face).

I will certainly write to Fréchet. But for now I am extremely short on time. I am even more deeply involved in school matters: in the ordinary school in Bolshevo with one collaborator I am trying to teach the elements of differential calculus in the ninth class and to introduce the elements of set theory to this class (on the theme "the geometrical meaning of equations and inequalities"). They have made me chairman of the mathematical section of the committee, which will really draw up the programs and order the corresponding textbooks. This is quite important, and it is not hopeless that something can be done.

Of the 101 first-year students of the boarding school[8] only 44 wanted to go into the Faculty of Mathematics and Mechanics, and 32 were accepted (about 70%, from the seventh school (Kronrod) about 60%, and much fewer of the rest). But then in the Physics and Engineering Institute *all* our candidates were accepted; evidently our training is more suitable there (the competition was no less there). In the Physics Faculty the pupils had rather less success (about 60%) because of the extreme formalism of the requirements

[8] *Editors' note*: A special mathematical boarding school, see the article by Sossinsky in this volume.

in "Novoselov-type" mathematics,[9] and possibly ill will towards our institution.

I have received the biography of Favard, and have still not made any use of it, but wish to convey my thanks, and hope that something about Favard will appear here.

In Brittany there are many places more attractive than Le Croisic. In late autumn it is everywhere empty, so you could set off to travel without booking places in hotels in advance. In fact I hope you'll visit Batz and Le Croisic. I lived there in the Hôtel de l'Océan on the very edge of the sea. But you have access also to various Maisons de Jeunesse if they are open out of season.[10]

I know this article on the complexity of algorithms about which you wrote. It is a whole small direction of research which, however, needs substantial improvement: the Turing machines are not a suitable apparatus here. It is possible to give a reasonable definition of "minimal possible complexity", which is unique to within a bounded multiplier under broad natural assumptions. Whereas Turing machines under true complexity of order T sometimes give T^2. Now we have succeeded in giving it sufficient simplicity...

... I would welcome any participation of yours in writing textbooks for the regular school, but I think the team of authors must be involved with the experimental teaching in the regular school. For algebra in classes 9–10 my team includes Shereshevskiĭ, who now teaches in the boarding school, but has quite a large experience in regular school, and a certain Suvorova, who tests written experimental sections in Bolshevo.

I was in I. G. Petrovskiĭ's office when someone[11] rang from Paris about your participation in the journal devoted to the new "Invesigational mathematics", which should in some way contrast with all the old journals. We explained that we are still not well informed about the existence of "invesigational mathematics" as a new sci-

---

[9] *Editors' note*: Allusion to the Novoselov textbook with specially chosen exercises based on long and senseless calculations.

[10] I went to Batz one October evening surprising even myself by boarding a train that had departed from the Gare de Montparnasse. The train arrived at Le Croisic at 1 a.m., the small town was deserted, the porter at the Hôtel de l'Océan did not let me in, not believing that I was alone, and not a representative of a band of gangsters. I ended up sheltering from the bright moon and a still rather cold wind, smelling of iodine under a thuya in the garden surrounding the pill-boxes of the Atlantic bank, partly transformed into villas (in any case the surrounding empty villas did not recall the remaining pill-boxes in their architecture). Next morning I went to Batz and found Mademoiselle Cornu in her tobacco kiosk, surrounded by numerous cats. The tomb of Uryson by the cemetery wall was thoughtfully kept. (Uryson drowned before the eyes of Mademoiselle Cornu in 1924).

[11] *Editors' note*: Presumably from the Soviet Embassy in Paris: a Soviet scientist was able to join the Editorial Board only with special permission, as difficult to obtain as the one for travelling abroad.

I. G. PETROVSKIĬ

ence, but we know Malgrange and Tits as excellent mathematicians...

This journal is Inventiones Mathematica. As to writing textbooks, I categorically refused to take part in this, both because of a desire to spend all my time in mathematics and because of serious differences with Andrei Nikolaevich (who was inclined to regard all schoolchildren as genial mathematicians, like himself).

I remember how one day (in the mid-1950s) Andrei Nikolaevich, having assembled his pupils (undergraduates and postgraduates) at his home for Christmas delivered a speech about mathematical talent. According to his theory, the earlier the stage of general human development at which a person stops the higher his mathematical talents. "Our most genial mathematician," said Andrei Nikolaevich, "stopped at the age of four or five, when people love to cut off the legs and wings of insects" (in the words of Andrei Nikolaevich, this mathematician regretted that he had not succeeded in catching Hitler alive—then, he thought, he would have put him in a cage and by a special apparatus pulled out his intestines by a centimeter each day and nipped them off publicly). Andrei Nikolaevich regarded himself as staying at the level of thirteen, when boys are very inquisitive and interested in everything in the world, but grown-up interests do not distract them (I remember he estimated P. S. Aleksandrov's level at sixteen, or even eighteen).

At any rate, Andrei Nikolaevich always assumed that the person he was talking to had an intellect equal to his own, not, probably, because he estimated reality incorrectly ("for the majority of students it is all the same what we teach them at lectures—they simply learn by heart the statements of a few theorems for the examination," he said about the students of the Faculty of Mathematics and Mechanics at Moscow State University), but because he was brought up in this way (and probably regarded this trust in his audience as wholesome and elevating). Probably because of this his remarkable lectures were not understandable to the majority of students (by the way, even formally his lectures were very far from the standard dull dictation, which predominated in the teaching of mathematics even then and which was well ridiculed by Feynman in "Surely you are joking, Mr. Feynman").

"The only person who can actually teach mathematics well," said Andrei Nikolaevich, "is one who has a passion for it himself and perceives it as a living and developing science." In this sense his lectures, despite all their technical shortcomings, were remarkably interesting for those who wished to understand ideas, and not to trace the signs and indices (among his lectures that I managed to hear were those on Galois fields, dynamical systems, Euler's summation formula, Markov chains, information theory, and so on).

Perhaps his free life as a student in Moscow in the 1920s, which he later recalled as a very happy time, influenced Andrei Nikolaevich's approach to teaching. A student was then supposed to take 14 examinations on 14 different branches of mathematics. But an examination could be replaced by an independent result in the corresponding field. Andrei Nikolaevich said that he did not take any of the examinations, writing in place of them 14 articles on different themes with new results. "One of the results," added Andrei Nikolaevich, "turned out to be false, but I realized this only after I had passed the examination."

Andrei Nikolaevich was a remarkable dean. He said that one should forgive talented people for their talent, and he rescued more than one of the new well-known mathematicians from exclusion from the university. Even when denying a grant to a wild student, this dean secretly helped him live through a difficult time. The Department has never reached, and will hardly ever reach, a level higher than then.

"It is nothing short of a miracle," wrote Einstein, recalling his student days, "that the modern methods of instruction have not yet entirely strangled the holy curiosity of inquiry: for this delicate little plant, aside from stimulation, stands mainly in need of freedom. It is a great mistake to think that a sense of duty and compulsion can be conducive to finding joy in seeking and learning. A healthy carnivore animal would have refused food if by cracks of a whip he were continually compelled to eat meat, particularly if the compulsively offered food were not chosen by him."

Andrei Nikolaevich's profound respect for the personality of a student differed from that of other professors known to me. From a student he always expected to hear something new and unexpected.

Andrei Nikolaevich definitely loved to teach and give lectures, even irrespective of visible results; in particular, he regarded with great misfortune the discontinuance of a reading-room for students (which existed at the Moscow Mathematical Society until the dean O. B. Lupanov removed the society from the leadership of the Moscow Mathematical Olympiad and school "*kruzhoks*"[12]).

The next letter gives an idea of the pedagogical load of Andrei Nikolaevich (just removed from the post of dean because of disturbances in the faculty connected with the Hungarian uprising).

> Kislovodsk, 31 May 1957.
>
> ... You have not replied to me about a *kruzhok* or seminar for first-year students. Without you I shall not start anything for first-year students, since my program as it is consists of
>
> 1) a more active leadership than in previous years of our regular

[12]*Editors' note*: "кружок" ("circle") is the Russian equivalent of "working group": there was a system of such working groups in which university students taught high school students one evening per week. They were mainly devoted to solving series of well-chosen problems leading to serious theories.

seminar in the department of probability theory;

2) regular meetings of research assistants and postgraduate students of the Steklov Institute and the department on various applied problems (we have taken from the present fifth-year students, apart from the postgraduate students, Aivazyan, Gladkov, Kolchin, and Leonov, as young scientific research assistants on specific applied themes);

3) a course on "stochastic processes"—this is a compulsory course for students in the fourth year of our speciality—I think it would do you no harm to hear it;

4) a seminar on dynamical systems and stochastic processes for the postgraduate students Alekseev, Meshalkin, Erokhin, Rozanov and you (there will be others taking part, but these are already prepared to work quite intensively and systematically);

5) a seminar with V. Tikhomirov for third- and fourth-year students on selected questions of probability theory and combinatorics, to which you are not forbidden to go, but which I wish to maintain in a quite accessible style.

Nevertheless, I will keep to my promise to be each week at a *kruzhok* or seminar for first-year students and to bring enough problems there, and also to protect you from a possible tendency to train small boys in irresponsible and idle chatter, as happens (despite its interest) in the *kruzhok* of [...], if the whole undertaking takes place... People write to me that all Moscow students in July will be obliged to help the Moscow militia (?!) in connection with the Festival,[13] but perhaps this is a kind of malicious libel?...

2 April 1958, Toussuire (Savoie).

... For the last two days my life in France has seemed extremely nonacademic. Yesterday there was a 24-hour strike on the railway and the Paris metro and buses. We—as suburban dwellers—were, however, given a huge quantity of military trucks that took us into Paris. But the public gave the equivalent of the cost of a ticket on the metro to the soldiers transporting us, who, because of this, were awfully cheerful and courteous. This morning the travel bureau somehow got me a ticket here. In view of the Easter holiday and yesterday's strike, train no. 609, on which I had to leave, was disconnected so that (calling itself one train!) it stood on three tracks, and the numbers of the coaches were allocated according to the rules of chance from 1 to 55. I got on my no. 17 only a minute

[13]The 1958 Moscow Festival was a very important and unique event, since it was the first (and only) large student international meeting in Moscow. It was a sign of the changes under Khrushchev's détante.

before the departure of that section, but thankfully I took up my reserved seat, although the corridors were full of passengers with no reservations. By some special (I don't know what accounted for it) courtesy of the official I even obtained at the second sitting (there were four altogether) an elegant tasty grand déjeuner in the restaurant car. The journey, which lasted for eight hours, was interesting: first the flat country of France with barges, which trucks hauled along the canals, then tunnels and rocky mountains, the Rhône, the exceptionally beautiful Lac du Bourget, and finally our valley with a small mountain river and mountains covered with snow. The train went on to Italy, and at the end of the trip my compartment consisted only of Italians, extremely dirty and even smelling of garlic.

In St.Jean-en-Maurienne I scandalously forced the Italians to unbarricade the door blocked by suitcases, managed to jump out, and at once found a small company searching for a bus to Toussuire. The bus turned up, but since there were only six of us it was replaced (at the same price) by a car, which quickly lifted us to a height of 1800 meters.

Toussuire consists of ten houses, of which 5–6 are small guest houses, each comprising 10–15 rooms. In some of them were "dormitories". The rooms were all locked up, and I put up at the dormitory until Monday. I hired skis for ten days and walked round. The weather is lovely, and it is now almost full moon. Tomorrow I shall take with me bread, cheese, butter, and bananas (in large quantities) and shall spend the whole day wandering on skis over small ridges at a height of 2200–2400 meters. Today a communist (there were a few of them among the employees, guides and instructors of the ski school) stopped me with a most friendly conversation, forcing me to drink two wine glasses of some very strong drug. At the pension, which includes a dormitory and very substantial food, I am paying 1400 francs, that is, a little more than in Paris, for my room with conveniences and a bath. I have still to get acquainted with the touring public. This consists of 1) the student kind of youth, 2) modest intelligentsia families with children from 4 to 17, who are all, of course, passionate skiers...

Now something about your attacks on me...

... I regard formal rigor as *obligatory* and think that in the end, after large-scale (and generally *useful* for a final understanding) work, it can always be combined (in the course of presentation of *important*, that is, in point of fact, *simple* results) with complete simplicity and naturalness. The only way to realize these ideals is the strict requirement of logical precision, even where it is, for the time being, burdensome.

... I have never had the time (or the energy) to write properly. The diversity of my mathematical and nonmathematical work (if one regards the latter, like the deanship, as something useful) somewhat excuses such a position, but I well understand how much I express everything badly and scrappily. I hope to bring back from this trip several model pedagogical writings for publication in French and Russian.

See, however (with regard to a summary presentation without proofs) my Amsterdam lecture...

15 April 1958, Paris (in my "office" in the Poincaré Institute).

... I stayed exactly eight nights in Toussuire (since the bus that I came on left on the ninth day). The weather was very capricious: each day it snowed a few times, each day the sun shone, and at intervals a blinding golden haze. Because of the haze and possible avalanches after 30–40 centimeters of new snow, it was not possible to go very far. Only on the last day did I discover on ski-tracks in the morning that a group of five people had gone on a six-hour journey to the highest point in the neighborhood (Pointe d'Ouillon, 2436 meters), from where there is a descent, well known to the guides, by the crest to another mountain (Mont Cartier, 2250 meters) and much further below Toussuire (to get to Toussuire directly it would have been necessary to climb rock-faces). Along the traces, I followed all this quite a light but rather complicated route (in the sense that any deviation from the road lead to rock-faces or slopes that were dangerous because of avalanches), much of it under the shining sun, but three times in the fog and under large flakes of snow (not, however, causing any desire to put on clothes over one's underpants). I did not catch up with my predecessors; completely in accordance with the rules they descended at the end to a bridge through a mountain stream; but I with Russian wildness threw my skis much higher and got across the stream over the stones. I returned to Toussuire at 14:50, ordered a splendid farewell lunch with more expensive fine wine (usually I take a standard 1/4 liter of red jug wine), drank coffee, settled with the hotel, presented the rest of the Téléski coupons to some boys, handed over the skis, and settled in the bus at exactly 16:30.

As a result of the sun all the skin peeled off my face, though rather painlessly, but my whole body became as sunburnt as is possible in eight days (without peeling).

Then there were two days visiting Favard in Grenoble, where we also went up into the mountains (by car), sat in a café and observed how Favard's ten-year-old daughter skied on a small hill by the café. Grenoble itself, the dense fir forests piled with snow in the neighboring hills, a castle that we visited, and an art museum in Grenoble, were also quite interesting, and our reception by Favard's family was really warm.

I have now given two lectures after the Easter break, tomorrow I am giving a lecture at the probability seminar, and now I am going to a "mathematical tea", which takes place each Wednesday at 16:45 . . .

—— • ——

Calcutta, 16 April 1962.

. . . Many servants, working in the garden of Professor Magalanobis, see that the room for honoured guests is occupied by a grey-haired sunburnt person who does not speak English, and who gets up at sunrise and strolls in the garden in silence. Furthermore, before my arrival they were ordered to clean out the pond, so that I could bathe in the mornings (from which nothing resulted). So a ten-year-old girl in a colorful shawl covered with beads persistently wished to obtain the answer that I am Indian, and my "I am Russian" did not convince her—she may have thought that this was some peculiar Indian tribe.

It's now 3:30. Still dark. Since it's cool in the morning, I've turned off the air-conditioning and opened the window. The birds are singing. I shall go to sleep again.

At 6:30 a handsome youth will come barefoot and in a pale blue shirt, and will place in my room (on the bed if I am lying there) a small table with tea and fruit. At 7:00 an American student will arrive, with whom we go to swim in the pool at the students' hostel.

At 8:00 we will have a real breakfast, to which Prof. Magalonabis will come as well as all those living in the guest-house. Since Madame Magalanobis is ill, an English geologist, Pamela Robinson, is the hostess at the table.

Then I shall go to our consulate and clear up the details of the return journey in the travel bureau.

In Bombay I saw terrible contrasts between the magnificent hotels in the center and the beggars in the settlements on the outskirts. Calcutta is more traditional and poorer in the middle, but now there is evidently a successful period, and I have not found any starving people. As for the poor people, inexperienced foreigners may exaggerate their numbers. In a sculpture museum there were barefooted men with wives, and these had children at their breasts. Servants only watched in order that the families did not sit on the floor to have lunch—by the outward appearance, a European naturally accepts all this society as consisting of beggars, but they are examining the gods assembled here.

9 August 69. The scientific research vessel "Dmitrii Mendeleev".

... Yesterday I spent the first half of the day with Maurice Peixoto (the "x" in Portugese is pronounced "sh"). We discussed the question of a symposium in Brazil in August 1971 with the participation of Smale, you, and Sinai. We even envisaged not a symposium, but something like a summer school for a whole month in a place which in Peixoto's words is more attractive than Rio. In Rio, Erlen Viktorovich Lenskii conducted me everywhere; he was there for a year and chatted freely with colleagues and simple people (which, as you know, is more difficult) in Portugese. I trained him to bathe in spite of the rain of the local "winter" (the temperature of the water was never less than 18°C). The four days in Rio were very interesting, but given to fussing, so the return to a measured existence on board ship gave pleasure. I am mostly engaged in the exposure of cases when the instruments instead of turbulence produce a spectrum of vibration of the cable by which we pull the thermoanemometer, and so on.

My part in the work of the expedition was reflected even in verses, composed for Neptune's festival[14] on crossing the equator:

Первыйкрестник — нету споров
Академик Колмогоров
Задает ученым взбучку,
Тянут те прибор за ручку
И спускают всех чертей
Для закона двух третей...
(The first to be baptised, no doubt,
Is Academician Kolmogorov.
He blames the scientists
For the way they pull their instruments
And put to work all devils
For the law of two thirds.)

The devils treated me comparatively lightly, and since I was dressed in a state-owned linen sheet like in a Roman toga, it was the only object that suffered from the soot on the machine oil, which they were smearing.

However, we did have more cultural entertainment. In the evenings the works of Vivaldi, Bach and Schumann were even accompanied by rotaprinted programs.

As well as in Rio, we were in Reykjavik and we made a great journey on buses to geysers and waterfalls (that is, it was great with regard to impressions, while it took only one whole day). Then unexpectedly (in order to send a sick person to Moscow in

[14]A person who crosses the equator for the first time should be "baptized" by Neptune and his devils by being put in a barrel of water.

an aeroplane) we called in at Conakry. We now had a month's supply of food and fuel, which we had obtained in Kaliningrad for the whole trip, and fresh water was obtained from distillers. But before our return we shall undoubtedly call in at Gibraltar, from where this letter will probably come...

—— • ——

The last decade of Andrei Nikolaevich's life was darkened by serious illness. First he began to complain about his eyesight, and the usual forty kilometers of cross-country skiing along the Vorya river had to be shortened to twenty kilometers along the Skalba river. However, even at the time of our last ski run the almost completely blind Andrei Nikolaevich jumped on skis from the bank to the ice of the Klyazma river, across a stream of open water. Later, in the summer, Andrei Nikolaevich began to struggle with difficulty with the sea waves, but in the autumn he still ran off beyond the fence of "Uzkoye" from the strict supervision of his wife Anna Dmitrievna and the doctor to swim in the pond (and taught me where it was convenient to climb through the fence to get to Uzkoye from Yasenevo; however, Andrei Nikolaevich was never too well-behaved, and not without pride did he talk about his fight with the militia on Yaroslavl station in Moscow).

In his last hears his life was very hard: sometimes he literally had to be carried in one's arms. Anna Dmitrievna, the nurse Asya Aleksandrovna Bukanova, the pupils of Andrei Nikolaevich and the graduates of the boarding school that he founded were in round-the-clock watches on him for some years.

Now and then Andrei Nikolaevich could utter only a few words an hour, but all the same he was always interesting—I remember how some months before this death Andrei Nikolaevich said how very slowly the tracer shells flew over Komarovka, and how he lived, upon his return on a summons from the Artillery Administration in 1942 to Moscow from Kazan, on a divan in the building of the Praesidium of the USSR Academy of Sciences in the Neskuchnii garden.

I remember his story about the winter ascent of Broken in the thirties: proudly descending on skis in swimming trunks Andrei Nikolaevich met two young people with a camera. They asked him to stop and go up to them. Instead of photographing him, as he expected, the young people asked *him* to photograph *them.*

—— • ——

In the Faculty of Mechanics and Mathematics of Moscow State University I think one can still see the picture where Mikhail Ivanovich Kalinin talks with the professors, lecturers and postgraduate students in the old building of the university on Mokhovaya Street. There it is easy to recognize A. N. Kolmogorov, S. A. Yanovskaya, V. V. Golubev, V. F. Kagan, P. S. Aleksandrov, and others. I reproduce here Andrei Nikolaevich's account of this event.

> At that time the daughter of Mikhail Ivanovich Kalinin was friendly with a postgraduate student in mechanics, and Mikhail Ivanovich arrived to meet the faculty. He made a short speech, and then asked everybody to speak about their troubles. Each one began to speak about his difficulty: the postgraduate students about the shortage of dormitories, especially for families, somebody about the necessity of sending graduates to provincial universities, and Pavel Sergeevich Aleksandrov about the leaking roof in the lavatory. In his concluding words the All-Union head said: Well, I see, you are all at sixes and sevens. As to the postgraduate students, they should first settle down to work and set up a dwelling place, and then get married. And as to the lavatory, that's why you have a people's commissar for education...

—•—

One more speech of Andrei Nikolaevich that I recall was about Hermann Weyl. In the words of Andrei Nikolaevich, Weyl loved the songs of Russian Cossacks. In the music room in his Göttingen flat, which occupied a whole floor, he sat close to the loud-speaker with his back to the guests and, having tuned in the radio, listened... There was a special room for ping-pong, and in general there was a feeling of excess compared to the level of a usual professor—it came from his wife, who belonged to the highest musical-artistic circle, somehow connected even back with Wagner...

From Andrei Nikolaevich's talk about Hadamard:

> Hadamard was a passionate collector of ferns. When he came to Moscow, Andrei Nikolaevich with Pavel Sergeevich Aleksandrov took him boating [apparently on the Obraztsov pond on the Klyazma—V. A.]. Suddenly Hadamard somehow looked at the bank and asked to put in urgently. He crossed to the prow of the boat, and when he approached the bank he was so worried as he rushed to the bank that he fell in the water. It turned out that a fern of an unusual kind grew there, which he had been seeking everywhere for many years. Hadamard was perfectly happy. But it was necessary to convey him quickly to a reception with the president [it seems that the president then was Komarov—V. A.] in the Praesidium of the USSR Academy of Sciences.
>
> They had to dress Hadamard in a costume of Pavel Sergeevich. But this was very noticeable (Hadamard was much taller). At the reception everybody said to Hadamard: "Mr. Professor, what has happened to you? You are not in your own costume—have you fallen in the water?" To which Hadamard proudly replied: "Why

do you think a professor of mathematics cannot have any *other* adventures?"

The last time Andrei Nikolaevich visited Hadamard was apparently when Hadamard was ninety. They talked, among other things, about the school Olympiads—in France there had long been a similar Olympiad, the Concours Général, in which the best took part (in each subject separately), the final-year students of the secondary schools in the whole of France at the same time. The questions were chosen from those composed by the best teachers in the whole of France—the teachers sent the problems to Paris, and according to the quality of these problems the ministry could judge the quality of the teachers (which would not be a bad thing for us to copy). The results of the competition determined the first mathematician among the students of that year, the second, the third,... the thousandth...

Hadamard vividly remembered the Concours Général in which he took part. "I was second," he said, "and the first, he happened to be a mathematician. But he was much weaker—he was always weaker." And it was clear that Hadamard still took his "defeat" at the Concours Général painfully!

For Andrei Nikolaevich mathematics was always partly a sport. But when at his jubilee (in a Lecture to the Moscow Mathematical Society) I compared Andrei Nikolaevich with an alpine mountain-climber, contrasting him with I. M. Gel′fand, whose activity I compared with constructing a highway, they were both offended: "... Why don't you regard me as capable of constructing general theories," said Andrei Nikolaevich. "Why do you think I'm not capable of solving difficult problems?" added Gelfand.

Andrei Nikolaevich himself passionately loved music, and was ready to listen to his favourite records endlessly; he had a set of these both in Komarovka and in Moscow. For me he always put on Schumann's quintet, and this turned into a holiday even my watches by his side, when Andrei Nikolaevich could hardly talk.

Some amusing events also took place.

Anna Dmitrievna's kitchen in the Kolmogorovs' flat in the professorial zone L of Moscow State University (the term "zone" has remained ever since the convicts constructed the building) was managed at that time by a middle-aged intelligent assistant Galina Ivanovna. She returned home late at night and did not have time to buy her food. She therefore asked Anna Dmitrievna to get her a pass into the university (which was guarded by the militia), where she could buy something for supper. Having consulted the governing body of the department, Anna Dmitrievna definitely refused: it is impossible to do anything for a person with *such* a surname. Galina Ivanovna asked me to help her. "My God," I asked Galina Ivanovna, "what *is* your surname?" "Marx,"[15] she answered.

---

[15]Jewish-sounding name.

Apparently it was assumed to be as difficult to obtain a pass to the university with such a surname as to enter it to study. (Fortunately she soon succeeded in finding a head of a department with a broader view on things.)

—— • ——

Sometimes his illness receded, and Andrei Nikolaevich could talk for longer. True, to understand his peculiar diction was not easy, even before his illness. It is said that the time when Andrei Nikolaevich's jubilee was celebrated Izrael Moiseevich Gel′fand referred to his visit to Komarovka. Pavel Sergeevich Aleksandrov immediately confirmed that Izrael Moiseevich was *indeed* in Komarovka, and even rescued a *cat* who was locked in the stove, which had begun to be stoked up. Legend (quite likely, however) asserts that Izrael Moiseevich thus commented on this: "Yes, I did indeed discover the cat in the stove, but by then I had been hearing meowing for half an hour, but had wrongly interpreted it."

—— • ——

More than his mathematical achievements, Andrei Nikolaevich was proud of his sporting achievements. "In 1939," he said, "being already an Academic secretary[16] I decided to test how far I could swim in the icy water of the Klyazma, and returned on skis to Komarovka with so high a temperature that in the hospital on Granovskii Street (where an Academic secretary was supposed to be treated) they feared for my life. So I realized that my possibilities were limited. But although I was already in my seventies, at the start of winter I ran from the university to swim in the Moscow River to the Neskuchnii garden. The embankment was so covered in ice that it was impossible to crawl out, and there was nobody near. I looked for a place to crawl out for longer than that time on the Klyazma—and I was not at all ill."

—— • ——

Andrei Nikolaevich remembered with pleasure his youthful journeys in the North, his longest being Vologda-Sukhona-Vychegda-Pechora-Schugor-Sos′va-Ob′-Biĭsk (and further, barefoot through the Altaĭ mountains. In his journeys along the Kuloy and Pinega he succeeded in setting sail, not yielding to the efforts of the local fishermen, after which Andrei Nikolaevich was recognized by them as their own (this showed in that they began to swear at him equally as at each other).

One of the last long conversations with Andrei Nikolaevich was about the future of mankind. Andrei Nikolaevich always referred with uncertainty to the list of former editors on the cover of Mathematische Annalen: "How will the cover look in 500 years' time?" he asked Hilbert. In addition, he was doubtful about the possibility of the existence of our culture for so long a time. First of all because of the demographic catastrophe forecast by Malthus. Andrei Nikolaevich dreamed of a new structure of society in which the wealth

[16]*Editors' note*: The Chairman of the Department of mathematics and physics of the Academy is called the Academic secretary.

of spiritual life overcomes instincts. Strange and naive as these ideas are, it is difficult to argue seriously with them: mankind is rather late in listening to the warnings of thinkers, and Andrei Nikolaevich regarded it as his duty to recall this at the end of his long and, in spite of everything, happy life.

Translated by E. PRIMROSE

Pages of a Mathematical Autobiography (1942–1953)

M. M. POSTNIKOV

I entered the first-year program in physics and mathematics at Perm University (then called Molotov University) in 1942 at the age of 15 (actually, 14 years and 10 months), after having completed only the eighth grade of the ten-grade school system. I was what was called a "book worm", and had spent the previous several years in a process of self education that in many aspects took me far beyond the confines of what was taught in school. According to the rules, youngsters had to complete ten years of schooling in order to be accepted at the university; however, because of the War it was possible to fill only a quarter of the quota of first-year students in physics and mathematics (only three of whom were men). Hence, the university administration was willing to overlook the fact that I did not have a high school diploma or other required papers, and they allowed me to take the entrance exams. This whole business was not concluded without some fuss, but, to make a long story short, in 1942 I became a regular student and started attending lectures.

At first I was completely satisfied with my university studies, which served to systematize my fragmented knowledge of mathematics. But soon, comparing the lectures with the textbooks, I was bitterly disillusioned: it turned out that the lecturers at Perm University were merely repeating what was in the textbooks, without adding anything new of their own. The only bright spot in the generally bleak picture was S. A. Yanovskaya. An old Bolshevik and participant in the Civil War and the Odessa underground, she worked on the philosophy of mathematics, had been a professor at Moscow University, and by a twist of fate had ended up in evacuation in Perm. Although not herself a research mathematician, for a long time she had circulated among the leading representatives of the pre-war Moscow mathematical school, and this could not help but impart to her a certain mathematical culture.

Along with her, there was a third-year mathematics student at Moscow University named E. B. Dynkin, now a well-known mathematician living in the United States. After associating with Yanovskaya and Dynkin, I realized that a proper mathematical education was possible only at Moscow University.

In the spring of 1943, Yanovskaya and Dynkin returned to Moscow, and I was faced with the problem: How could I arrange to follow them there? There were three main obstacles:

(1) It was wartime, Moscow was under siege, and a special pass was required to enter the city.

(2) To obtain a pass one needed an invitation from an authorized Moscow organization.

(3) One also needed a passport, which I was still too young to have been issued.

This is not the place to relate how I overcame these obstacles (and a whole array of other petty obstacles—for example, the virtual impossibility of a private individual buying a train ticket to Moscow). Suffice it to say that my efforts were crowned with success, and one fine evening in early November 1943, I got off the train in Kursk Station in Moscow and headed out to find Yanovskaya—the only living soul I knew in Moscow. I found her home fairly quickly, but not immediately, and so it was late at night, around 11, when I rang her door bell. The surprised Yanovskya did not send me away, she allowed me to spend the night on her sofa, made tea for me in the morning, took me to the university with her, and presented me to the dean of mathematics at the time, I. G. Petrovskiĭ.

I shall not describe how I spent the next month in wartime Moscow, with no place to stay, with essentially no papers, and with little right to be there. At the beginning of December, Petrovskiĭ signed an authorization for me to enroll as a second-year student of mathematics, I obtained a dormitory room, and my stay in Moscow finally became legal.

During the next two years I studied mathematics intensively. I lived in a dormitory on the outskirts of Moscow, near the Timiriazev Academy. I would wake up at 7 a.m., take a cup of tea with bread, and leave for the university. It would take an hour and a half to get to the campus. From 9 a.m. until 1 p.m. I attended lectures, and then I had lunch. Lunch would consist of three courses—a bowl of watery soup, some mashed potatoes or noodles with a tiny piece of meat or fish, and glass of a sweet liquid made by boiling dried fruit. A half pound of bread was allotted for the meal. Naturally, such a meal would not keep away hunger for long. To take my attention away from the hunger I would either go to the library (called the "Mathematical Room") or else attend more lectures.

As it happened, in October 1941 the main classroom building of Moscow University had been destroyed by German bombardment. Because of the resulting shortage of classroom space, upper-level undergraduate lectures were being given in the afternoon and early evening. The lectures ended at about 9 p.m., I would then return home, grab some tea with a piece of bread, and go to sleep.

Thus, it was not so much out of a particular desire on my part as out of a need to distract myself from the nagging hunger that in 1943–1944 I took not only the usual second-year courses, but also the third-year courses and several more advanced courses on a range of subjects (Bessel functions, homological groups, calculus of variations, Riemannian geometry, Markov processes, etc.). After attending the lectures, it was natural for me to take the exams and then, after passing the exams, move on to the next year of study. Thus, in autumn of 1944 I was considered a fourth-year student, and in spring of 1945 I was a fifth-year student. By summer of 1945 I finished the university, receiving a degree with honors (I had received only two B's—in political economy and in the history of the Communist Party).

In addition to the required courses, I passed exams in 17 electives (the norm was 3)—this was something of a record. As they say, I became noticed by the faculty, and this had a long-term impact on my mathematical life.

In the spring of 1945, when I was starting my fifth year, the assistant dean in charge of programs of study called me in to say, "Postnikov, you need to have an advisor for your senior thesis. Whom do you want, Lazar′ Aronovich [Lyusternik] or Pavel Sergeevich [Aleksandrov]?" He knew that I was an active participant in Lyusternik's seminar on topological methods in the calculus of variations. I cannot remember why he suggested Aleksandrov.

Of course, my answer to this question would be decisive for my entire future career. And at this point a very strange thing happened. Without a moment's hesitation I blurted out, "I want Lev Semenovich Pontryagin!" The reason why this was so strange is that I did not know Pontryagin, and had only taken his very boring, formalistic course on homological algebra (which later became the basis for Pontryagin's book [**1**]). I cannot even recall ever before having spoken with Pontryagin. (Of course, since I had taken his course, this meant that I had taken an oral exam from him, but that had left no trace in my memory.) Yet my spontaneous choice turned out to be an unusually good one: if it weren't for Pontryagin, it's hard to say what would have become of me.

In my life it has frequently happened that I have had to make just such a spontaneous decision affecting my entire future—with no conscious information and with virtually no careful thought. As a rule, these decisions have turned out to be the right ones. (The only exception—and this is subject to debate—was my refusal in 1956 to have my name submitted as a corresponding member of the Soviet Academy of Sciences—but that is another story.)

The day after my conversation with the assistant dean, seeing Pontryagin in the hall, I said to him, "Lev Semenovich! My name is Postnikov, I am a fifth-year student, and I told the dean that you are my advisor!" Pontryagin was startled, but he answered, "Fine. Come to see me at home tomorrow, and we'll talk." That is how I became Pontryagin's student.

In 1945 it was still not known whether any three-dimensional manifold

L. S. Pontryagin

that is a boundary can be obtained from the sphere using Dehn surgery. Pontryagin suggested that I investigate how the intersection ring of a three-dimensional manifold over $\mathbf{Z}/2$ changes under Dehn surgery, and, based on this, come up with a characterization of all rings which occur as intersection rings of three-dimensional manifolds which are boundaries. I found that in such a ring, in addition to the obvious algebraic conditions implied by Poincaré duality, the following identity is satisfied by any two elements $x, y \in H_2$:

$$x^2y + xy^2 = xyu, \tag{1}$$

where u is the so-called "cutting homology class" of the manifold (equal to zero if and only if the manifold is orientable). This aroused a great hope: perhaps I had found an invariant property which distinguishes boundary manifolds from other manifolds. But alas! I soon proved that condition (1) holds for any three-dimensional manifold. (We now understand that it could not be otherwise; but at the time this was—at least for me—a big disappointment. It seems that Pontryagin, however, was not surprised.) This work was the subject of my senior thesis, which I defended in the spring of 1945.

At this time in Moscow, Aleksandrov's school was very excited about the so-called "duality laws" which relate the homology of various subsets $X \subset S^n$ to that of the complement $Y = S^n \backslash X$. In the case when X is closed, this connection is given by classical Pontryagin duality; so the problem is to carry over this duality to more general X. I came up with some (rather vague) ideas about how this can be done, by approximating X from the inside by closed sets and approximating Y from the outside by open sets. I hurried to share these thoughts with Aleksandrov and Pontryagin, who reacted in completely different ways, as was characteristic of them.

Aleksandrov made an appointment to see me early in the morning of May 9, 1945, which was Victory Day. He discussed my ideas with me in detail—the conversation lasted over an hour—and generally gave me his blessing to develop them farther, although without offering any new constructive ideas.

Pontryagin, on the other hand, spent only a few minutes talking with me, during which time he pointed out that, in the first place, the whole topic was not of great interest; in the second place, I had taken a false path, since the homology I had introduced is not invariant, and depends upon the embedding of Y in S^n; and, in the third place, in a few months G. S. Chogoshvili was going to defend his doctoral dissertation, one chapter of which develops similar ideas (but in much greater depth!), and if it became known that I was trying to do something related, then this could inadvertently get in the way of his thesis defense. Thus, he firmly advised me to leave this topic, and in any case not to discuss it with anyone in the near future. Naturally, I followed his advice. However, some information about what I had been doing leaked out anyway—apparently though Aleksandrov—and so this was not the end of the story.

Six or seven years later, by which time Chogoshvili was working in Tbilisi, he invited me to Georgia for several months to lecture on topology. Chogoshvili's idea, which in the end he was able to carry out brilliantly, was to create a school of topology in Tbilisi, to which visiting mathematicians from Moscow and elsewhere would be attracted by the natural beauty and the famous hospitality of Georgia. Over thirty years have passed since then, but to this day I have especially warm feelings toward Georgia and its people, and I have often visited Tbilisi feeling as if I were returning to my native city.

However, my own role in the establishment of the Tbilisi school of topology was minimal. Chogoshvili entrusted his graduate student Tamaz Larishvili to my care, and I spent most of the summer teaching him the foundations of homotopy theory. When I left Moscow, as an exercise I asked Tamaz to construct a theory of obstructions to the existence of fibre spaces with no assumptions about the base or fibre being simply connected. Tamaz was supposed to construct this theory using as a model the obstruction theory that was already known in the case of simply-connected base and fibre and the obstruction theory for mappings of nonsimply-connected spaces (where, unlike in Eilenberg's classical obstruction theory, one uses cohomology with local coefficients). A little while later, Chogoshvili asked, with some embarrassment, whether this problem could be the theme of Lazrishvili's Candidate's Degree dissertation. I had no objection, since the work had stretched out into a dissertation; but I felt that one could not expect anything far-reaching here. And in fact, in due time Tamaz successfully defended his thesis, and went on to teach mathematics for many years at a technical school, no longer working in topology. He is completely satisfied with his life, and when I visit Tbilisi it is always a great pleasure to see him.

But let us return to duality laws. Before my departure for Moscow, after I had finished working with Lazrishvili, Chogoshvili invited me to take a vacation at his summer home in Tsagveri, in the mountains between Borzhomi and Bakuriani. While walking in the forest, as if making a casual remark he said, "You know, we have gotten pretty deep into the forest—not a soul around for several kilometers. If you shout, no one will hear..." Then, after a moment's silence, he suddenly asked, "And is it true that you claim that in 1945 you thought up everything in my doctoral dissertation?" "Georgiĭ Sever′yanovich", I answered "What are you saying?! I have never claimed to have had even a theorem from your dissertation, let alone the whole dissertation." The incident was over, and we returned home.

After finishing undergraduate studies in 1945, I was accepted as a graduate student in the Mechanics-Mathematics Department of Moscow University with Pontryagin as my advisor. At the time graduate students had to pass four examinations in mathematics. I decided to make use of this opportunity to systematically summarize my knowledge, by writing detailed notes on the topic of each exam. The first was in associative algebras. My notes on this theme covered everything from the basics to the theory of central simple algebras (crossed products). Ten years later, graduate students were still using my notes to prepare for their qualifying exams.

The second exam was in number theory, encompassing Hecke's book, Hilbert's survey, and Chevalley's class field theory. The third exam, on Lie groups, caused me the most trouble, because Dynkin's work in this area had only just begun, and I was not yet aware of the simplifications he achieved (the explicit form of the Campbell–Hausdorff formula, the theory of simple roots, etc.). I ended up not having to take the fourth exam, on algebraic geometry, and this explains why to this day I understand that field less than, say, Lie groups.

While working on the exam preparation, I was also studying the topology literature, and every week I would meet with Pontryagin to discuss topological themes. It must be said that Pontryagin was a very thoughtful and conscientious advisor, not hesitating to devote time and energy to his students. However, his concern went perhaps too far, since he would always give his graduate students problems to which he already knew the solution. This gave him the following advantages: (1) The graduate student would always finish with a completed dissertation. In the eyes of the administration this demonstrated Pontryagin's great success as an advisor. (2) The dissertation would be of sufficiently high quality, and the advisor could be completely sure of the correctness of all of its theorems.

The drawback of this approach was that the graduate student did not develop independent creativity and persistence in overcoming difficulties. In short, Pontryagin won the battle, but lost the war.

Of course, I did not understand all of this right away. When I was writing my senior thesis, I has sensed something of the problem, but I did not completely realize the situation until some time in my second year of graduate school.

In any case, back in 1945 I had decided not to prepare my senior thesis for publication as an article, thinking that my personal contribution to the topic was too small. However, in 1948, when my graduate studies were nearing an end, Pontryagin formally ordered me to write such an article, since it would be improper not to have a publication before finishing graduate studies. I wrote something, but without any special feeling of inspiration. Pontryagin was unhappy with the text, and he arranged for V. A. Rokhlin, who was then working as his assistant, to edit it. Rokhlin discovered that it was simpler for him to rewrite everything from the beginning. It was in this way that "my" first paper appeared [2]. The main idea of the paper and a large part of the implementation were due to Pontryagin, and the entire text was rewritten by Rokhlin.

In the winter of 1945/46, Pontryagin gave me a problem (which I no longer remember). I started working on it with enthusiasm, but during a meeting with Pontryagin I suddenly realized—because of a careless remark of his—

that he knew the complete solution, and was merely giving me hints to lead me to it. I was quite shocked, and decided to work on something completely different.

At this time I read an article by Whitehead [3], in which he studied the question of what happens to the homotopy groups π_r when an n-dimensional cell is adjoined. Whitehead considered the case $r \leq n+2$, and I decided to look at the case $r = n+3$. Whitehead's method—which was more or less standard at the time—consisted in considering the preimages of points and reducing the problem to a certain classification problem for maps of $(r-n)$-dimensional polyhedra to the given space. As a first approximation to my problem—and I never did make any further progress in this direction—I studied the general problem of classifying an arbitrary three-dimensional polyhedron X in an arbitrary simply-connected space Y.

Let X and Y be topological spaces. The homotopy classification problem for maps $X \to Y$ asks us to find necessary and sufficient conditions for any two continuous maps $f, g\colon X \to Y$ to be homotopic, and, based on this, to describe the set $[X, Y]$ of all homotopy classes of continuous maps $X \to Y$. Here it is natural to suppose that X and Y are connected spaces. In 1945 it was also assumed that the space X is a simplicial polyhedron (it can be partitioned into simplices). Somewhat later, after a long series of attempts, Whitehead finally gave a satisfactory definition of a cell-space (called a $\widetilde{\text{CW}}$-complex); then the requirement that X be a simplicial polyhedron in homotopy classification questions was replaced by the requirement that X be a $\widetilde{\text{CW}}$-complex. This change did not require any new ideas or constructions, and could be carried out in a completely automatic way. Hence, in what follows I will suppose that X is a $\widetilde{\text{CW}}$-complex, although, of course, there is a certain anachronism is doing this.

The first result on the homotopy classification problem was obtained in 1932 by Hopf [4], who studied maps from an n-dimensional $\widetilde{\text{CW}}$-complex (actually, a simplicial space, of course) to the n-dimensional sphere S^n. Hopf used homology groups, but when cohomology appeared in 1935, almost immediately it was noticed (by Whitney [5]) that it was cohomology rather than homology that provided a suitable apparatus for the classification problem.

It is clear that any map $X \to S^n$ is homotopic to a map which takes the $(n-1)$-dimensional skeleton X^{n-1} of the space X to a certain point (fixed once and for all) of the sphere S^n (Whitney used the term normal for such a map $X \to S^n$). Thus, the classification problem reduces to the search for conditions for normal maps to be homotopic.

Let $f\colon X \to S^n$ be an arbitrary normal map. Then, given any n-dimensional oriented cell $e^n \in X$, the restriction $f|_{\overline{e^n}}$ has a well-defined degree (we are assuming an orientation of the sphere S^n), and the function $d_f\colon e^n \mapsto \deg(f|_{\overline{e^n}})$ is an n-dimensional cocycle on X (with coefficients in the group of integers $\mathbf{Z}$). This cocycle is called the characteristic cocycle of the normal map f, and the cohomology class d_f is called the characteristic class.

The Hopf theorem (in Whitney's formulation) states that for $\dim X = n$ two normal maps $f, g: X \to S^n$ are homotopic if and only if their characteristic cocycles are cohomologous: $d_f \sim d_g$. Since one easily sees that any n-dimensional cocycle is the characteristic cocycle of some normal map, this means that the formula $[f] \mapsto d_f$ gives a one-to-one correspondence

$$[X, S^n] \Leftrightarrow H^n(X; \mathbf{Z}), \qquad \dim X = n, \tag{2}$$

between the set of homotopy classes $[X, Y]$ and the cohomology group $H^n(X; \mathbf{Z})$.

Whitney also noticed that this result (and its proof) immediately carries over to the case of an arbitrary $(n-1)$-connected space Y (i.e., a space such that $\pi_r Y = 0$ for $r \leq n-1$). Here d_f turns out to be a cocycle with coefficients in the group $\pi_n Y$, and the one-to-one correspondence (2) takes the form

$$[X, Y] \Leftrightarrow H^n(X; \pi_n Y). \tag{2'}$$

The proof of the Hopf-Whitney theorem proceeds by reducing the question of homotopy of maps $f, g : X \to Y$ to the question of extending to all of $X \times I$ the map from $X \times \{0\} \cup X \times \{1\}$ to Y given by $(x, 0) \mapsto f(x)$, $(x, 1) \mapsto g(x)$. The general form of the extension problem relates to a pair (X, A) and an arbitrary (normal) map $f : A \to Y$; one seeks conditions which ensure that f can be extended to all of X. In the case of the Hopf-Whitney theorem one must do this in the case $A = X^n$ and $\dim X = n+1$. Of course, in order for f to extend it is necessary that the cochain d_f (which is a cocycle on A) be part of a cocycle on X. The Whitney extension theorem (of which the Hopf-Whitney classification theorem is an immediate corollary) states that when $\dim X = n+1$ this condition is also sufficient.

The next step in the classification problem was to study the case $\dim X = n+1$ and $Y = S^n$. In 1938, Pontryagin completed this step for $n = 2$ **[6, 7]**.

The general principles (the so-called theory of obstructions) for solving the classification and extension problems were developed by Eilenberg in 1940 **[8]**. In 1938 Pontryagin had already mastered this technique, but he had not gone to the trouble of pulling it out of his proofs and stating the method explicitly. Here we see a clear example of two types of mathematician: the first solves a problem and is uninterested in (and unable to) formulate the results in complete generality, while the second type of mathematician is continually thinking about putting things in a general form.

Let $f: X^n \to Y$ be a continuous map from the n-dimensional skeleton X^n of a cell complex X to the space Y (which is assumed to be simply connected). The restriction of f to the boundary of each $(n+1)$-dimensional oriented cell $e^{n+1} \in X$ determines an element $c_f(e^{n+1}) \in \pi_n Y$. The resulting

$(n+1)$-dimensional cochain $c_f\colon e^{n+1} \mapsto c_f(e^{n+1})$ on X over the group $\pi_n Y$ turns out to be a cocycle. It is zero if and only if f extends to X^{n+1}, and for this reason it is called the obstruction to extending f. If $c_f = 0$, in which case f has an extension $f'\colon X^{n+1} \to Y$, then the obstruction $c_{f'}$ to extending f' (this is an $(n+2)$-dimensional cocycle over the group $\pi_{n+1} Y$) is called the second obstruction to extending f; it is denoted z_f.

Suppose that the maps $f, g : X^n \to Y$ coincide on X^{n-1}, and let $e^n \in X$. We regard the sphere S^n as two copies of the cell e^n glued together, and we define a map $S^n \to Y$ by setting it equal to f on one copy of the cell and equal to g on the other copy. Let $d_{f,g}(e^n) \in \pi_n Y$ be the homotopy class of this map, and let $d_{f,g}$ denote the cochain $e^n \mapsto d_{f,g}(e^n)$. This is called the difference cochain of the maps f and g; it vanishes if and only if f and g are homotopic relative to X^{n-1}. In addition, we have

(3) $$\delta d_{f,g} = c_g - c_f.$$

If the space Y is $(n-1)$-connected and the map $f : X^n \to Y$ is normal, then $d_f = d_{*,f}$ where $*$ is the constant map; hence, by (3), we have $c_f = \delta d_f$. This equality is called the first obstruction formula. The Whitney extension theorem is a direct consequence of this formula.

Thus, the proof of the Hopf-Whitney theorem follows the scheme

(4)
$$\begin{gathered}\text{Obstruction formula}\\ \Downarrow \\ \text{Extension theorem} \\ \Downarrow \\ \text{Classification theorem.}\end{gathered}$$

If we apply (3) to two extensions to X^{n+1} of the same map $f : X^2 \to Y$ (and replace n by $n+1$), we immediately find that for any extension of $f\colon X^n \to Y$ to X^{n+1} the second obstruction cohomology class z_f is uniquely determined by f (it does not depend on the choice of extension). Here $z_f = 0$ if and only if f can be extended to X^{n+2}.

If $n = 2$ and $Y = S^n$, the groups $\pi_n Y = \pi_2 S^2$ and $\pi_{n+1} Y = \pi_3 S^2$ are isomorphic to $\mathbf{Z}$; hence, for any normal map $f : X^2 \to S^2$ which extends to X^3, the cohomology classes d_f and z_f (of dimension 2 and 4, respectively) may be regarded as classes with coefficients in $\mathbf{Z}$. In particular, it makes sense to speak of the product $d_f \cup d_f$ of d_f with itself.

The Pontryagin formula for the second obstruction gives the following equality for $Y = S^2$:

(5) $$z_f = 2d_f \cup d_f.$$

This gives the Pontryagin extension theorem: a map $f : X^2 \to S^2$ extends to X^4 if and only if $\delta d_f = 0$ and $2d_f \cup d_f = 0$. The corresponding Pontryagin classification theorem states that two normal maps $f, g : X^3 \to S^2$ which coincide on X^2 (this is the only interesting case, by the Hopf-Whitney theorem) are homotopic if and only if there exists a cohomology class $\overline{e} \in H^1(X; \mathbf{Z})$ such that

$$d_{f,g} = 2d \cup \overline{e}, \quad \text{where } d = d_f = d_g. \tag{6}$$

This leads to the following description of the set $[X, S^2]$ when $\dim X = 3$:

(a) $[X, S^2]$ is a union of the disjoint subsets $O(d)$ which are the preimages of the cohomology classes $d \in H^2(X, \mathbf{Z})$ under the map $[X, S^2] \to H^2(X; \mathbf{Z})$, $[f] \mapsto d_f$;

(b) each set $O(d)$ is in one-to-one correspondence with the cokernel of the homomorphism $H^1(X; \mathbf{Z}) \to H^3(X; \mathbf{Z})$, $e \mapsto 2d \cup e$. One could not ask for a better description.

For $n > 2$ the groups $\pi_{n+1}S^n$ are groups of order two, and because of the 2 in the formula (5), Pontryagin had the mistaken impression (see **[6]**) that the second obstruction z_f always vanishes for $n > 2$. In fact, however, for $n > 2$ Steenrod in 1946 **[10]** proved the second obstruction formula

$$z_f = \mathrm{Sq}^2 d_f \tag{7}$$

and the corresponding extension and classification theorems (obtained by the same general scheme (4)). Here Sq^2 is the famous Steenrod square, which is defined for cohomology classes over $\mathbf{Z}$ and itself is a cohomology class over $\mathbf{Z}/2$.

The arrival of the journal with Steenrod's article was a big blow to Pontryagin, and not only because it revealed his error. Already in 1942 (see **[11]**), while solving the problem of computing π_3, Pontryagin had essentially introduced the Steenrod operations. However, as in the case of obstruction theory, he did not undertake a general development of the theory, and simply passed it by (although he did write down the basic Steenrod formula for $\delta(u \cup_i v)$ in the case $i = 1$).

Naturally, Pontryagin quickly mastered Steenrod's paper. He immediately armed himself with Steenrod squares, and used them in **[12, 13]** to finish the development of the ideas in **[11]**. As Pontryagin's student, I, too, studied Steenrod's paper thoroughly. I also knew Pontryagin's work [7] on the classification of maps $X \to S^2$ for $\dim X = 3$. Hence, for me it was completely natural to try to carry over Pontryagin's result to the case of maps $X \to Y$ where $\dim X = 3$ and Y is arbitrary (but simply connected). (This was not a question for Pontryagin himself, because of his general tendency to work with concrete situations.)

The main difficulty—and essentially the only one—consisted in explaining the 2 in Pontryagin's formula (6). The answer occurred to me while studying the Whitehead product (which, I should mention, was introduced in the same paper [3]): I noticed that the Whitehead product of a generator of $\pi_2 S^2$ by itself is equal precisely to twice a generator of the group $\pi_3 S^2$. Hence, if we define $\cup$ to be the multiplication in $H^2(X; \pi_2 Y)$ with values in $H^4(X; \pi_3 Y)$ that one obtains from the Whitehead product $\pi_2 Y \otimes \pi_2 Y \to \pi_3 Y$, then Pontryagin's formula (5) takes the form

$$z_f = d_f \cup d_f, \tag{8}$$

where the "mysterious" 2 has disappeared, and the formula now makes sense for any simply connected Y.

It was then a question of proving (8) (the derivation of the classification theorem from (8) is routine). This I did, by suitably modifying Steenrod's argument. It turned out that the formula (8) holds only when the group $\pi_2 Y$ has no 2-torsion. In the general case, the operation $d \mapsto d \cup d$ must be adjusted, so as to obtain a new operation denoted Кв^2. In the case $\pi_2 Y = \mathbf{Z}/2$, this modification had already been proposed by Pontryagin—for other purposes—in the paper **[11]** that I mentioned before. For this reason, I called the operation Кв^2 the Pontryagin square. Thus, the correct version of (8) has the form

$$z_f = \text{Кв}^2 d_f. \tag{9}$$

In the corresponding theorem, instead of the operation Кв^2 we need a so-called suspension over Кв^2. The computation of this suspension amounts to a rather long calculation with Steenrod operations $\cup_i$. There was one place in this computation where at first I made an error of sign and obtained zero. Rokhlin noticed the mistake, and then I had to rewrite what had already become a complicated manuscript. The final formula involved a new operation Кв^1, which has the property that $2\text{Кв}^1 d = 0$ for any d. Whitehead later referred to this as the Postnikov square **[14]**.

I published these results without proof in a short note **[9]**, with the idea of giving a full exposition later. However, as it happened I never got around to doing this. In general, mathematical work consists of three stages. The first stage is the most pleasant: thinking up the theorem and an idea for proving it. The second step—less pleasant, but still with an element of satisfaction—is to organize the proof, fill the gaps, check and simplify the computations, derive the necessary technical lemmas. The distasteful third stage—which, alas!, takes up 90% of one's time—consists in writing out the proof on paper in an understandable form. It is not surprising that many mathematicians shortchange this last stage, as a result of which their work becomes a mysterious puzzle, and one often finds it easier to think up a proof oneself than to figure out the author's text.

Pontryagin was very demanding when it came to mathematical exposition. He trained me in this—for which I am grateful. Nevertheless, whenever possible in these situations I have always preferred not to spend much time writing articles, but rather to stick to short notes and preprints.

But despite my dislike of writing things up—and perhaps even because of it—my results did not go unnoticed. In particular, Whitehead **[14]** proved them again (also for $n > 2$; see below) using his theory of the secondary boundary operator.

Simultaneously with me, Whitney **[6]** also obtained the classification of maps $X^3 \to Y$. However, his methods were so unusual that I never was able to understand them thoroughly.

At this time—summer of 1947—a change occurred in my living situation. There was a conflict in the Moscow University dormitory, where I was then living, for which I was held responsible. This is not the place to discuss whether this was fair. In any case, when the conflict became public knowledge the administration reacted strongly: I was both evicted from the dormitory and expelled from graduate school. I was saved by Pontryagin, who immediately took me in as a graduate student at the Mathematical Institute of the Academy of Sciences. At that time this was not an exceptional response. There were very few graduate students at the Mathematical Institute—only two or three—and they were students who for one reason or another were not eligible for regular graduate studies at Moscow University.

My move to the Mathematical Institute turned out to be very fortunate for me—"every cloud has a silver lining", as the saying goes. I did not have to look for work after graduate studies, since—again by Pontryagin's initiative—I was kept on at the Institute.

This was an example of my general good luck. In Russia we have a half-joking superstition to the effect that a $2n$-digit number whose last n digits are the same as the first n digits brings good luck. I was born on October 27, 1927, i.e., 10/27/1927, which is almost a lucky number. According to experts in numerology, this causes me to be lucky most of the time; but because of incomplete symmetry of my date of birth $(0 \neq 9)$ I never became a darling of good fortune, and my "successes" always contained a fly in the ointment.

An example of this general principle was my Candidate's dissertation defense in February of 1949. According to the rules, every dissertation is evaluated by two official opponents, who are always very diligent about studying the work. Although Pontryagin had heard all the details of my dissertation, naturally he had not read it carefully. For this reason, he insisted—for the first time ever, and probably the last, in the history of Candidate's dissertation defenses—that I have a third opponent, whom he could trust to actually read my dissertation. This third opponent was Rokhlin (the "regular" opponents were Aleksandrov and M. F. Bokshteĭn). At the defense, Rokhlin, in his typical sarcastic eloquence, spent a good twenty minutes ridiculing the text of my dissertation, pointing out various sorts of real and imagined linguistic slips and grammatical lapses. (By the way, after Rokhlin's tirade I started writing much more slowly, carefully choosing my words and rereading the text several times.) Petrovskiĭ, who was attending the defense and heard Rokhlin's criticisms, proposed that such an unpolished and poorly written dissertation should not be accepted until it was rewritten. Rokhlin jumped up, and spent another twenty minutes explaining that there was no need to do this, that the dissertation was very good, that it represented a major scientific advance, etcetera, etcetera. Although his words of praise were even less well-founded than his earlier criticisms, they accomplished their purpose, Petrovskiĭ took back his suggestion, and the defense was concluded successfully.

As I already mentioned, right after the defense I started work in Pontryagin's department of the Mathematical Institute. The next two years were full of intensive work, as I enthusiastically developed the area of research that had opened before me.

In the first place, I carried over the results of my dissertation to the case $n > 2$ (see **[15]**), i.e., to the case of maps from an $(n+1)$-dimensional cell-space X to an $(n-1)$-connected space Y. This case turned out to be less interesting, mainly because, instead of the Whitehead product, one encounters an uninteresting product $\pi_n Y \otimes \pi_n Y \to \pi_{n+1} Y$. Somewhat later (referring to my work), Shimada and Uehara **[17]** re-proved this result, as they were obliged to do, since I had not yet published a detailed version of the work.

I then took up the case when $\dim X = n$ and $\pi_i Y = 0$ for $1 < i < n$. In the first place, one has to carry over Eilenberg's general obstruction theory to the case of maps to nonsimply-connected spaces. Although this is trivial to do—just replace the usual cohomology by cohomology with local coefficients (which was invented by Steenrod several years after Eilenberg's work appeared)—I still had to write a special note about this (see [18]) in order to have a reference for the result.[1] When $n = 2$, then the problem of classifying maps $X^n \to Y$ was solved by Robbins in 1942 **[19]**. By analyzing his proof, I realized what has to be done in the case of arbitrary n. The result (see **[20]**), though simple in principle, is rather complicated to state.[2] The key role is played by a certain cohomology class $k^{n+1}(Y) \in H^{n+1}(\pi_1 Y, \pi_n Y)$, which was introduced at this time (for a completely different reason, see below) by Eilenberg and Mac Lane (and also, in essence, by Robbins for $n = 2$).

Enraptured by the vistas opening before me, I did not always take the trouble to publish my results, even as short notes (in those days one could not even think of preprints—at least not in Moscow), and quite a few of

[1] I later learned from a review in Mathematical Reviews that this had already been done—in a different form—by A. Komatu (Japan J. Math. **17** (1941), 201–228).

[2] Again in Mathematical Reviews, I learned that this classification problem was also solved by—A. Komatu (Japan J. Math. **17** (1941), 201–228).

these results remained buried in my desk drawer. Here is a typical example.

At some point in 1951, Aleksandrov asked his student O. V. Lokutsievskiĭ to speak on the 1933 work of Hopf and Pannwitz [**21**] at our topology club, which was still functioning at the time. This paper raised the question of conditions on a pure n-dimensional simplicial polyhedron K which ensure that there does not exist a map $K \to K$ onto a proper subset $L \subset K$ (actually, onto a subset of the form $K\backslash\sigma^n$, where σ^n is an n-dimensional simplex) which is homotopic to the identity map id: $K \to K$. Hopf and Pannwitz proved that

(a) a sufficient condition is that for any n-dimensional simplex σ^n of K there exist $m \geq 0$ and a homlogically nontrivial cycle z modulo m in which the simplex σ^n occurs with nonzero coefficient;

(b) if K is a simply-connected polyhedron, then this condition is also necessary;

(c) in the general case this condition is not necessary.

In his introductory remarks before Lokutsievskiĭ's presentation, Aleksandrov commented on the geometrical beauty and elegance of these theorems, and expressed his regret that the paper [**21**] had been largely forgotten—perhaps because its results had such a final form. Lokutsievskiĭ spoke twice—at the first meeting he explained the results, and at the second meeting he gave the proofs. As I listened to Lokutsievskiĭ, I set about restating the Hopf-Pannwitz condition in cohomological language, at which point I realized that to make it into a necessary and sufficient condition one need only introduce cohomology with local coefficients. When Lokutsievskiĭ's first talk ended, I announced this to the group. Aleksandrov was shocked, and apparently was insulted as well: some young upstart announces that right in the course of the talk he was able to solve a problem that had stumped Hopf himself! He's obviously lying, and must be punished.

"Very good, Mikhail Mikhailovich!" he said with exaggerated solicitude, "Perhaps you would be so kind as to describe your 'solution' to us right here at our next meeting?" "I would be glad to," I answered. And so it was decided.

While preparing my presentation for the following week, I realized that I could actually solve a much more general problem. Let X be an arbitrary n-dimensional polyhedron (cell-space), and let (Y, B) be a pair such that $\pi_r(Y, B) = 0$ for any $r \leq n-1$. We shall say that a map $f: X \to Y$ is normal if $f(X^{n-1}) \subset B$. It is clear that any map is homotopic to a normal map. On the other hand, every normal map $f: X \to Y$ determines an n-dimensional cocycle d_f of the space X over the local system of groups $\pi_n(Y, B, f(e^0))$, $e^0 \in X$, whose value at a cell $e^n \in X$ with prescribed vertex e^0 is equal to the homotopy class of the map

$$f \circ \chi : (B^n, S^{n-1}, s_0) \to (Y, B, f(e_0)),$$

where χ is the characteristic map of e^n. When $B = \text{pt}$, this is precisely the characteristic cocycle d_f in Hopf's theorem. It turns out that there is an analogue of Hopf's theorem in this more general situation; in particular, the map f is homotopic to a map to B if and only if $d_f \sim 0$ (in cohomology with local coefficients). A generalized Hopf-Pannwitz theorem is an immediate corollary of this statement. Another corollary is Pontryagin's theorem in **[22]**, which states that a map $f : X^n \to Y$ from an n-dimensional cell-space X to a (simply-connected) cell-space Y is homotopic to a map to the $(n-1)$-skeleton Y^{n-1} if and only if the map induces the zero homomorphism on n-dimensional homotopy for any coefficient group.

All of this I explained at the next meeting. The situation in topology in Moscow at this time is shown up clearly by the fact that, after hearing my presentation and accepting it without any objections, the audience continued listening to the end of Lokutsievskiĭ's talk with undiminished interest.

Nevertheless, when the seminar ended, Aleksandrov came up to me and said, "Everything you explained is very interesting. Please write a detailed exposition, and give it to me. I'll read it through, and we'll have it published." I wrote the paper he asked for, and gave it to him. The manuscript remained in his hands, and I did not see it again. At that time I would have considered it awkward to remind Aleksandrov—and, to tell the truth, I myself simply forgot about it, since I had taken up a completely different subject.

When Whitehead was in Moscow (see below), I reported on this work in his presence. Whitehead gave no reaction—I assumed at the time that this was because he thought both the problem and the solution were trivial. I later learned that he had just submitted for publication a joint paper with Spanier **[23]**, in which they had given a completely different—in some sense a dual—solution of the same problem. (Their result applied to a wider range of dimensions, but only in the simply-connected case.) I do not know why he said nothing about this in Moscow. Most likely, he had simply not listened very carefully to my report.

The question of publishing my work later became moot, since I. Berstein obtained the same result completely independently (see **[24]**).

It was already known to Poincaré that the abelianization of the fundamental group $\pi_1 X$ is isomorphic to the homology group $H_1(X; \mathbf{Z})$. In 1942, Hopf showed that if $\pi_2 X = 0$, then $\pi_1 X$ also determines the group $H_2(X; \mathbf{Z})$, and by 1945 he had generalized this result to arbitrary aspherical spaces (see **[25]**). Hopf's results were greatly simplified and generalized in a series of papers by Eilenberg and Mac Lane (see, for example, Eilenberg's survey **[26]**). The success of Eilenberg and Mac Lane is explained, first of all, by their decision to use cohomology rather than homology. Their papers laid the foundation for the study of group cohomology and the general development of homological algebra. The basic results they obtained before 1950 can be described (in a somewhat modernized form) as follows.

Let (X, x_0) be a pointed topological space. A simplicial subset $M =$

$M(X)$ of the singular simplicial set $S(X)$ is said to be minimal if $S(\{x_0\}) \subset M$, and if for any singular simplex $\sigma \in S(X)$ all of whose faces are in M, there exists a unique simplex in M which is homotopic to σ (relative to the boundary). It is easy to see that minimal subsets exist, and that any two minimal subsets are isomorphic. In addition, the imbedding $M \subset S(X)$ induces an isomorphism of (co)homology groups (it gives a chain equivalence). If X is a $K(\pi, n)$-space (i.e., $\pi_r X = 0$ for $r \neq n$, and $\pi_n X = \pi$), the subset M is isomorphic to the simplicial set $H(\pi, n)$ whose r-dimensional simplices are the n-cocycles of the standard r-simplex Δ^r and whose boundary operators are induced by the inclusions of Δ^{r-1} in Δ^r as a face (and the degeneracy operators are induced by the projections $\Delta^{r+1} \to \Delta^r$). To construct this isomorphism, we define the n-cochain d^n on M (an n-cocycle, as we easily see) whose value $d^n(\sigma)$ on an n-simplex σ is the element of the group $\pi = \pi_n X$ that is determined by σ (namely, the map $(\Delta^n, \dot{\Delta}^n) \to (X, x_0))$. Then the isomorphism $\varphi : M \to H(\pi, n)$ is given by the formula $\varphi(\sigma) = \chi_\sigma^* d^n$, where $\sigma : \Delta^r \to X$ is an arbitrary simplex in M, and $\chi_0 : S(\Delta^n) \to M$ is its characteristic map.

Moreover, the cocycle d^n and hence also the map φ are defined for any $(n-1)$–connected space X; and if $\pi_r X = 0$ for $n < r < m$, then φ is an isomorphism in dimensions $< m$ (and induces an isomorphism on homology in these dimensions). If $n = 1$, this give Hopf's results.

In dimension m, the map φ has a section (i.e., a right inverse ψ). Suppose we choose such a section. Then for any $(m+1)$-simplex $\sigma \in H(\pi_n X, n)$ the simplices $\psi\sigma^{(i)}$, $0 \leq i \leq m+1$, form a simplicial m-sphere in M, and so determine an element $k^{m+1}(\sigma)$ of the group $\pi_m X$. The resulting function $k^{m+1} : \sigma \mapsto k^{m+1}(\sigma)$ is a cocycle whose cohomology class does not depend on the choice of section ψ. Eilenberg and Mac Lane call this cocycle the k-invariant of X (with the property that $\pi_r X = 0$ for $r < m$ and $r \neq n$). If $n = 1$, then this is precisely the above cocycle k^{m+1} (with $m = n$).

Let G be a group, and let $Z(k^{m+1}, G)$ be the group of all pairs of the form (ρ, c), where ρ is a homomorphism $\pi_m X \to G$ for which $\rho k^{m+1} \sim 0$, and c is an m-cochain in $C^m(H(\pi_n X, n); G)$ such that $\delta c = \rho k^{m+1}$. Further, let $E(k^{m+1}, G)$ be the quotient of $Z(k^{m+1}, G)$ by the subgroup of all pairs of the form $(\rho, \delta d)$, where $d \in C^{m+1}(H(\pi_n X, n); G)$. The final theorem of Eilenberg and Mac Lane says that $E(k^{m+1}, G)$ is isomorphic to the group $H^m(X; G)$ (see **[26]**).

After becoming aware of this theorem, I soon realized that

(a) the group $E(k^{m+1}, G)$ introduced by Eilenberg and Mac Lane is actually the m-dimensional cohomology of a certain simplicial set which is constructed from $H(\pi_n X, n)$ and k^{m+1};

(b) the construction of this set can be iterated, leading to an algebraic description of the minimal set M of any connected topological space X.

At first, I iterated the construction in a completely straightforward way, working with spaces X whose homotopy groups are nonzero only in dimensions $n < m_1 < m_2 < \cdots$. But when I explained this to Pontryagin, he immediately observed that there was no need for this assumption, and one should take $n = 1$, $m_1 = 2$, $m_2 = 3$, etc., i.e., X is a completely arbitrary (path connected) space. To this day I remember my feeling when I heard his words—as if a fog had lifted from my eyes. I quickly wrote the note [27], and submitted it for publication.

In this note the role of the simplicial sets was played by a special device which I thought up for this purpose. But after the article [28] arrived in Moscow, introducing simplicial sets for the first time (under the name of semi-simplicial complexes), it became clear to me that it is simplicial sets that should form the conceptual basis of the whole theory.

The interactive description of the set M goes as follows (under the assumption that X is simply connected; the general case requires local coefficients). Suppose that K is an arbitrary simplicial set, π is an abelian group, and k^{m+1} is a cocycle in K over π. For any simplex $\sigma \in K$, we can define its characteristic map $\chi_\sigma : \Delta^r \to K$, $r = \dim \sigma$, and hence the cocycle $\chi_\sigma^* k^{m+1} \in Z^{m+1}(\Delta^r; \pi)$. We consider the set K' of all pairs (σ, c), where $\sigma \in K$ and c is a cochain in $C^m(\Delta^r; \pi)$, $r = \dim \sigma$, such that $\delta c = \chi_\sigma^* k^{m+1}$. K' is a simplicial set with the natural boundary and degeneracy operators, and the canonical projection $\gamma\colon K' \to K$, $(\sigma, c) \mapsto \sigma$, is a simplicial map. Here γ is an isomorphism in dimensions $< m$, and in dimension m it has a section $\sigma \mapsto (\sigma, c)$.

Now to every simply-connected space X we can associate a sequence (tower) of simplicial sets

$$K_2 \xleftarrow{\gamma_2} K_3 \longleftarrow \cdots \longleftarrow K_n \xleftarrow{\gamma_n} K_{n+1} \longleftarrow \cdots \tag{10}$$

with the following properties:

(a) $K_2 = H(\pi_2 X; 2)$;

(b) for any $n \geq 2$ there exits a simplicial map $\alpha_n\colon M(X) \to K_n$ which is an isomorphism in dimension $\leq n$ and satisfies the relation $\alpha_n = \gamma_n \circ \alpha_{n+1}$;

(c) for any $n \geq 2$ the simplicial set K_{n+1} is the set K' constructed for the simplicial set $K = K_n$, the group $\pi = \pi_{n+1} X$, and a certain cocycle $k_n \in Z^{n+2}(K_n; \pi_{n+1} X)$ (for $n = 2$ this is the Eilenberg-MacLane k-invariant, and for $n > 2$ it is constructed in an analogous way using the isomorphism α_n and the section of the previous map γ_{n-1}).

It is natural to define the "limit" K_∞ of the tower (10). By condition (b), it is isomorphic to the minimal set $M(X)$, and hence it has the property that for any group G, $H^*(K_\infty; G) = H^*(X; G)$.

The results of Eilenberg and Mac Lane can be obtained from this by looking at the first two nontrivial levels in the tower (10).

At this time there was not a soul in Moscow who was interested in this sort of problem. It was hard enough for me even to find someone who would agree to listen to me. Pontryagin did this more or less out of a feeling that it was his official responsibility—he had no special interest in it. So I decided to turn to A.G. Kurosh, who was the chair of the department of algebra, since I could argue that the construction was purely algebraic. Kurosh received me very respectfully, and agreed to hear me out.

Of course, all of this was new to him, and he could understand the construction only on a formula level. However, he was an experienced mathematician, and after hearing my report he asked me, "Can't you use the tower (10) to compute other invariants of the space X besides the cohomology groups?" My face lit up, and I cried out, "But of course! You can compute everything! This tower determines the homotopy type!" The next day I wrote **[29]**, and submitted it for publication.

By now the path I followed has been forgotten, and no one regards the tower (10) as a way to describe the cohomology groups.

Of course, the transition to the homotopy type was not completely automatic. Here one needed the notion of a geometrical realization, which at that time was in the air, and was simultaneously proposed by Giever **[30]**.[3] (Nowadays one uses a more efficient geometrical realization that was introduced somewhat later by Milnor; see **[31]**.) But once this realization was introduced, the proof did not present any major difficulties.

Since the tower (10) determines the homotopy type of X, it was not hard, based on ideas known at the time, to obtain from (10) a homotopy classification in terms of group cohomology for the maps $K \to X$ for any cell-space K (see [32]). In particular, from this point of view the classification theorems in my Candidate's dissertation were simply the computation of the first two levels of the tower.

This type of construction was in complete contradiction to Pontryagin's philosophy of life. According to his belief, what is fundamental in mathematics is solving a clearly stated concrete problem. Any general construction or principles that arise in the course of the solution are a sort of diversion—often useful, but by no means the main point. Since Pontryagin then was the only authority on algebraic topology in Moscow, this meant there was a general negative reaction to my results. At the All-Union Topology Conference (in 1952, it seems), after my talk I recall how Pontryagin bombarded me with questions of the type, "What's this good for?", "What does it lead to?", "What new problems can you solve with this?" Flustered, I could not come up with satisfactory answers, and I had to take my seat surrounded by mocking laughter.

However, one must give Pontryagin his due—he did not take any administrative measures against me, and, moreover, he made a sincere effort to determine what was going on. First of all, he told me to write a detailed,

[3]See also Hu (Pacific J. Math. **1** (1951), 583–602), an article which escaped my attention.

clear and self-contained exposition (see [**39**, **40**]). In the second place, several times he arranged for me to give talks at the topology club. After the second or third talk he announced, "Perhaps there is something here, but the problem is that you are unable to convey it to others. Explain it all to Boltyanskiĭ, and let him give the report."

At the time V. G. Boltyanskiĭ was Pontryagin's graduate student. His training and predilections made him what we would now call a geometrical topologist. He was famous for having constructed a two-dimensional compactum whose square is three-dimensional. He knew nothing of homotopic topology. Over the course of several months, meeting four or five times a week, I told him everything I knew and had thought up. I would start talking at the institute, and continue on the way to Boltyanskiĭ's apartment, where, combining business with pleasure, we would sit down to a card game of preference. Boltyanskiĭ not only was able to understand everything—he himself began to do research in algebraic topology. My mathematical solitude came to an end.

Every continuous map $f : X \to Y$ can be identified with a section $X \to X \times Y$, $x \mapsto (x, f(x))$, of the trivial bundle $X \times Y \to X$. This leads to the problem of carrying over results on continuous maps to the case of sections of arbitrary (locally trivial) bundles. There is no difficulty in carrying over the definitions and general properties of obstructions, or in proving an analogue of the first obstruction formula (see [**33**]; note that the well-known Stiefel-Whitney, Chern and Pontryagin characteristic classes can be interpreted most naturally as various kinds of first obstructions). But the second obstruction formula requires new ideas; it was proved by Boltyanskiĭ, first in the case of vector fields [**34**], and then in the general case [**35**].

In his monograph [**36**], Boltyanskiĭ, along with his own results, gave a detailed exposition of the related results in my Candidate's dissertation. In this way I was finally relieved of the necessity of writing them up for publication.

Boltyanskiĭ's results became the content of his Candidate's and Doctoral dissertations. I recall the first thing he said at his Doctoral defense: "Our formula for the second obstruction consists of three terms: the first is Steenrod's term, the second is Postnikov's term, and the third and most important—is mine!" In Russian the world "term" has another [obscene] meaning and so the room broke out in laughter. To this day I do not know whether or not the pun was intentional.

In the spring of 1953, after hearing Boltyanskiĭ's report to the topology club, Pontryagin said of my work: "Well, all right! You have a Doctoral thesis!" The opponents were appointed (Kurosh, Aleksandrov and Bokshteĭn), and in December of 1953 the defense was held. I remember part of Aleksandrov's presentation:

> Reading Hopf's papers is not an easy task. It can be compared with clearing a path in the rain forest, through the lianas, wind-

> fallen trees and swamp land. But when you finish the paper, you find yourself at the edge of the forest. The beautiful landscape of a new country opens before you, and you understand that all your efforts were not in vain. Reading Postnikov's dissertation is not less difficult. However, after reading it, you find yourself not at the edge of the forest but rather still in the woods, up to your waist in the swamp. Instead of a new landscape before your eyes, you face the same overgrown trees through which nothing can be seen.

This passage gives a clear picture of the atmosphere surrounding me then in Moscow.

Two weeks after the defense, by chance I happened to be looking through V. F. Kagan's biography of N. I. Lobachevskiĭ **[37]**. To my surprise, in this biography I found exactly the same characterization—delivered 80 years before by Bachmann in reference to Lobachevskiĭ! It is fortunate that I did not know this at the time of my defense: I undoubtedly would have been unable to resist responding to Aleksandrov, with unpredictable results for the outcome of the defense. As it was, I kept silent, and the defense was concluded without any mishaps.

Several months before, for the first time since World War II a representative delegation of British scientists, headed by Bernal, visited Moscow. Among the delegates was the unforgettable J. H. C. Whitehead. I was what was called the "designate" for him, and I accompanied him everywhere. But I fear that I was not of much use, since at that time I knew very little English.

Naturally, Whitehead was invited to give a lecture. He began his talk by apologizing for "bringing coal to Newcastle", and he then proceeded to give an exposition of his paper **[38]**, in which the tower (10) was constructed anew in the language of cell-spaces. (Somewhat later the construction was reformulated—I think by Serre—in the language of fibre bundles; that is the generally accepted version today.) It is sometimes said that it was after this that my work became accepted in Moscow, and the way was cleared for my Doctoral defense. But I want to emphasize that this is not the case—the decision to go ahead with the defense was made by Pontryagin and Aleksandrov several months earlier.

After Whitehead's report, Professor (now Academician) S. M. Nikol′skiĭ, a specialist in approximation theory who attended the lecture, asked me, "Is Whitehead really a good mathematician?" After my enthusiastic "Yes!", Nikol′skiĭ continued, "Well, of course! He praises you, and you praise him!" And Nikol′skiĭ was someone who was friendly towards me. One can only imagine what the general view was.

I must say that, because of my poor knowledge of English, I did not understand Whitehead all that well, and so did not really have any reaction to his talk. Whitehead was disappointed; however, his rapport with me did not change.

This was my first encounter with a foreigner. Some of the incidents stand out in my mind to this day.

During Aleksandrov's report, which took place in the old building of the Institute of Crystallography on the Ordynka street, Whitehead suddenly jumped up and rushed from the room. The speaker's mouth fell open, and no one knew what to say or do. Several minutes later, the door opened a little, and the somewhat embarrassed face of Whitehead appeared with the question, "Where can one find paper?" It turned out that he had suddenly felt the call of nature, and the Institute's bathroom had no toilet paper.

Yes, indeed, the toilet paper problem! Later I went to the bathroom, and was horrified. The walls were covered with peeling dark paint and splotches of dirt; and the cement floor had an inch-deep layer of some watery substance, the composition of which we need not discuss. Of course *post hoc non propter hoc;* but several months after this incident the bathrooms in all of the Academy institutes underwent major repairs and were put in reasonable shape.

At the time of his departure, Whitehead had been imbibing a bit freely, and in the airport he loudly exclaimed, "I'm an old communist!", at which point another member of the delegation (a historian) remarked, "All mathematicians are mad!" The departure from Vnukovo airport took place in a completely informal atmosphere. All who were seeing him off—including me—were allowed aboard the plane. No customs people or border guards were in evidence. Whitehead had received quite a lot of souvenirs, which he had not taken the trouble to pack properly. He walked out to the plane assisted by three porters, loaded down with more than a dozen bundles and packages tied up with string. In the plane—an American D2—we needed fifteen minutes to find places for everything in the luggage compartments.

Nowadays such patriarchal atmosphere seems like something out of a storybook. But that is how it was.

The writing of my dissertation and the struggle to have my work recognized interrupted the flow of my thought, and I was never again able to work so intensively and prolifically. It was not only a question of the interruption—I became occupied with other matters. The dissertation defense thus brought an entire period of my life to a close, and so this is a good place to end.

References

1. L. S. Pontryagin, *Foundations of combinatorial topology*, OGIZ, Moscow-Leningrad, 1947; English transl., Graylock Press, Rochester, N.Y., 1952.

2. M. M. Postnikov, *The structure of the ring of intersections of three-dimensional manifolds*, Dokl. Akad. Nauk SSSR (N.S.) **61** (1948), 795–797.

3. J. H. C. Whitehead, *On adding relations to homotopy groups*, Ann. of Math. (2) **42** (1941), 409–428.

4. H. Hopf, *Die Klassen der Abbildungen der n-dimensionalen Polyeder auf die n-dimensionale Sphäre*, Comment. Math. Helv. **5** (1933), 39–54.

5. H. Whitney, *The maps of an n-complex into an n-sphere*, Duke Math. J. **3** (1937), 51–55.

6. L. S. Pontryagin, *A classification of continuous transformations of a complex into a sphere*, I. Dokl. Akad. Nauk SSSR **19** (1938), 361–363.

7. ——, *A classification of mappings of a three-dimensional complex into the two-dimensional sphere*, Mat. Sbornik (9) **51** (1941), 331–363.

8. S. Eilenberg, *Cohomology and continuous mappings*, Ann. of Math. (2) **41** (1940), 231–251.

9. M. M. Postnikov, *The classification of continuous mappings of a three-dimensional polyhedron into a simply connected polyhedron of arbitrary dimension*, Dokl. Akad. Nauk SSSR (N.S.) **64** (1949), 461–462.

10. N. Steenrod, *Products of cocycles and extensions of mappings*, Ann. of Math. (2) **48** (1947), 290–320.

11. L. S. Pontryagin, *Mappings of the three-dimensional sphere into an n-dimensional complex*, Dokl. Akad. Nauk SSSR **34** (1942), 35–37.

12. ——, *The homotopy group $\pi^{n+1}(K_n)$ $(n \geq 2)$ of dimension $n+1$ of a connected finite polyhedron K_n of arbitrary dimension, whose fundamental group and Betti groups of dimensions $2, \dots, n-1$ are trivial*, Dokl. Akad. Nauk SSSR **65** (1949), 797–800.

13. ——, *Classification of the mappings of an $(n+1)$-dimensional sphere into a polyhedron K of arbitrary dimension, whose fundamental group and Betti groups of dimensions $2, \dots, n-1$ are trivial*, Izv. Akad. Nauk SSSR. Ser. Mat. **14** (1950), 7–44.

14. J. H. C. Whitehead, *On the theory of obstructions*, Ann. of Math. (2) **54** (1951), 68–84.

15. M. M. Postnikov, *Classification of continuous mappings of an $(n+1)$-dimensional complex into a connected topological space which is aspherical in dimensions less than n*, Dokl. Akad. Nauk SSSR (N.S.) **71** (1950), 1027–1028.

16. H. Whitney, *Classification of the mappings of a 3-complex into a simply connected space*, Ann. of Math. (2) **50** (1949), 270–284.

17. N. Shimada and H. Uehara, *On a homotopy classification of mappings of an $(n+1)$-dimensional complex into an arcwise connected topological space which is aspherical in dimensions less than n $(n > 2)$*, Nagoya Math. J. **3** (1951), 67–72.

18. M. M. Postnikov, *Homology invariants of continuous mappings*, Dokl. Akad. Nauk SSSR (N.S.) **66** (1949), 161–164.

19. H. Robbins, *On the classification of the mappings of a 2-complex*, Trans. Amer. Math. Soc. **49** (1941), 308–324.

20. M. M. Postnikov, *Classification of the continuous mappings of an arbitrary n-dimensional polyhedron into a connected topological space which is aspherical in dimensions greater than unity and less than n*, Dokl. Akad. Nauk SSSR (N.S.) **67** (1949), 427–430.

21. H. Hopf und E. Pannwitz, *Über stetige Deformationen von Komplexen in sich*, Math. Ann. **108** (1933), 433–465.

22. L. S. Pontryagin, *On a connection between homology and homotopy*, Izv. Akad. Nauk SSSR. Ser. Mat. **13** (1949), 193–200.

23. E. Spanier and J. H. C. Whitehead, *Obstructions to compression*, Quart. J. Math. Oxford (2) **6** (1955), 91–100.

24. I. Berstein, *Essential and inessential complexes*, Comment. Math. Helv. **33** (1959), 70–80.

25. H. Hopf, *Über die Bettischen Gruppen, die zu einer beliebigen Gruppe gehören*, Comment. Math. Helv. **17** (1945), 39–79.

26. S. Eilenberg, *Topological methods in abstract algebra. Cohomology theory of groups*, Bull. Amer. Math. Soc. **55** (1949), 3–37.

27. M. M. Postnikov, *Determination of the homology groups of a space by means of the homotopy invariants*, Dokl. Akad. Nauk SSSR (N.S.) **76** (1951), 359–362.

28. S. Eilenberg and J. A. Zilber, *Semi-simplicial complexes and singular homology*, Ann. of Math. (2) **51** (1950), 499–513.

29. M. M. Postnikov, *On the homotopy type of polyhedra*, Dokl. Akad. Nauk SSSR (N.S.) **76** (1951), 789–791.

30. J. B. Giever, *On the equivalence of two singular homology theories*, Ann. of Math. (2) **51** (1950), 178–191.

31. J. Milnor, *The geometric realization of a semi-simplicial complex*, Ann. of Math. (2) **65** (1957), 357–362.

32. M. M. Postnikov, *On the classification of continuous mappings*, Dokl. Akad. Nauk SSSR (N.S.) **79** (1951), 573–576.

33. N. Steenrod, *The Topology of Fibre Bundles*, Princeton Univ. Press, Princeton, N.J., 1951.

34. V. G. Boltyanskiĭ, *Vector fields on a manifold*, Dokl. Akad. Nauk SSSR (N.S.) **80** (1951), 305–307.

35. ——, *Cross-sections of fibre bundles*, Dokl. Akad. Nauk SSSR (N.S.) **85** (1952), 17–20.

36. ——, *Homotopy theory of continuous mappings and of vector fields*, Trudy Mat. Inst. Steklov. no. 47. (Izdat. Akad. Nauk SSSR, Moscow, 1955).

37. V. F. Kagan (Editor), *Collected Works of N. I. Lobachevskii*, Vol. 4, Gosudarstv. Izdat. Tehn.-Teor. Lit. Moscow-Leningrad, 1948.

38. J. H. C. Whitehead, *The G-dual of a semi-exact couple*, Proc. London Math. Soc. (3) **3** (1953), 385–416.

39. M. M. Postnikov, *Investigations in homotopy theory of continuous mappings.* I, *The algebraic theory of systems.* II, *The natural system and homotopy type*, Trudy Mat. Inst. Steklov., No. 46 (Izdat. Akad. Nauk SSSR, Moscow, 1955).

40. ——, *Investigations in homotopy theory of continuous mappings.* III, *General theorems of extension and classification*, Mat. Sb. N.S. **40** (82) (1956), 415–452.

Translated by NEIL KOBLITZ

Markov and Bishop: An Essay in Memory of A. A. Markov (1903–1979) and E. Bishop (1928–1983)

BORIS A. KUSHNER

Amicus Plato, sed magis amica veritas

"Plato is my friend, but the Truth is more important"—
How sad the sound of this Latin...
For only Thou, Almighty God,
Hast all the Truth.
I ask so little—
Let us to be Thy image and likeness.
Keep the truth
And give us the capacity to love.
I know that knowledge is gained through struggle,
And it is difficult to live in another way,
But a friend crying before me
Is more important than any truth...

1. It is difficult for me to characterize these remarks. It is probably best to describe them as an essay about two of the most eminent creators of Constructive Mathematics. One of them was my teacher, Andreĭ Andreevich Markov and the other was E. Bishop. The years of my association with Markov were the happiest years of my life and he remains for me an unforgettable and unique person. I was not acquainted with Bishop, but now that fate has brought me to the United States, the country where Bishop did his creative work and founded a flourishing school, I find myself thinking about him and about the culture of his country, and I am beginning to understand better his originality, at once so American and so universal. I was very much aware of his creative concerns when I was still in Moscow, but now that I live in his land I have gained a sense of closeness to him without which writing is devoid of life.

This work was supported by a research grant from the University of Pittsburgh at Johnstown.

A. A. Markov, Jr.

2. Luckily for all, the Great Constructive Revolution in mathematics did not take place. Now the passions are subsiding, the issues and their creators are slowly entering the history of our science, and time brings a new understanding of the scale and grandeur of their victories and the sorrow of their defeats.

The cold cruelty of the Latin saying in the title of the small epigraph-poem has always shocked me. How many Platos have been destroyed by champions of the Truth! What lofty words have accompanied shocking deeds! How readily and gladly have the words been enacted and what release have they brought from all things humane! History keeps spelling out the consequences of this cruel saying but sways no one. Mathematics is no exception. Here, too Truth was made into an idol and was offered suitable sacrifices. Of course, these sacrifices have not been as dreadful as the sacrifices offered on the altar of Social Justice, but they have been often painful and sometimes fatal. And now when I hear ever more clearly the voice of my personal history it tells me that Plato is more important than the truth...

3. I fell under Andreĭ Andreevich's spell the moment I first came into his presence. That was in 1960 or 1961.[1] We students of the School of Mathematics (Mekh-Mat) at Moscow's Lomonosov University (Moscow State University) had to choose an area of specialization. This meant choosing a department. A meeting was held in one of the two amphitheaters on the sixteenth floor (the room number was 16–10 or 16–24), where department representatives addressed the students. (I still remember the calm and convincing presentation of Academician G. I. Petrov, the only mechanics specialist in the room). A grey-haired, mysteriously handsome man immediately attracted my attention. He listened to the speaker and smiled sarcastically. When his turn came, he was introduced as Andreĭ Andreevich Markov,[2] Chairman of the Department of Mathematical Logic. Until that moment I had been able to see him only from the side, but now I was struck by his dazzling blue eyes and veritable halo of grey hair. His speech was unusual. It had the quality of a recitation. Later I realized that, when writing, Andreĭ Andreevich literally drew each letter and endowed it with special significance, so that not only his speech but also his writing had the quality of a recitation. This solemn quality was not a mannerism of a public speaker; Andreĭ Andreevich maintained it in private conversation and even when speaking on the phone, so that the very process of speaking acquired the status of a religious rite. But back to the meeting in the amphitheater.

[1] I sometimes omit names for understandable reasons. Some details and dates may not be entirely correct.

[2] A more accurate English version of my teacher's name is Andreĭ Andreevich Markov, Junior. His father was the great mathematician Andreĭ Andreevich Markov, Senior (1856–1922), famous, in particular, for Markov chains. Andreĭ Andreevich Markov, the third, and Andreĭ Andreevich Markov, the fourth—son and grandson, respectively, of A. A. Markov, Junior—live in Moscow.

A. A. MARKOV AND B. A. KUSHNER: PHOTOGRAPH SHOT IN MARKOV'S MOSCOW STUDY, MARCH 1979

Andreĭ Andreevich spoke about the new young science of mathematical logic and the even younger department that intended to develop it in all possible ways. He mentioned logical connectives and emphasized disjunction, which he illustrated with a phrase like "either he loved her or she loved him", taken (he said) from a defining dictionary. There followed a solemn analysis of this "love phrase" which greatly amused the audience.[3] Only the speaker remained straightfaced. Markov then spoke with great warmth about S. A. Yanovskaya and Vladimir Andreevich Uspenskiĭ. "But Vladimir Andreevich is a classicist," he said of the latter. I took this remark quite literally and was surprised when I met Uspenskiĭ, who turned out to be an energetic young man without a large beard and grey hair. Markov continued to talk to the attentive audience about his department. Touching on special courses, he mentioned his own course, Constructive Mathematical Logic (here his voice grew solemn and his pauses within words endowed them with special significance, attended by "stu-dents, gra-du-ate stu-dents, assis-tants, do-cents, profes-sors". At this point a senior professor was heard muttering: "Only the Rector does not attend." Andreĭ Andreevich turned to him in a questioning manner. "You just said that your lectures are attended by students, gradu-

[3]Later, I heard similar analyses of encyclopedic and other articles in Markov's public presentations. His sharp glance located the almost inevitable flaws in definitions, and the contrast between the elementary nature of such flaws and the solemn tone of the texts in which they occurred never failed to amuse his audiences.

ate students, assistants, docents and professors. Only the Rector does not attend," said the senior professor. "But the Rector is a pro-fes-sor!" rejoined Markov. The audience burst out laughing. My fate was sealed. I left algebra for mathematical logic.

Much has happened since that day. Many of my youthful aspirations have ended in disappointment and I have sustained many painful losses. But the fact that I had met my Teacher has remained a source of abiding happiness.

The key to Markov's personality was his artistic approach to everything. He was an artist in the broadest and loftiest sense of the word. He perceived life itself as an artistic performance. Even now I recall this funny and lofty scene: Markov has just finished his lecture and is walking along the sixteenth floor corridor to wash his hands. His hands are majestically stretched out and he looks like a surgeon approaching the operating table. He seems to be sailing in the sea of students like a battleship at a parade, not caring if there is any vacant space before him. By some miracle, there always is. Yes, Moscow University was a remarkable place in my student days.

Most people were attracted to Markov by his artistic disposition, although it could at times be disturbing. Not everybody was comfortable with his unusual sense of humor and his permanent readiness to mystify.[4] Sometimes there was a Mephistophelian touch behind it. Experience notwithstanding, his students would inevitably fall into the same trap. Markov would begin with "Last night I left my house" and would continue with a tale of varied and progressively more incredible events. We listened spellbound. Then he would surprise us utterly with "And then...I woke up" and smile. One of those incredible worlds which he created for us on so many occasions was described in the only prose work of Markov's that I know of. It was titled "The case of Professor Ivanov" or something to this effect.

The same peculiar perfection could also be found in Markov's poems. It was in the mid-1960s, in a *dacha* near Moscow, that I first heard these poems. We—A. A. Markov, I. G. Bashmakova, A. S. Kuzichev and I—visited S. A. Yanovskaya and she asked Markov to read his poems. The poems' architectural harmony, their sonic expressiveness, enhanced by the author's conscious use of the beauty of Russian vowels, their sense of novelty resulting from his use of mystification to convert the ordinary to the mysterious and exciting, and his remarkable reciting ability had a breathtaking effect.[5]

[4]Markov appreciated such inclinations in other people. At one time, some of my mischievous friends amused themselves by making phone calls and asking to speak with Beethoven. Of course, the usual reaction was one of indignation. Markov told the caller that Beethoven couldn't come to the phone because he wasn't living in the apartment. In response to the more impudent request to pass greetings to Beethoven, Markov imperturbably told the caller that he was sorry but couldn't do that either because Beethoven had died in Vienna in 1827. In response to the caller's "Sorry, I seem to have got the wrong number," Markov replied "That's okay. All the best." The caller couldn't have tried a more suitable number...

[5]The number of poems was relatively small. Unlike many professional poets, Markov wrote poetry only out of absolute inner necessity.

Markov's performance was marked by a subtle feeling for the sound and syntactic structures of the Russian language and for their interactions. Later, in the summer of 1971, I managed to tape one of his performances. Copies of this tape are owned by some of Markov's students and keep alive the memory of his artistry.

Another manifestation of Markov's tendency to mystify was his manner of reading, or, rather, reciting, bureaucratic documents. Many of the members and guests of the Scientific Council of Mekh-Mat undoubtedly remember the spectacle: Markov holds a bureaucratic document in his outstretched hands in the manner of an oriental potentate about to peruse a missive from a fellow ruler (I always felt on such occasions that all that was missing was a servant who would hold the document and another a fan). The voice of Markov is full of ringing and imposing overtones. Banal bureaucratic formulas that hardly enter our consciousness regain their initial fatuous significance. The audience laughs uproariously and even the dissertator is relieved. I recall one such Council Meeting at which one of Pavel Sergeevich Aleksandrov's students defended his dissertation. It was a rather tragic case. To this day I do not know who was right when it came to the so-called essence.

Undoubtedly, the opponents were acting in full accord with the notorious Latin formula. I think that now, with Pavel Sergeevich gone, at least some of these young, energetic and talented scientists remember this incident with sadness. At the height of the scandalous situation Markov behaved in a seemingly paradoxical manner. One of the designated opponents, a famous Russian mathematician, unaccountably linked the dissertant's abstract topological constructions with the probable future solution of the problem of delivery of meat and milk (at this time it was not yet proper to speak of the production of these goods). Markov engaged him in a discussion. Treating the other's baffling statement like a bureaucratic pronouncement, Markov would return to the "meat and milk delivery" over and over again, provoking repeated explosions of laughter. I remember Shafarevich laughing to tears. Most likely, Markov behaved in this way to impart a humane touch to the depressing situation.

In the early seventies Markov enjoyed a kind of second youth. He got closer to his pupils and was often the soul of our informal meetings and parties. A visiting scientist from *Gruzia* (Georgia), a nice, kind person, brought with him some marvelous homemade wine, far superior to store wine, and this made our meetings even more exciting and informal. It is difficult to forget Markov's sunny mood in Dilijan (Armenia) and in Obninsk (near Moscow) in the summer of 1970. The two conferences followed one another with practically no break and the intimacy between the Teacher and the members of his School increased remarkably. Once, in Obninsk, after a crowded party in a small hotel room with wine, songs, discussions, and a poetry reading, we set off, long after midnight, for a walk in the neighboring forest. It was a delightful walk. I managed to lose all my ID cards (but they were returned to me at 6 a.m. in the hotel lobby by a gloomy operative from one of the local "classified" institutions). When we returned to the hotel at dawn the chambermaids reproached Markov: "Aren't you ashamed to keep such company, an old, grey, respectable man like you?" The next night they used their unlimited power (whose mysterious source I could never fathom) to prevent Markov from joining us by locking all the doors on his floor (after suitable announcements). To make up for this we brought him in the morning bucketfuls of lilies-of-the-valley. (We hope the "Greens" everywhere will forgive us.)

At about that time Markov began to visit me at home. His visits were informal. They began and ended late at night. Markov always walked home. He liked long walks. Once the two of us walked from the South-West underground station in Moscow to the Vnukovo Airport—a distance of 11.5 miles. Along the way we discussed general inductive definitions. We had coffee at the terminal, regretted that it was too late for a return walk, went by bus to Markov's apartment, and continued our discussion.

One summer evening we were awaiting Andreĭ Andreevich in my apartment. N. M. Nagornyĭ arrived with his wife, we took the excellent Georgian wine from its hiding place and waited for Andreĭ Andreevich. Suddenly there was a tropical downpour. A pondlike puddle formed at the intersection of Vernadsky and Lobachevsky Avenues and disrupted traffic. We were sure that Markov would not come. Suddenly the bell rang and a smiling, soaking-wet, Andreĭ Andreevich entered. We gave him some dry clothes and he related his odyssey. "I left my apartment and reached the corner of Lenin Avenue when it began to rain. I was determined to go on but was soon stopped by a huge puddle. What to do? I decided to get around the puddle underground and went to the nearest subway station ("Vernadsky Avenue"). Once there I discovered that I had no money. I went up to a lady, tipped my hat (Markov had arrived with a soaking wet straw hat on his head) and said: "Madam, would you kindly give me five *kopeeks*?" The lady got scared and gave me the required cash without delay." Imagine the scene...

Markov liked—and was able—to feel young. He joked that he was young before his seventieth birthday and would be old after it. His seventieth birthday was celebrated in one of the big halls on the sixteenth floor of Moscow University and Markov delivered a speech in his usual unusual manner and concluded with the words: "It is usual to consider a man in his seventies as old. But I am not old at all. In fact, I have just submitted for publication eight articles!" To the delight of his audience, Markov raised his voice when uttering the last words and practically shouted the word "eight". (Incidentally, the editors of *Doklady Akademii Nauk* at first refused to publish so many articles at one time and argued that this was really one article in eight parts. The matter was settled by the intervention of Academician A. N. Tikhonov.) Markov was young and merry during the celebration party. Among the guests of honor was Andreĭ Kolmogorov.

It is necessary to say a few words about the relationship between these two outstanding personalities. There are mathematical folklore stories about their bickering. Their relationship was anything but simple. Nevertheless, Markov had great respect for Kolmogorov's personality and achievements. He was undoubtedly hurt when the honor of teaching for the first time the newly introduced undergraduate course in mathematical logic, required of all mathematics students at Moscow University, went to Kolmogorov. Nevertheless, he attended all the lectures and found them interesting. On the other hand, his presence at the lectures made even Kolmogorov somewhat nervous. Once, as if in answer to a question I did not ask, Markov said to me: "Perhaps you are surprised that I did not protest, did not fight. The point is that there are two kingdoms in mathematics—the kingdom of Light and the kingdom of Darkness. Andreĭ Nikolaevich belongs to the first, and that is more important than anything else." The respect of these two men was mutual: I could sense it in all my contacts with Kolmogorov. Another relevant detail—one that I consider to be of key importance—is that when Markov died, Kolmogorov was the one person who may be said to have shielded his orphaned department.

Markov was an unusually wide-ranging scientist.[6] Much has been written about his mathematical works and more will be written about them in the future. I was near him during the last period of his life when his interests centered on the constructive trend in mathematics that he gave rise to. I will not write now in detail about this creation of his. He wrote a number of excellent expository works on constructive mathematics (see Markov **[9–13]** and Markov and Nagornyĭ **[14]**. I wrote a special (as yet unpublished) article on the subject (see also the Introduction to my monograph **[6]** and my article in the Mathematical Encyclopedia [7]).

There are characteristics of Markov's constructive mathematics and of his poems that remind one of their creator.[7] The author and his works brought to mind a classical temple—a concatenation of straight lines that fit harmoniously with one another. I think that the concept of constructive process and constructive object, the keystones of Markov's constructive world outlook, derived from his "discrete" style of writing—the most impressive construc-

[6]Besides his famous works in the theory of algorithms, mathematical logic, foundations of mathematics and constructive mathematics, he had essential achievements in such various topics as theoretical physics and theoretical astronomy, theory of elasticity and axiomatic set theory, topology and computer science, geometry and dynamical systems, geophysics, etc. His early student-years publications were in chemistry.

[7]The same was true of Markov's mathematical style. He disliked talk about "ideas of proofs", and sometimes denied their very existence—a touch of his usual attempts at mystification. All details, including minute ones, had to be stated in full. He would often single out gaps and even mistakes in details, and counter "but that's obvious" with "then it's very easy to prove". Sometimes he gave the impression of napping in a seminar—after all, details are hardly exciting. But he was certain to wake up at the first sign of a mistake.

tive process I have ever witnessed. His "majestic slowness"[8] notwithstanding, Markov accomplished a great deal.

After Markov's death I began to edit an unfinished manuscript of his. It was later splendidly developed by N. M. Nagornyĭ and became the Markov and Nagornyĭ monograph [**14**] mentioned earlier. The manuscript dealt mostly with semiotics. Some of its initial ideas provoked in me a sense of protest (in particular, I found no essential difference between things proved and things accepted as self-evident), but I was deeply impressed by the clarity, power, and consistency with which Markov developed his ideas. Alas! I could recognize in his very handwriting signs of the painful disease which finally separated us.

Markov had devoted many years of his life to the creation of constructive logic. That was a difficult and sometimes painful experience, perhaps an inevitable cross of all creators of constructive mathematics.[9] Finally, he managed to develop the remarkable concept of a ramified semantic system designed to explain, among other things, the phenomenon of implication in constructive mathematics. This theory, which began at the point where many other attempts to explain the constructive meaning of mathematical sentences had halted, is the best memorial to Markov's self-sacrificing activity in mathematics.

I must admit that Markov's use of general inductive definitions (similar to some famous Brouwer constructions) as a semantic tool came to me (and not only to me) as a complete surprise. Many of the participants in the seminar of the Computing Center of the Academy of Sciences of the USSR are certain to remember the fervent discussions about these problems. And here I must recall an endearing trait of Andreĭ Andreevich: in those discussions he never let me feel in any way the difference in our ages, experience, and social position. He never used his enormous authority as a tool of persuasion. He was always prepared to hear me out and to try to understand my arguments. Sometimes he got irritated and even angry, and our relationship deteriorated momentarily, but in the end his inexhaustible sense of humor would prevail, a smile would return to his face, and both of us would laugh when his newest attempt at mystification was a success.[10] For him Plato was more important than the Truth.

The last years of Markov's life were marked by a constant struggle with a painful disease and by tragedies in his family. His courage was incredible. Regardless of all difficulties, he was invariably polite and attentive, and an

[8]See Nagornyĭ's recollections in the Preface to the monograph by Markov and Nagornyĭ [**14**].

[9]It is interesting that while Bishop initially rejected philosophy and favored concrete mathematical activity, he ended by immersing himself in profound reflection on the nature of implication (cf. Bishop [**2**] and Bishop and Bridges [**4**, p. 13]).

[10]It is important to note that, in spite of Markov's outward orthodoxy and revolutionary constructive declaration, representatives of very different trends and schools got along well and could work undisturbed in his department.

outside observer could never guess how merciless fate had been to him in those years. He kept on working to the last moment, when the world seemed to be falling around him.[11]

At the end of September 1979 I visited him in the hospital. I was about to leave for Warsaw and I sensed that I would not see him again. I am glad that the last time I saw him, I saw him alive. The image of the radiant Markov from the sunny University of my youth, and the image of his farewell smile in the grey room of the Hospital of the Academy of Sciences are like two photos from a family album. They enclose the essence of a life.

4. In 1966 an International Mathematical Congress was held in Moscow. I well remember that mild summer and the unusual atmosphere in the main building of Moscow University. The arrival of a large number of foreign colleagues and the possibility of personal contacts with them—all that was entirely exceptional for everybody and especially for us graduate students. Names that hitherto had just symbolic significance turned out to be names of live and friendly individuals. My friends from the graduate school and I were delighted to meet such legendary scientists as Church, Kleene, Heyting, Tarski, Kalmar, and A. Robinson. When I looked over the program of the Congress I came across the plenary lecture of the American mathematician E. Bishop, devoted to the constructivization of mathematical analysis. For about 15 years before the Congress, a school of constructive mathematics, with its own specific philosophy and techniques, had been developing and growing around Markov. Many of the results obtained in those years belonged to constructive analysis. From the traditional standpoint, constructive analysis could be regarded as a consistently algorithmic version of classical analysis. I was unaware of Bishop's name, let alone his constructive ideas.

To come back to that day of the Congress, a friend of mine—then a graduate student and now a well-known Soviet mathematician—told me with amazement that Bishop was a first-rate analyst, and that his colleagues regarded his "conversion" to constructivism as a mathematical disaster. My friend's teacher—also an accomplished analyst—tried to persuade Bishop to "return to the fold" but to no avail; it seems that the "disease" of constructivism does not respond to homemade remedies. Bishop's one-hour lecture was received by the audience with the restrained politeness not unusual on such occasions. My spoken English was at the time rudimentary and I could only vaguely understand his lecture. Nor did I dare introduce myself and speak with him with the help of my bilingual friends. To this day I regret the missed opportunity. I had another chance encounter with Bishop that I am certain to remember. The incident took place on the same sixteenth floor of the main university building near the office of the department of mathematical logic. I was about to enter the office to pick up some papers but

[11] He once said to me: "I feel that I am falling down and that everything around me is falling down." The disease had badly damaged his vestibular apparatus.

E. BISHOP

stopped when I heard voices inside. A discussion was in progress in the office. Suddenly the door opened, Bishop rushed out, and practically ran along the corridor toward the elevators. He looked tired and aloof. Then Markov appeared with the usual mystifying smile on his face. He was followed by one of his closest associates who repeated excitedly "But he has no standpoint, he has no standpoint!" His excitement was entirely understandable. Bishop's "standpoint" was clear: he preferred "live" mathematical activity to the searching for a standpoint. On the other hand, Markov's associate was forever engaged in searching for a reliable standpoint, and at the end of each round of his search he would arrive at a new stage of "nonunderstanding of mathematical sentences".[12]

The reader will have guessed that I witnessed the first—and, as far as I know, the only—meeting of the leaders of two main constructive trends of our time. At the time I was not close enough to Markov to dare ask him what had transpired during the discussion. Nor did I explicitly broach the topic later, and now all that remains is regret.

I believe that this unfortunate incident, probably due to personality differences, excitement, and the participants' devotion to truth (*Amicus Plato...*), resulted in the subsequent complete divergence of the Bishop and Markov schools. And to think how much they had in common, how many problems they could have usefully discussed, and how much they could have learned from one another!

Shortly after the Congress, the editors of the magazine *Novye knigi za rubezhom* (New books abroad) asked me to review Bishop's monograph **[1]**. I was amazed by his creative power and energy. It was unprecedented that a thinker who had just stated a new and original concept would immediately confirm its fruitfulness by using it to develop concrete mathematical disciplines, including areas of mathematics presumably too abstract for constructivization. The book opened with what can only be described as an inspired "Constructivist Manifesto" in which the author presented his credo in a concise, clear, and elegant way. I strongly recommended that the monograph be translated into Russian. This recommendation was repeated by the well-known logician Vl. Lifschitz—all to no avail. Meanwhile Bishop, in cooperation with D. Bridges, undertook an essential and thorough revision of his book. It is safe to say that they created a new monograph. It was published in 1985 after Bishop's premature death. I can only hope that this monograph (Bishop and Bridges **[4]**) will be translated into Russian.

Now, as I write these words in the gentle hills of Pennsylvania, I have on my desk another—and probably the last—version of Bishop's Constructivist Manifesto with the incredible title *Schizophrenia in contemporary mathematics.* This is the printed version of the lecture Bishop presented at a 1973

[12] Of course, this casual remark is basically flawed. I sincerely respect the deep honesty and self-sacrifice of this scientist and recognize his valuable contributions. But I suspect that the Truth dwells quite far from the formal roads where he hopes to find it.

Meeting of the AMS. It was later re-edited (Rosenblatt, Ed. [**16**]) and published by the AMS together with recollections of Bishop's colleagues and other materials, as a memorial issue.[13] How sad and exciting it is to read Bishop's lecture, probably his last attempt to persuade the "disbelievers"! How clear his despair, the despair of a prophet faced by trivial and indifferent incomprehension! And yet the Truth is so close, so obvious, that it seems enough to say a few words—nay, make a few signs—to communicate it to all these nice and intelligent people in the audience. What he regarded as the irrationality of their resistance must have created in his mind an apocalyptic image of a world that had lost its reason and induced him to give his lecture its unprecedented title. (I am convinced that Bishop's "diagnosis" of schizophrenia was directed not so much against contemporary mathematics as against certain contemporary mathematicians. His emotional response was more likely the result of his interaction with some of his colleagues than of his reading of their mathematical works.) Already in the initial lines Bishop refers to the many years during which he made unsuccessful attempts to convey to his colleagues a feeling for the key philosophical problems of mathematics. His efforts and ideas were not taken seriously. Worse than that, the publication of his monograph, now recognized by these very colleagues as outstanding, encountered serious obstacles. One of the referees stated boldly that publication of the monograph would be a disservice to Mathematics. Bishop's pupils and collaborators were faced with serious problems in their professional careers, which is why he would not take on new students.[14] One can easily appreciate how deeply wounded this generous and inspired man must have been by the indifferent and even hostile reception of his ideas!

I do not intend to go into details of Bishop's constructive concepts. This he did himself in his original works with great technical skill and consummate artistry. What I propose to do here is outline the psychologically interesting relationships among three ideologies: the constructivism of Bishop, the constructivism of Markov, and the intuitionism of Brouwer.

In his *Schizophrenia in contemporary mathematics* Bishop formulated four main principles of the constructive world outlook.

(A) Mathematics is common sense.
(B) Do not ask whether a statement is true until you know what it means.
(C) A proof is any completely convincing argument.
(D) Meaningful distinctions deserve to be maintained.

[13]E. Bishop died on April 14, 1983. On September 24 of the same year the University of California (San Diego) where Bishop was a professor for many years, held a Memorial Meeting in his memory. The proceedings of this meeting were published in the issue mentioned earlier. (I find it depressing to think that, to the best of my knowledge, neither the Moscow nor the Leningrad Mathematical Society ever held a Markov Memorial Meeting.)

[14]Cf. Nerode's recollections in the memorial issue. Nerode mentions that Bishop turned down an invitation to deliver a lecture at the Summer Institute on Recursive Theory in 1982. The reason he gave was that the bitter disappointment he had experienced during his lecture tour a decade earlier had contributed to his heart attack. All assurances and persuasions were of no avail.

Of course, such lists should never be taken too literally and too seriously. They are attempts to outline in a few words a circle of ideas and concepts, and as such are invariably vulnerable. Sometimes one can sense the special circumstances that shaped them, such as the author's heightened emotional state, a crusading mood aroused by intense discussions, and so on.

Take Bishop's four principles. I find (B) attractive, (A) and (C) too vague, and (D) rather tautological. It is difficult to avoid the impression that the overly emphatic nature of (A) and (C) is due to Bishop's radical rejection of the formalistic conception adopted by the extreme followers of Hilbert (but not, I think, by Hilbert himself), which denied almost completely the role of intuition and common sense in mathematics.[15] The more moderate mathematical majority would, most likely, readily accept (A) and (C) with the proviso that everyday common sense, or—to use more refined terminology—initial intuition, is sufficient only as a basis of mathematical activity. In the course of this activity we develop a multileveled mathematical intuition which enables (at least some of) us to feel at ease in multidimensional spaces and to "see" in them geometric images, or to perceive the reality of large transfinites with as much—or more—acuity as that associated with our perception of the physical world. The possibility of developing such additional intuitions (physical, chemical, mathematical, technical, medical, musical, and so on) is one of the most fruitful and exciting abilities of *homo sapiens*. Nobody but a dyed-in-the-wool formalist would deny (C). The only trouble here is that the possibility of objectivization of the completely subjective category of what is "convincing" can only be envisaged by optimists (in this connection see the interesting article of V. A. Uspenskiĭ [**1**]); one of the lessons of history is that the most honest and sophisticated individuals may hold very different, even completely opposite, views of what is true. The crises in the Foundations of mathematics—the last of which is still with us—supply us with many relevant examples. For example, is it obvious who was right in the mathematically fruitful but humanly tragic debate between Brouwer and Hilbert?

Any reasonable person would endorse in principle statement (D). But we are not always prepared to analyze every meaningful distinction. For example, the classical treatment of existential statements does not take into consideration the undoubtedly essential distinction between constructive and nonconstructive methods. The beauty, power, and perfection of the resulting theory are a reward for the loss of nuances, for the abstraction from the real diversity of the mathematical world. A similar situation prevails in the

[15] I believe that when inveighing against formalists Bishop is really attacking not formalism as a philosophical and mathematical trend but its use as a psychological shelter, a kind of "truth of last resort", by practical mathematicians, often largely ignorant of this doctrine, who do not care to deal with subtle issues such as "the nature of mathematics, its concerns, the status of mathematical objects, and so on". Bishop blamed "logicians" for this state of affairs and was suspicious of them (see Nerode's recollections).

natural sciences. For example, physics operates with concepts such as material point, absolutely rigid body, incompressible fluid, and so on, and, more generally, works with mathematical models of a pointwise continuum—an undeniable idealization from the viewpoint of our sensory perception of reality.[16]

Charged with ignoring the meaningful distinction between constructive and nonconstructive existence theorems, a traditional mathematician who approves of (D) might say what a physicist would in a similar situation: "Of course, these distinctions are important, and we will try to develop theories which take them into account, but—sad to admit—we are still unable to create a harmonious and comprehensive picture of the world that would take into account the endless variety of all phenomena." He might add that the study of effective existence theorems is possible within the framework of classical mathematics provided it was equipped with all the necessary tools. Such an argument involves a great deal of common sense in (A). But we can hear the whispering of a higher common sense that suggests the profound and intrinsic importance of concepts of the constructive world outlook that refined our understanding of the nature of mathematics and enriched us by a number of conceptual mathematical achievements, including intuitionistic logic, the theory of choice sequence and constructive models of the continuum.

This discussion demonstrates the limitations of lists like (A)–(D) above. On the other hand, it is inadequate in that it does not cover all the directions in which Bishop developed his theses.

I cannot refrain from commenting on the important and tragic difference between the *Constructive manifesto* of 1967 (see Bishop **[1]** and Bishop and Bridges **[4]**) and the *Schizophrenia in contemporary mathematics.* The first work is full of optimism and radiance. It has a Mozartian quality. There are no obstacles and the polemical side of the work seems to be of secondary importance. The essence of Bishop's constructive mathematics is expressed in aphoristic form at the very beginning: "... every mathematical statement ultimately expresses the fact that if we perform certain computations within the set of positive integers, we shall get certain results." The key charge a finite being could make against classical mathematics is the lack of a numerical sense.[17] The optimistic and revolutionary tone is characteristic as well for the Prologue to the 1967 monograph (Bishop **[1]**, reprinted in Bishop and Bridges **[4]**): "The book has a threefold purpose: to present the constructive point of view, to show that the constructive program can succeed, and to lay

[16]For example, it is enough to ask whether the velocity of light (in, say, kilometers per second) is a rational number, or not to realize how far from reality the traditional concept of real number takes (or abstracts) us. Brouwer's "developing" numbers are in better agreement with the nature of physical constants. However, to the best of my knowledge, no successful physical theories that operate with such numbers have been created thus far.

[17]One could consider the situation from a different viewpoint and ask if being capable of creating classical mathematics is truly "finitary". The thesis of the "finitarity" of a human being is viewed as self-evident by virtually all constructive schools.

a foundation for further work. These immediate ends tend to an ultimate goal—to hasten the inevitable day when constructive mathematics will be the accepted norm." Alas (or perhaps not "alas"), in spite of all efforts of all constructivists of all orientations, this day is as remote today as it was in 1967 (and its theoretical reality is questionable). Bishop's growing sense of isolation in the mathematical world and the indifference of his colleagues (which must have hurt him more than their harshest objections) are clearly responsible for the bitter polemical tone of the 1973 lecture.

What can be said about Bishop's attitude toward other constructivists? Let us quote from Nerode's recollections (M. Rosenblatt, Ed. [**16**, pp. 81–82]): "We asked him how much he had been influenced by Brouwer's writings. He said that he had been influenced by Weyl's book, but had looked only briefly at Brouwer's works for fear of being led away from his own natural lines of development." This is an understandable position for an eminently creative personality such as Bishop's. But in view of this position, it is especially sad to note his almost undisguised disregard for intuitionism. While recognizing the importance of Brouwer's critique of classical concepts (in particular, of the universal validity of the principle of the excluded middle) and of his discovery of the constructive meaning of logical operators, Bishop utterly rejects Brouwer's concepts and constructions aimed at the creation of a nonpointwise theory of the continuum and based on the idea of freely developing sequences (choice sequences). Here it is important to bear in mind that the relation between the discrete and the continuous has had a long history and that "atomistic" concepts of the continuum (including Bishop's) are not conceptually irreproachable. Also, it is not obvious which type of intuition—that of continuous extension or that of the integers (the latter being the keystone of most constructive trends)—is more strongly developed in *homo sapiens.* (I recall one of Kolmogorov's lectures in which he definitely assigned primacy to the intuition of continuity.) In a word, it would be a good idea to apply (D) to the distinction between the discrete and the continuous instead of dismissing it from the outset. As for choice sequences (a concept that goes back to E. Borel), it is sage to say that, the well-known difficulties of interpretation of Brouwer's ideas notwithstanding, the progress of science has fully confirmed their originality and fruitfulness. This remarkable theory of uncompleted objects is another proof of Brouwer's perspicacity. (For an updated version of Brouwer's ideas see, for example, the recently published two-volume monograph of Troelstra and van Dalen [**15**]; a historical and philosophical treatment of Brouwer's theory of the continuum can be found in my work [**8**].)

Bishop is even more dismissive of Markov's constructive school. In fact, he ignores its very existence in his 1967 monograph and devotes to it just a few sentences in his 1973 lecture. It seems that Truth was not born of the discussion in Moscow. It is an intellectual tragedy that these two schools lived their best years without the joy of friendly contacts."[18] And yet, they have much in common! For example, when Bishop speaks of the classical existential quantifier I literally hear Markov. The same can be said of Bishop's remarks about the esoteric nature of classical mathematics (except that Markov, with his mocking attitude toward lofty terminology, would use simpler language).

The main difference between the two schools was probably the Original Sin—committed in Markov's school—of identification of the intuitive concept of algorithm with one of the contemporary precise definitions. There are a number of serious arguments in favor of this approach. I admit that I am not greatly impressed by Bishop's antirecursive arguments or convinced of the fatal consequences of Markov's choice. For a long time the intuitive concepts of reals, functions, limits, continuity, and so on, sufficed for the successful development of mathematics. On the other hand, the (relative) rigorization of these concepts in the nineteenth century did it no harm. Nor can it be denied that the theory of recursive functions exerted an enormous influence on the depth and exactness of our ideas about computability, effectiveness, and algorithms. As for the charge that we have not yet obtained absolute precision, it is necessary to point out that this reproach echoes the old superstition that endowed mathematics with the special status of a science that was ideal, eternal, and uniquely precise."[19] Absolute Truth is not ours, but we carry in us its image and likeness and are able to sense our progress toward it. Their disagreements and differences notwithstanding, the two great personalities to whom I now say farewell were endowed with this ability to the highest degree. That is all that matters. Everything else passes in Time.

Let them rest in peace.

ACKNOWLEDGMENT

I am grateful to Dr. Sergeĭ Demidov (Moscow) for suggesting that I write this essay. I am deeply indebted to Ms. Naomi Yoran for her hearty and invaluable support, specifically, for useful discussions and suggestions, and for proofreading. I would like to thank Ms. Bonnie F. Morris for her help in improving the language of my translation. Dr. Katalin Balla (Budapest, Hungary) contributed essentially by offering valuable suggestions and spotting a number of defects in my English manuscript. I am deeply thankful to her. I owe a special debt of gratitude to Professors H. Grant and A. Shenitzer of York University (in Canada) who were most helpful in the preparation of

[18]This makes the recent joint publication by Bridges and Demuth **[5]** all the more valuable. Unfortunately, Dr. Osvald Demuth, a pupil of Markov's and the founder of an important constructive school in Prague, a fine person and talented scientist, passed away in September of 1988. I mourn the passing of this dear friend.

[19]See the interesting discussion in Uspenskiĭ **[17]**.

the final version of the paper. Finally I would like to thank my colleagues at the University of Pittsburgh at Johnstown, Dr. Mark Mehlman and Dr. Alexander Wilce, for their friendship—the most valuable thing one person can offer to another. It helped me tremendously. I goes without saying that I am solely responsible for the contents, appearance and possible defects of this essay.

References

1. Errett Bishop, *Foundations of Constructive Analysis*. McGraw-Hill, New York, 1967.

2. ——, *Mathematics as a numerical language*, Intuitionism and Proof Theory, (Proc. Conf., Buffalo, N.Y.). North-Holland, Amsterdam, 1970, pp. 53–71.

3. ——, *Schizophrenia in Contemporary Mathematic*, Errett Bishop: *Reflections on Him and His Research* (San Diego, Calif., 1983) (M. Rosenblatt, Editor), Contemp. Math., vol. 36, Amer. Math. Soc., Providence, RI, 1984, pp. 1–32.

4. Errett Bishop and Douglas Bridges, *Constructive Analysis*, Springer-Verlag, Berlin-Heidelberg-New York-Tokyo, 1985.

5. Douglas Bridges and Osvald Demuth, *On the Lebesgue measurability of continuous functions in constructive analysis*, Bull. Amer. Math. Soc. **24** (1991), 259–276.

6*. B. A. Kushner, *Lektsii po konstruktivnomu matematicheskomu analizu*, Nauka, Moskva, 1973 (Russian); English Transl.: B. A. Kushner, *Lectures on constructive mathematical analysis*, Transl. Math. Mono., Vol. 60, Amer. Math. Soc., Providence RI, 1984.

7*. ——, *Konstruktivnaya matematika*. Matematicheskaya Entsiklopediya, t. 2, Sovetskaya Entsiklopediya, Moskva, 1979, pp. 1051–1053. (Russian).

8. ——, *Printsip bar-induktsii i teoriya kontinuuma u Brauera*, Zakonomernosti Razvitiya Sovremennoĭ Matematiki, (M. I. Panov, Editor), Nauka, Moskva, 1987, pp. 230–250 (Russian).

9*. A. A. Markov, *O konstruktivnoĭ matematike*, Trudy Matem. Inst. Akademii Nauk SSSR (Steklov Institute), t. 67, 1962, pp. 8–14 (Russian).

10. ——, Comments of the editor of the translation of the book by A. Heyting, *Intuitionism* (translation from the English), Mir, Moskva, 1965. (Russian)

11. ——, *O logike konstruktivnoi matematiki*, Znanie, Moskva, 1972. (Russian)

12*. ——, *Konstruktivnaya matematika*, Bol′shaya Sovetskaya Entsiklopediya, 3-e izdanie, t. 13, Sovetskaya Entsiklopediya, Moskva, 1973, pp. 148–151. (Russian)

13*. ——, *Konstruktivnoe napravlenie v matematike*, ibid., pp. 151–152. (Russian)

14*. A. A. Markov and N. M. Nagornyi, *Teoriya algorifmov*, Nauka, Moskva, 1984 (Russian); Engl. Transl.: *The Theory of Algorithms*, Kluwer Academic Publishers, Dordrecht-Boston-London, 1988.

15. Anne Troelstra, and Dirk van Dalen, *Constructivism in mathematics*: An introduction. Vol. 1–2, North-Holland, Amsterdam-New York-Oxford-Tokyo, 1988.

16. Murry Rosenblatt, Editor, *Errett Bishop*: *Reflections on him and his research*. (San Diego, Calif. 1983), Contemp. Math., Vol. 39, Amer. Math. Soc., Providence, RI, 1985.

17. V. A. Uspenskiĭ, *Sem' razmyshlenii na temy filosofii matematiki*, Zakonomernosti Razvitiya Sovremennoĭ Matematiki (M. I. Panov, Editor), Nauka, Moskva, 1987, 106–155. (Russian)

(An asterik marks Russian publications available in English translations)

JUNE 1991, JOHNSTOWN, PA, USA

Translated from the Russian by the author
with the editorial assistance of H. GRANT and A. SHENITZER

Étude on Life and Automorphic Forms in the Soviet Union

ILYA PIATETSKI-SHAPIRO

This is a short account of my life and work in the Soviet Union. It is an incomplete account, omitting, for example, my contacts with D. Kazhdan, J. Bernstein, and S. I. Gelfand, which were very important for my future work on L-functions. But I hope that my story will be interesting for people who follow the mathematical developments in the USSR. This paper would not have been possible without the efforts and patience of my friends Jay Jorgenson and Tohru Uzawa. I thank them for their efforts to force me to clearly dictate the text and for the patience with which they accepted my seemingly endless changes. I thank my friend Roger Howe for his encouragement, as well as Atle Selberg and Jim Cogdell, who read the entire text and suggested many changes. I also thank my son Gregory, who edited the final version of this text, and many other members of the mathematical community for helpful remarks and encouragement.

I was born in 1929 in Moscow, the capital of the Soviet Union. My parents were both from traditional Jewish families. My father came from Berdichev, a small, heavily Jewish city in the Ukraine, and my mother came from Gomel, another small city with a large Jewish population. Both were from middle-class families, who became poor after the October revolution of 1917.

My father was an engineer who worked in research related to shoe production. His specialty was synthetic soles. He was not very successful, but I believe that he was a good engineer who knew his specialty very well. After World War II, when he was 50 years old, he defended a Ph.D. thesis. He was motivated by the salary increase for the Ph.D. degree, given to workers in research institutes. That increase was instituted by Stalin, who recognized the importance of science when the atomic bomb destroyed Hiroshima. However, the Ph.D. did not help my father very much, since soon after defending his thesis he was demoted and sent to work in a factory where having a Ph.D. degree produced only a little addition to his salary.

My father suffered from his lack of success. He considered that his failure was the result of his not being a member of the Communist Party. When I was in my twenties, he strongly advised me to join the Party. We had many arguments about it. Even now I am sad that I was not able to explain to him why I refused to join the Communist Party. I am happy, however, that I followed my way. My mother did not get a higher education and worked as a typist most of her life. She usually worked at home, on a small typewriter she kept there. I remember how excited she was when she helped type my translation of Siegel's lecture notes. Later, I dictated my thesis to her.

My parents basically forgot, or were afraid to follow Jewish traditions. However, they were not against them. Sometimes at Passover we did have matza. I remember that in 1940 my grandfather, then 80 years old, moved to our apartment, where he stayed till the end of his life. He kept Jewish traditions. He died in our apartment in October of 1941, when Hitler's armies were close to taking Moscow. After emigrating to Israel, I learned about his cousin A. Shenkar, who emigrated with his wife to Israel in the early 1920s. He was a successful businessman in Palestine, and was the first chairman of an organization of businessmen in Palestine. Not having any children, he donated a lot of money to various universities in Israel. I was happy to learn that one of the buildings at Tel Aviv University is named after Shenkar.

I remember myself in 1939, reading newspaper stories about great public trials. It was clear to me that the stories were fabricated, but I was afraid to talk about this, even with my parents. The purges of the 1930s touched my family as well. My uncle, a brother of my father, was one of the leading lawyers in Moscow when he was arrested in 1937 and sent to a labor camp. He died there but we never knew exactly when and how.

On June 22, 1941, Hitler attacked the Soviet Union. Even now I remember Molotov's speech in which he announced that Hitler had moved his army into the Soviet Union without a declaration of war. In 1942, our family was evacuated from Moscow to the town of Kirov, in northern Russia not far from the Ural mountains. The local population was hostile to relocated people. They had good reasons for this, since they had to divide scarce food and housing with the newcomers. The food supplies that were never very good became worse in the wartime. The local children were very anti-Semitic and often picked on me. Fortunately, after one year, we went back to Moscow, where things were a little better. I was too young for the army and so I went to a high school.

It was a difficult time for me. Food was rationed during the war and we never had enough of it. To get better rations, I went to a special technical high school which prepared engineers for the railways. In addition to the usual subjects we studied technical topics such as drafting. Despite better rations, I did not like the school and was not a good student. I was dismissed from the school and sent to work at the railway station. However, my father talked to the director of the school and persuaded him to excuse me from that work in order to continue my education. I went to an evening high school and graduated in 1946.

My interest in mathematics started when I was about ten, with my father showing me the negative numbers. I remember that I was struck by their charm and the feeling of something unusually beautiful. Later, while still in high school, I rediscovered the binomial coefficients. My interest in mathematics was developing mostly at home until the later years of high school, when I started to attend the math seminars for teenagers organized by students of Moscow University. I also took part in several Moscow Mathematical Olympiads, with moderate success. There, the participants were given several hours to solve difficult problems. Whoever could solve the most problems was the winner. I remember that I won one third prize and a couple of honorable mentions. In high school, I also became interested in number theory after learning about Fermat's Last Theorem. I remember that several times I tried to prove it (not quite successfully).

In 1946, after graduating from high school, I entered Moscow University. I was still mainly interested in number theory. Soon I started to attend the number theory seminar organized by A. O. Gelfond, who was famous for his results on transcendental numbers α^{β}. His seminar was attended by his students and ex-students, some of whom were already professors. I remember N. M. Korobov and A. G. Postnikov. At the seminar, I met my friend G. Freiman. He was also a student, two years ahead of me. I also met U. V. Linnik, a friend of Gelfond, and already a professor in Leningrad.

The seminar was devoted to diverse topics in analytic number theory, including additive number theory, the Riemann zeta function and so on. The center of interest in this seminar was the method of trigonometric sums and applications. Occasionally we had talks about transcendental numbers. We had already realized that it was impossible to solve fundamental problems like the Riemann hypothesis using old methods in analytic number theory. We started to explore other topics such as uniform distribution of fractional parts, where progress looked more achievable. Gelfond was very interested in the theory of the zeta function. When he died in 1966, I was present in the hospital. I remember that he was trying to write some formula and tell me something which was clearly related to the zeta function. He could not because he was already paralyzed.

Later on, when I was a third year student, I attended courses of N. K. Bari on problems of trigonometric series. I started to work on one of the problems, the uniqueness problem. A subset X of S^1 is called a set of uniqueness if any trigonometric series which converges outside of X to zero has zero Fourier coefficients. It is easy to prove that any countable set is a set of uniqueness. Later on, Salem proved that the standard Cantor set is a set of uniqueness. He constructed certain of closed sets of cardinality equal to the continuum called H-sets. He conjectured that all sets of uniqueness are H-sets. I found a new construction of sets of uniqueness which are not H-sets. I also found a criterion for a closed set to be a set of uniqueness.

For this result in 1952 I received a Moscow Mathematical Society award for young mathematicians. Let me recall that 1952 was a year of great anti-Semitism in the Soviet Union. So it was a great surprise that I received the award. I shared the prize with Prokhorov, a specialist in probability theory, who is now a member of the Soviet Academy of Sciences.

I graduated from Moscow University in 1951. My advisor at Moscow University was A. O. Gelfond. He was a very warm person, very humane and sensitive to me and to the other students. He was a member of the Communist Party. His father was personally acquainted with Lenin. This was widely known, since Lenin, in his only philosophical book (required reading for all students), criticized the philosophical beliefs of Gelfond's father. When I asked Gelfond about this, he said that his father and Lenin had disagreements in public life, but in private they were friends. Being a member of the Communist Party, Gelfond felt that he had some influence, and recommended that the Moscow University accept me as his graduate student. However, that year was one of great anti-Semitism in Russia. Let me recall that anti-Semitism became very strong in Russia after the end of World War II. It was a strange inheritance that Stalin got from Hitler.

And so, the recommendation from Gelfond did not help me enter the graduate school of Moscow University. I was denied admission by the party committee of the mathematics department. The reason given was that my grade in military training was only "C". I did not take military training seriously and used to play chess during class (something I never did in math classes). In fact, military training was not an important subject and my "C" grade was just an excuse to keep me out of graduate school.

A. O. Gelfond still wanted me to continue with mathematics and he suggested to his friend A. D. Buchstab to try to take me in the Moscow Pedagogical Institute. My friend A. Lavut (later a well-known dissident) made the joke that Moscow University and Moscow Pedagogical Institute are fighting for Piatetski-Shapiro. The University wanted the Pedagogical Institute to take him, and the Pedagogical Institute wanted the University to take him. Since Moscow University was stronger, I entered the graduate school of the Moscow Pedagogical Institute.

At that time, the department of mathematics of the Moscow Pedagogical Institute was very strong. It included the famous logician P. S. Novikov. My advisor A. D. Buchstab was known for his work in sieve methods. He was a very good advisor for me since he let me do what I wanted. At the same time, he helped me very much to enter the graduate school. The main obstacle was the oral examination in Marxism and Leninism. A. D. Buchstab was present at the exam. It was a funny examination. It was well known that for Jews there were only two grades: satisfactory ("C"), and unsatisfactory ("F"). I think that I answered all the of the questions; however, the person conducting the exam refused to give me a grade. He said that only the chairman of the department of Marxism and Leninism could give me a grade, and the chairman was not present for the exam. In the end, I got the highest grade for a Jew which was "C". The director of the Moscow Pedagogical Institute knew about the style of the exam. Sometimes he accepted Jews who scored "C" on that exam, something he never did for non-Jewish students. This was a funny reverse discrimination.

Graduate school was organized differently in the Soviet Union. Everyone accepted received a stipend for three years with no teaching obligation. However, there was no way to extend the stipend beyond three years. The stipend was very small, but our needs as graduate students were very modest.

Accepted with me was my friend Yu. Sorkin, also Jewish. He was working in general algebra, with his advisor A. Dizman. He was also from Moscow University. Together we went through a strange, Kafka-style, adventure during our first year in graduate school. After we were both accepted to the graduate school we both received letters informing us that we were supposed to go to teach in a high school in Karaganda, in the middle of Asia. The letter said that if we did not go, we would be ordered by the courts to go. My parents panicked and advised me to go. They said it would be better to go to Karaganda than to go to a camp. My reaction to the letter was not so strong and my friend Sorkin agreed with me. At that time, after a student graduated from college, he was supposed to be assigned by a special commission to work somewhere. The justification of this was that since the state educated us for free, the state in return could send us to work where it wants. Since both Sorkin and I were recommended but not accepted to the graduate school of Moscow University, we both were given this assignment. In the meantime, we were accepted to the graduate school of the Moscow Pedagogical Institute. We did not sign the agreement to accept the assignment of the commission. However, the decision of the assignment commission was compulsory, and, theoretically, we could have been forced to go by the court. In reality, such cases seldom went to court. Fortunately, Sorkin had a friend in Karaganda who wrote that they did not need us. It was true, the school did not need us. We went through this adventure safely, but it took about one year.

In 1954, after I defended my Ph.D thesis, I went to Kaluga for three years. Kaluga is about 100 miles away from Moscow; by train it took a few hours. The apartment of my parents in Moscow was very close to the train station, so it was very easy to commute to Moscow for weekends.

My work in the theory of automorphic functions started when I. R. Shafarevich suggested to me to translate Siegel's lecture notes from the Institute for Advanced Study on automorphic functions. As far as I remember, I met Shafarevich for the first time in 1949 when I was a student in my third year of undergraduate studies. I attended a course of B. N. Delone on the geometry of numbers. The course included some elementary material such as

Minkowski's Lemma and introduced algebraic numbers as a lattice in multi-dimensional Euclidean space. He talked about various problems including the famous tower problem of Hilbert. (If you construct a tower of unramified maximal abelian extensions, is it finite or not?) He was trying to say something about this problem using his geometric approach. He said he had a student named Shafarevich, a genius, who felt that this problem could not be solved in such an elementary way. Later on, Shafarevich and E. Golod solved this problem in a more sophisticated way. I do not remember any serious interaction with Shafarevich until later, when I became a graduate student and started to participate in his seminar.

Shafarevich was very interested in Siegel's theory of modular functions. He conducted a seminar to understand Siegel's lecture notes on automorphic forms.[1] He invited me to translate the lecture notes from English to Russian; I was happy to do this since I could earn a small amount of money. I remember that my mother, who was a typist, helped me very much since I dictated the text to her and she typed it right away. The lecture notes of Siegel were published in Russian—their only publication.

I remember the moment when we learned that not every complex torus has an algebraic structure. Shafarevich was very surprised by this and went to discuss it with Gelfand. It was later understood by several mathematicians that in higher dimensions, complex manifolds are not algebraic in general. The general criterion for Kähler manifolds to be algebraic was given by the celebrated theorem of K. Kodaira. By that time, I had started to work on the theory of automorphic functions. The seminar of Shafarevich was very inspiring for me. About that time, I obtained my first results about automorphic functions. I remember going to Shafarevich's apartment and discussing the difficulties I had encountered. That was always very helpful. He had a very good grasp of the general picture and of all technical details. At that time, it was typical to go to the apartment of your professor because in the universities there was not enough space to work.

I remember that our conversations were not restricted to mathematics, and after finishing our mathematical discussions we frequently turned to politics. Shafarevich, a son of a professor, was a well-educated man who knew French and German. Even then, he made it clear that he disliked the October Revolution. Of course, he did not say that explicitly, which would have been dangerous. At that time, during Stalin's rule, no one could dream of being a dissident. However, it was clear to me that Shafarevich had negative feelings for Communism. Of course, he never was a member of the Communist Party. More interestingly, he was against all revolutionary movements in principle. At that time, Dostoevskii was not easily available in Russian,

[1]These lectures were given at the Institute for Advanced Study in Princeton during World War II.

but Shafarevich quoted the very negative depiction of revolution from the famous novel "*Devils*".

Returning to mathematics, the central result of Siegel's book was the so-called theorem of algebraic relations for Siegel modular functions. In my Ph.D. thesis, I proved a generalization of this theorem for the case of Siegel-Hilbert modular functions. Actually, I introduced the terminology "Siegel-Hilbert modular functions" in my thesis, and it was accepted.

Siegel's book also contained explicit descriptions of symmetric domains; at that time, I did not know the general theory of Lie groups. So this description was very important for me. The realization of bounded domains as unbounded domains played a very important role in Siegel's methods. For an important class of bounded symmetric domains, these unbounded domains are called the Siegel half-plane. The general notion of Siegel half-plane can be described as follows. Let V be a convex cone in $\mathbb{R}^n$ which does not contain any line. Consider the following set of points:

$$H = \{x + iy ;\ y \in V ,\ x , y \in \mathbb{R}^n\} \subset \mathbb{C}^n$$

This is what I called a Siegel domain of the first kind. For example, the Siegel half-plane is given by taking V to be the set of positive definite $n \times n$ real symmetric matrices. The natural problem was to extend this description to other symmetric domains. The natural question was how to do this for other types of symmetric domains. The difficulties of this problem were manifest in the two-dimensional complex ball B^2. The two-dimensional complex ball is given by

$$\{(z_1 , z_2) ; |z_1|^2 + |z_2|^2 < 1\}.$$

It is easy to prove that there is no realization of B^2 as a Siegel domain of the first kind. I found a realization in the following form:

$$\{(z , u) ; \Im z - |u|^2 > 0\} \subset \mathbb{C}^2.$$

This example led me to the general definition of a Siegel domain of the second kind. Let $V \subset \mathbb{R}^n$ be a convex cone with no lines inside. Let W be a complex vector space and let F be a map satisfying the following conditions.

(1) $F : W \otimes W \to \mathbb{C}^n$.
(2) F is linear in u and antilinear in v.
(3) $F(u , v) = \overline{F(v , u)}$.
(4) $F(u , u) \in \overline{V}$ and $F(u , u) = 0$ if and only if $u = 0$.

Then the Siegel domain associated to (F , V) is given by

$$H = \{(z , u) ;\ \Im z - F(u , u) \in V\}.$$

It was proved later in collaboration with S. G. Gindikin and E. Vinberg that any bounded homogeneous domain has a realization as a Siegel domain of the second kind with transitive action of linear transformations. There is a very nice exposition of the story of Siegel domains by S. G. Gindikin [1].

In those days, I was able to check by hand that all of Siegel's examples can be written in this form. Also by guessing, I could find realizations for two other exceptional symmetric domains. However, the most unexpected application was the discovery of nonsymmetric homogeneous domains in $\mathbb{C}^4$. In hindsight, you just have to realize that such examples should exist. It is a simple exercise that you have to take $n > 3$, because you have the first nontrivial cone in dimension 3:

$$y_1y_2 - y_3^2 > 0,$$
$$y_1 > 0,$$

which is equivalent to

$$\begin{pmatrix} y_1 & y_3 \\ y_3 & y_2 \end{pmatrix} > 0,$$

where $A > 0$ for a symmetric matrix A means that A is positive definite. Then you consider

$$F(u, v) = \begin{pmatrix} u\bar{v} & 0 \\ 0 & 0 \end{pmatrix}.$$

Then the corresponding Siegel domain is

$$\begin{pmatrix} \Im z_1 - |u|^2 & \Im z_3 \\ \Im z_3 & \Im z_2 \end{pmatrix} > 0.$$

I remember that I published the definition of Siegel domain of the second kind a year before I realized that it can be used in order to construct examples. It is an interesting situation: if you knew the definition of Siegel domain of the second kind and knew that this definition led to an example of a nonsymmetric domains, then it would take at most an hour to find an example and the essential idea of the proof for showing that it is not symmetric. I believe that this situation is typical even now; nobody, even the author, reads articles with proper attention. I know other examples of this from my American life as well!

Let me explain how I came to the idea that nonsymmetric homogeneous domains should exist. I knew about this problem, but was not interested in it. I was under the general impression that such domains did not exist, but it would be difficult to prove the nonexistence. However, the study of automorphic forms naturally led me to the study of the geometry of symmetric domains. The following fibration is important.

Let D be a bounded domain. To each boundary component B_i, we construct a fibering by looking at all the geodesics that end in B_i and associating the end point to every point on the geodesic. In a typical situation, the fiber is not a symmetric domain. Thus I learned from the geometry of Siegel domains about the existence of nonsymmetric domains.

This fibering was very important for understanding Satake compactifications. It was also very important for generalizing Satake's construction to arbitrary arithmetic groups. At that time, the very important result of A. Borel and Harish-Chandra on the structure of fundamental domains for arbitrary arithmetic groups appeared. I remember that Shafarevich was invited to attend the International Congress of Mathematicians in Stockholm in 1962. I was invited to this Congress also. As usual, I was refused permission to go. I do not remember the formal explanation, but it was clear that I was refused because I was Jewish. Since I was invited to give a talk, I wrote the talk and asked Shafarevich to read it because I felt he was closer to the work than other people. He was willing to do this.

When Shafarevich returned, he told me that many people attended the presentation of my talk. I was very excited to hear this. Shafarevich passed to me the question of A. Borel whether I could prove the theorem of algebraic relations and the theorem of normal compactification of arbitrary symmetric domains and arithmetic subgroups. It was clear to me that we had the necessary tools: the generalized reduction theory of A. Borel and Harish-Chandra, and the geometry of symmetric domains by myself. I started to work on this topic and soon obtained the results. At about the same time, A. Borel and W. Baily proved the same results. I must confess that their exposition was more thorough than mine; everyone (including me) now calls it the Baily-Borel compactification.

Shafarevich introduced me to automorphic forms, the topic which became the main focus of my work. We shared a strong interest in number theory and later wrote a few papers together. One paper is on the tower of fields of automorphic forms. Let H denote the upper half-plane and let $\Gamma(p^n)$ be the modular group of level p^n. Let

$$X_{p^\infty} = \operatorname{proj\,lim} H/\Gamma(p^n).$$

Then $\widetilde{\mathrm{SL}}_2(\mathbb{Q}_p)$ acts on X_{p^∞} considered as a complex manifold, but if you also take into consideration the fact that X_{p^∞} is defined over $\mathbb{Q}(\mu_{p^\infty})$ (you add all the p^n-roots of unity), then $\widetilde{\mathrm{GL}}_2(\mathbb{Q}_p)$ acts on X_{p^∞}. It is clear that this fact lies behind arithmetic applications of automorphic forms. We started to work on it to understand it better. Later on, I found that the Eichler-Shimura congruence relation has a very natural interpretation in this language. In my article **[2]** I explained the interrelation of the action of the group of adèles and the Eichler-Shimura congruence relation. I remember that V. Drinfeld helped me very much in the preparation of this article. Soon after, V. Drinfeld came to his wonderful discovery on Shtuka; he told me that he was influenced by my article. This was, as far as I remember, my first article to be written first in English and then translated into Russian to obtain permission to be sent abroad. This article was published later in the Proceedings of the International Conference on Modular Forms. I was invited to this conference. But I could not obtain permission of the authorities to participate in this conference. Another paper, quite well known, was on the Torelli theorem for K3 surfaces. We started to work on this topic entirely on Shafarevich's suggestion. Shafarevich was an ideal collaborator; he would look into all technical details. It was possible to discuss everything with him.

When I decided to leave Russia, I went to Shafarevich and told him about it. He was very negative about this. He was not openly anti-Semitic at that time. But, by that time, I had heard many remarks from him that sounded strange to me. For instance, he was very critical of the Jackson Amendment which denied the most favorable status to countries which restricted emigration. He criticized Jackson and the American Congress even more than the Soviet media. He tried to persuade me not to emigrate. Earlier, Shafarevich was tried to persuade B. Moishezon not to emigrate, also to no avail. Shafarevich gave me many different arguments against emigration, the most funny of which was the following. He said that I would never learn how to correctly pronounce the Hebrew letter “ain”. He was right. I still cannot pronounce it correctly, but my daughter Shlomit can do it without a problem.

Recently, Shafarevich published an essay on “Russophobia”, placing himself on the extreme right of anti-Semitic Russian literature. It was very unpleasant to see a man I respected so much become a leader of anti-Semitism. I am not going to discuss the contents of this book, but I must say that I completely disagree with his basic statement that Jews were responsible for the October revolution and the evil that came from it. It is true that some Jews participated in the October revolution, but it is clear that the October revolution was not a plot of Jews against the Russian people. The Jews, like the other participants, honestly believed that this revolution would bring a good life for all people. I quite understand Shafarevich’s worries about the fate of the Russian people, and I agree with him that the October revolution was a catastrophe for the Russian people. But in the history of any people, there are similar or worse catastrophes. I think that if people would critically understand the reason for a catastrophe, then they could overcome it. Trying to make Jews the scapegoats for this catastrophe is very bad for the Russians themselves.

In 1958, at the end of my stay at Kaluga, Gelfand invited me to come to the Institute of Applied Mathematics in Moscow, which at that time was considered secret. Its director was the late M. V. Keldysh, one of the most important figures in Russian science, not only for the position he held (for a long time, the president of the Soviet Academy of Sciences), but also for his theoretical leadership of the Sputnik program. Let me digress for a moment and add some personal stories about Keldysh. In his memoir, A. D. Sakharov mentions that Keldysh was asked by Brezhnev how much time one needed to make Russian science Jew-free (Judenrein) and Keldysh replied that probably 15 to 20 years would be enough. In the same memoir, Sakharov writes that at the institutes Keldysh directed, M. V. Keldysh was not anti-Semitic. I worked at the Institute of Applied Mathematics for about 15 years and I share the same feeling that Keldysh was not anti-Semitic. After I resigned from the Institute in 1974 and applied for an exit-visa, Keldysh, as the director of the Institute, had to write a letter about me stating whether I was involved in classified research. After my application was refused, I asked him to confirm that I was not involved in any classified activity. Keldysh told me that he had already confirmed that. I believed him, but when I told my refusenik friends, they all laughed at my naïvete and one respected refusenik suggested that I tell American journalists that Keldysh was a hypocrite. I refused to follow this suggestion, and in fact, in less than a year I got permission to leave the Soviet Union. I consider this as proof that Keldysh was not lying.

Gelfand was the head of a theoretical department that was not involved in classified research. At that time, there were many mathematical departments involved in classified research, but researchers started to realize that they would be better off by staying away from classified research.

I remember that I had a chance to attend Gelfand's course in 1949, when Gelfand worked together with Naimark on unitary representation theory. About the same time I started to attend the Gelfand Seminar. The Gelfand Seminar was unusual in its breadth of topics covered—there could be talks on representation theory, functional analysis, hydrodynamics, sheaves, etc. It was Gelfand's intention to understand mathematics as a whole; no problem in mathematics was irrelevant to his seminar. The seminar was also flexible in its time schedule. Seminars started at 6 or 7 p.m. on Monday and went on to 10 p.m. or even midnight. One thing was certain: the seminar never started or ended on time.

Gelfand was very active. He would ask many questions and at the end replace the talker and present the talk in much better form. I remember giving a talk myself on representation theory of $\widetilde{\mathrm{GL}}(2, \mathbb{Q}_p)$. This was the very beginning of this work. The notion of smooth representation was not common at that time. Gelfand, together with M. I. Graev, already started to work on the classification of representations of $\widetilde{\mathrm{GL}}(2)$ over p-adic fields, but he only considered unitary representations. However, for a person working in automorphic forms, the natural notion was a smooth representation. My talk was about the Jacquet-Langlands theory of $\widetilde{\mathrm{GL}}(2)$. It covered smooth representations and Kirillov models, etc. Gelfand was not familiar with this theory but he immediately understood the importance of this notion. The point is that he was concerned with the notion of equivalence of representations in infinite-dimensional spaces. If you say that two Banach representations V_1, V_2 of a topological group G are equivalent if there exists a continuous isomorphism of these two spaces which commutes with the action of G, then you are in trouble. This is because representations which are naturally equivalent will not be equivalent under this definition. For example, consider the representation of $\widetilde{\mathrm{SL}}(2, \mathbb{R})$ on the space of holomorphic functions on the upper half plane. One can give two different norms on this space so that the two Banach spaces are not equivalent under the naïve definition. It is clear that the two representations should be called equivalent. Hence Gelfand was interested in the correct notion of equivalence. In the p-adic theory, if one considers smooth representations, then one can use the naïve definition. In the Archimedean case one can also define the notion of smooth representations, which were established much later by W. Casselman and N. Wallach.

Gelfand had a broad interest in representation theory. To him, representation theory is at the center of the whole of mathematics. He was not a specialist of the theory of automorphic forms at the beginning. The first important thing that started our collaboration was the notion of cuspidal automorphic forms. The definition of cuspidal automorphic forms was well known by that time. The definition of cuspidal Maaßwave form was also known; reformulation in terms of representation theory was very easy. When I discussed this notion with Gelfand, he became very enthusiastic. He said many times at that time that it was a very important definition. In the course of our discussions, Gelfand realized the connection between scattering theory and asymptotic properties of Eisenstein series (later on, L. D. Faddeev wrote a paper on this). Gelfand and I proved that the spectrum of cusp forms is discrete. Gelfand many times underlined the importance of this set of irreducible representations of the adèle group which can be realized as cuspidal. He said that the set should play an important role in arithmetic. We also introduced an important operator which is in some sense similar to the S-matrix. Unfortunately, I did not learn enough of scattering theory. At the end of our cooperation, we wrote the book *Representation theory and automorphic functions* with M. I. Graev.

Cooperation with Gelfand was very unusual. Typically, people who cooperate divide any type of work among themselves. But with Gelfand, it was different. One has to appreciate his deep understanding and knowledge of mathematics and wonderful ability to find unexpected relations. For instance, in the case of the theorem of discrete spectrum for cuspidal automorphic forms, Gelfand, from analogy with scattering theory, formulated the theorem, explained to me why it is true, and left me to work out the rest. Later on, cuspidal representations became known to Western mathematicians and they enthusiastically developed the ideas. The theorem of restricted tensor products, which describes representations of $G(\mathbb{A})$, was also first proved in our book. I remember that Gelfand explained to me the problem: how to relate representations of the adèle group $G(\mathbb{A})$ to representations of local groups. We started to work on this problem and soon we found the theorem about restricted tensor products.

I remember that the three of us, Gelfand, M. I. Graev and I, spent one vacation on the river Volga and worked on the book. Graev was also my friend, who started to work in the theoretical department of Gelfand at about the same time. Due to a shortage of office space in the Institute, we had to share our offices. Graev was a quiet person strongly devoted to mathematics, capable of doing a lot of hard mathematics. We owe it to him that the book finally came out. Unfortunately, after I left the Soviet Union, Gelfand and Graev stopped working on automorphic forms. I was very happy to meet

Gelfand very recently in the USA Dr. Klaus Peters suggested that we reprint our book as it is. Gelfand and I agreed immediately with this idea. We decided only to add a new introduction signed by the three of us.

Let me conclude these notes by mentioning my relation with Y. I. Manin. He was younger than me. He was a student of Shafarevich, but soon became completely independent. Our relationship was always warm and friendly. I remember especially the last few months in the Soviet Union, when I attended Manin's Seminar at Moscow State University about gp-adic L-functions. We had a number of mathematical discussions at that time. I remember one of them, where Manin told me that he expected the Mordell conjecture to be solved soon despite the fact that some very important tools were still not available. He said that he would not be surprised if an extremely talented and powerful mathematician came and solved this problem in a few years. Recently, I met Manin in the United States and in France and I found him the same warm, friendly person.

Even though I left the Soviet Union, I do not harbor ill feeling towards it. I recall my years there without regret. There was, and still is, an excellent mathematical school there. I am always happy to meet my friends from Moscow.

REFERENCES

1. S. G. Gindikin, *Seigel domains, in Festshrift in honor of I. I. Piatetski-Shapiro*. I, II (Tel Aviv, 1989), Amer. Math. Soc., Providence, RI, 1991, pp. 5–19.

2. I. I. Piatetski-Shapiro, *Zeta functions of modular forms, in modular functions of one variable*. II (Proc. Internat. Summer School, Univ. Antwerp, Antwerp, 1972), Lectures Notes in Math., Vol. 349, Springer, Berlin, 1973, pp. 317–360.

On Soviet Mathematics of the 1950s and 1960s

D. B. FUCHS

When I received an invitation to write a kind of short mathematical autobiography for this volume, I had a strong impulse to refuse. But then I realized that it was a good occasion to describe some events of which I became partly a participant and partly a witness and which definitely deserve to be recorded. In what follows I shall try my best to avoid personal reminiscences.

Some words to begin with. I was brought up in a mathematical family. My father, B. A. Fuchs, was a well-known mathematician, S. Bergman's student and the author of numerous books and articles in several complex variables. Some famous Russian mathematicians, including N. V. Efimov and V. A. Rokhlin, were friends of our family. So it was quite natural that at an appropriate age I had a firm intention to become a mathematician, which was more or less equivalent[1] to the intention to enter Moscow State University (MSU). Nevertheless my father, while supporting the first of the two intentions, was strongly opposed to the second. He tried to convince me that I could not be admitted at MSU just because of my name. (I should explain here that the name Fuchs sounds Jewish in Russia; and my father actually was a Jew, at least biologically, while my mother originated from Russian peasants of the Middle Volga.) But I was not then of an age to listen to wise warnings. I applied to enter MSU, and was admitted there without any noticeable problems. It was the autumn of 1955.

I could not appreciate the overwhelming significance of the events at that time, but I can now.

The Jewish problem: does it exist?[2] I know that the question is discussed quite seriously in the West. I can hardly convince those who do not want

[1]The pyramidal structure of our political organization reflects itself in pyramidal structures in many other aspects of our life. In particular, the level of Moscow University is uncomparably higher than that of Leningrad, Kiev, Novosibirsk, etc. Universities (and their level is uncomparably higher than that of, say, Saratov, Kazan, Odessa, etc. Universities).

[2]In what follows I discuss exclusively the situation in the Department of Mathematics of Moscow State University. The situation in the whole deserves more serious analysis; but basically it is the same everywhere in our country.

to believe, the more so because I have no reliable statistics. I simply want to say some trivial things known to everybody in our country. Before doing so I must stress that my own experience of this kind is far from dramatic. The three determining events of my life, namely entering MSU, becoming a graduate student at MSU, and getting a job there, coincided with three local minima of the anti-Semitic campaign; and the first of these minima was a global one, at least for the last 40 years.

It began with the struggle against the "cosmopolites" at the end of 1949. This struggle embraced all aspects of life—not only the entrance examinations. For example, my father had lost his job and was in various degrees of unemployment until the autumn of 1953, when he obtained a chair at a second-rate technical college. But this struggle did influence University entrance examinations—in a drastic way. Almost no Jews were taken into the Department of Mechanics and Mathematics (the usual abbreviation: Mekh-Mat) in 1950–1952 (though there were exceptions; for example, Ya. Sinai entered Mekh-Mat in 1952). I know of some ten Jews (or half-Jews) who entered in 1953 (some future superb mathematicians were among them, such as Yu. Manin, E. Golod, B. Mityagin and A. Dynin). I was told once that none of them was admitted to MSU according to the normal examination procedure (some were transferred from other Universities, while others were given the semilegal status of "a candidate to students"); probably this is only partially true. But in 1954, and particularly in 1955, the problem suddenly ceased to exist (or, rather, interrupted its existence). The about-turn was so sharp that many people (including my father) simply did not know of it, or could not believe it. During the next 15 years the situation took a definite turn for the worse. Nevertheless I believe that most of the people who examined prospective students in mathematics did it honestly (some mathematicians volunteered their efforts to support the racial purity of Mekh-Mat, but they were not numerous). The crucial job was done in those years by physicists (there were four examinations: written and oral mathematics, physics and a "composition"). But in the late 1960s and early 1970s the situation changed dramatically (I shall return to this period in my notes). There began a real war against Jews. All people with Jewish-sounding last names (some Germans and Estonians suffered without any guilt) or with Jewish patronymics or even with a parent with a Jewish patronymic were considered to be Jews, and every possible barrier was erected on their way to Mekh-Mat. The most effective weapons were special "Jewish problems" which were offered to "Jews" at the oral examinations in mathematics. Sometimes they were too difficult, sometimes they required a large amount of complicated calculations. Bad marks for the composition could be given to those who managed to escape failure at the oral examination. I know of many highly talented people who were not admitted to Mekh-Mat in the 1970s and later. It is enough to mention A. Beilinson, A. Givental and B. Tsygan (Beilinson was transferred to Mekh-Mat from the Pedagogical Institute when he was a third-year student,

Givental graduated from the Oil Institute, Tsygan studied in Kiev). I could name many less famous people who suffered the same fate. And how to count those who gave up?

As a rule in the years 1970–1988 there were some three to five Jews or half-Jews among the 500 students of Mekh-Mat each year. And usually they had to go through various Appellation Committees, complaints to the Ministry, etc.

I have the pleasant opportunity to confirm that in 1989 things were much better than in previous years; but the real trouble is that the examination commission consists mainly of the same people as before.

It would be interesting to understand the reason for this beastly unfair and pragmatically unwise campaign. I doubt that the Heroes of the Great Anti-Jew War were inspired by their pure Anti-Semitic feelings. According to my father, I. M. Vinogradov, who was always considered as the leader of Anti-Semitism in mathematics, surrounded himself with Jews in the late 1930s; in particular he never took any important decision as the Director of the Steklov Institute without a consultation with the local Party leader B. Segal; the last was actually Vinogradov's creature and protegé. Once (probably it was in 1968) I spoke to I. R. Shafarevich, and mentioned without any particular motive the year 1950 "when all the Jewish professors were driven from Mekh-Mat." "Who told you this?" Shafarevich asked with irritation. "Possibly it was Gelfand." I did not know what to say. Certainly, it was not Gelfand who had told me this, at least, for the first time; but it seemed to me that everybody knew it. "It is true that Gelfand had to leave Mekh-Mat then," Shafarevich continued, "but I had to do the same without being a Jew. They simply fired all the good mathematicians, Jews or not Jews." Now it seems to me that Shafarevich was more right than might seem at first glance. Mekh-Mat would never have been driven to its present miserable state if the policy of its authorities had been directed only against Jews. Russia is a very big country, and if you were not to admit the Jews to Mekh-Mat, or, say, left-handed, or blue-eyed ones, or apply some other arbitrary criterion, but were to honestly choose the best ones from the rest, then this would be highly unjust and immoral but would not have had such a terrible effect. The only way to deprive Russian mathematics of talent is to struggle against the talent, and this is exactly the struggle we were involved in. For example, those who graduated from the bet Moscow mathematical high schools were always regarded by Mekh-Mat's authorities as Jews, irrespectively of their actual origin. And this is a great (probably undeserved) honor to Jews that they were *a priori* included in the category of talented people.

To finish this subject I want to stress that in post-war Soviet Russia the notion of a Jew was by definition purely genetic, and Jewish genes were supposed absolutely dominant; you may be a quarter Jew and three quarters Russian (as Beilinson is), but you are regarded as a full-fledged Jew.

Mekh-Mat of the late 1950s. The "thaw" in the country gave rise to a

I. M. VINOGRADOV

"thaw" at the University. The level of (the best part of) the Mekh-Mat students was extremely high in those years. Anosov, Arnold, Golod, Kirillov, Manin, Novikov, Palamodov, Sinai, Vinberg, and many other first-class mathematicians studied there more or less simultaneously with me. Bari, Dynkin, Gelfand, Gelfond, Khinchin, Kolmogorov, Kurosh, Markov, Petrovskiĭ, Pontryagin, Postnikov, and Shafarevich were among the Professors. The Mekh-Mat schedule was oversaturated with first-rate lectures and seminars. Many undergraduate students wrote good scientific works (which is in general more common in the USSR than in the West), and Dima Arnold even solved Hilbert's Thirteenth Problem. According to Mekh-Mat's rules each second-year student had to choose an advisor from the staff of the Department of Mathematics (or outside of this staff: the times were liberal). My first adviser was B. N. Delone (he used to spell his name Delaunay). Many of us adored him, partly because of his independent and rather extravagant behavior. He organized a small seminar in Diophantine approximations which attracted a small group of second-year students. The seminar was quite unusual. You would just open your mouth, and Delone would exclaim, "How clever! We shall publish it in Uspekhi! No, better in Doklady! No, let it be Uspekhi after all." After several months there remained only two of us: Sasha Vinogradov (no relation to I. M. Vinogradov!) and I. And we did write a note in Doklady (jointly with Delone). I do not think anybody ever read it; in any case two lines were missing in the printed version of the main statement. Certainly we were pleased and proud (both Vinogradov and I were in our teens) but even then we did not take it seriously. In any case, besides Delone's seminar we attended Dynkin's seminar on Lie groups and homogeneous spaces. And after a year we decided to find an advisor in this seminar, but not Dynkin himself. This led us almost forcibly to A. S. Schwartz, who was then himself a last-year graduate student. Schwartz gave his okay and thus we became topologists.

Topology of late 1950s and 1960s in Moscow. In the 1930s and 1940s topology flourished in Moscow, and the center of it was L. S. Pontryagin. According to V. A. Rokhlin, Pontryagin knew much more than he published, and no article in topology published by anyone in the world contained much new for him. But in early 1950s the "French works" appeared. Many beautiful geometric results of Pontryagin's era were understood as mere trivialities. And Pontryagin abondoned topology for optimization theory. Different opinions on this dramatic episode were expressed by different people (including Pontryagin himself—see his autobiography in Uspekhi, 1981); I will not touch on this subject. But I cannot avoid reflecting on its terrible influence on topology at Mekh-Mat and consequently in the USSR.

The official head of Mekh-Mat's topology was P. S. Aleksandrov, whose interests were then restricted exclusively to "set-theoretical topology", or "general topology"; that is, the domain of mathematics which studies the notion of a topological space in maximal generality with special attention to various pathologies. The staff of the Chair of Topology consisted almost completely of Aleksandrov's students (the only exception was made for me in 1963–1967). In addition to this there existed a strong public opinion (cultivated by Pontryagin and his colleagues) that topology had exhausted itself and was no longer a worthy subject to work in. Nevertheless a small company of topologists has formed around A. S. Schwartz and M. M. Postnikov. The seminar had seven permanent participants. These were S. Novikov, B. Averbukh (he calculated the groups of oriented bordisms independently of J. Milnor, but after Milnor published his work), L. Ivanovskiĭ (who did an enormous amount of calculations of the homotopy groups of spheres, and never changed the subject), Vinogradov, myself, and two more mathematicians who never were topologists but wisely studied topology as if it were their field: D. Anosov and G. Tyurina. (The latter became my wife in 1961 and died in an accident in 1970.)

Our seminar had no formal leader. Schwartz received his Ph.D. and had to leave Moscow. (He had no Moscow "*propiska*"; I suppose here that the reader is familiar with this notorious notion. Aleksandrov possibly could have obtained permission for him to stay, but he definitely did not want to.) Postnikov did not like teaching, and he had some strong interests outside mathematics. So we were left to ourselves. Unfortunately, Novikov was of our age group, and it was embarassing for him to assume the usual functions of a seminar leader (to tell us what to read, to offer problems, etc.). The same difficulty, by the way, has always been present in my relations with Arnold. I could not be his student, but nevertheless some years later I became a kind of "in-house topologist" in Arnold's seminar. I even published two articles with solutions of his problems; both of them are still referred to, especially "Cohomology of the braid groups modulo 2".

Our small group was under hard pressure from two sides. Firstly, our fellow students reproached us that we pursued some homology and homotopy groups instead of doing something really worthwhile, say, in PDE or classical function theory. Our reaction was fully inadequate: it resulted in a sheer ignorance of analysis (at least on my part). Later I tried to study some analysis but the time was lost. Unfortunately, even in topology we tried to avoid everything which did not involve homological algebra. I learned the notion of a spectral sequence a year earlier than that of a barycentric subdivision. And do not ask me when I learned that the Grassmannian $G_+(4, 2)$ was diffeomorphic to $S^2 \times S^2$.

Secondly, there was a bitter struggle between set-theoretic topology and algebraic topology. Aleksandrov hated algebraic topology (of which he was one of the creators) and tried to stop its development by any means. Good-humored Anosov invented comic "Rules for the Chair of Topology" which prohibited studying "commutative diagrams and exact sequences which actually were neither commutative, nor diagrams, nor sequences, and the more so not exact, but were in fact senseless combinations of zeroes and arrows".

Once Schwartz wrote in an advertisement of his lectures that "the main difference between set-theoretic and algebraic topology is that while the latter solves difficult problems with simple methods, the former solves simple problems with difficult methods". Such jokes were among the reasons that after getting his Ph.D., Schwartz had to leave Moscow for Voronezh; and we had lost our potential leader. The main battle of the war took place some years later when Novikov and Shafarevich tried (in vain) to prevent three major students of Aleksandrov from getting their Doctoral degrees (much higher than the American Ph.D.). All this seemed very important to us until 1968 when both armies found themselves on the same side in the front line in the "Esenin-Volpin case" (see below).

In the meantime the general attitude toward topology changed, and the main reason was the Atiyah-Singer index formula. The index problem was actually formulated by Gelfand, and our mathematicians were always interested in it. The trouble was not in the mere fact that the solution was found by somebody from the outside; but our PDE experts could not even understand what was written on the right-hand side of the formula! They became eager to study topology. The modest lectures which I used to give to several bored students became one of the most popular courses at Mekh-Mat. The lecture halls were overcrowded. people wanted to learn everything in topology, but especially the ingredients of the Atiyah-Singer formula: K, ch and Todd.

Certainly, my topology was useless to all these people because the Atiyah-Singer formula was as far from me as from those who were listening to me. At the time I thought that the most important problem in mathematics was the calculation of the homotopy groups of spheres, and the most useful device was the Adams spectral sequence. This spectral sequence was the ultimate purpose of my lectures. But gradually everything became more or less normal. The ignorance of most of our mathematicians in topology became less dramatic, the popularity of topology decreased to a reasonable level, and my lectures in topology became better.

The only one of us who did really bright work in topology was Sergei Novikov. When he was an undergraduate student he worked mainly in homological algebra; he invented a kind of Steenrod square operations in the cohomology of Steenrod-like algebras and applied it to the Adams spectral sequence. Then he began to work more geometrically and wrote a paper on cobordisms entitled "Some homotopy properties of Thom spaces". The most outstanding among his later works were about smooth structures on manifolds, codimension 1 foliations on 3-manifolds, the topological invariance of the rational Pontryagin classes, and cohomology operations in cobordisms (the Adams-Novikov spectral sequence). Novikov became a theoretical physicist in the late 1960s, but he still writes in topology from time to time.

My own scientific achievements in topology were modest. I did some reasonable things under the guidance of Schwartz, but then I had to choose subjects myself. I invented a kind of a formalization of Eckmann-Hilton duality. Some people, including Peter Hilton, approved of it. Probably the work was not bad, but I could not follow the temptation to continue it. As a result of this, at the age of 27 or 28 I simply did not know what to do next. But then I made Gelfand's acquaintance, and it was a godsend for me.

Gelfand-Fuchs. I am sure that most readers who happen to know my name, know it in this combination. In 1967–1970 I worked jointly with Gelfand, and we published some 15 articles; some 5 articles more were published later. I had no idea that Gelfand knew about my existence until somebody (it was probably M. Agranovich) phoned and told me that Gelfand wanted to meet me. I came to his apartment where there were many people (I remember D. Kazhdan and G. Margulis), and the telephone rang every minute. Gelfand asked me what I was doing, and happily enough listed to my answer not very attentively. He said then that he always wanted to work in topology (*this was the truth*), and made me an offer to work jointly with him. And it turned out that he was quite serious; we began to work. Our first observation was that if one takes the quotient of $\mathbb{R}^\infty - 0$, where $\mathbb{R}^\infty$ is the Tychonov space, by the group of homotheties, then one gets a smooth ∞-dimensional manifold S (a "sphere") on which all globally defined continuous functions are constant. And the projection $\mathbb{R}^\infty - 0 \to S$ is a nontrivial principal $\mathbb{R}$-bundle (which was impossible over a paracompact base). Moreover, this $\mathbb{R}$-bundle turns out to be universal in some sense. We tried to generalize this construction to a general noncompact Lie group, and rediscovered the Van Est cohomology groups as groups of certain characteristic classes. Then we approached infinite-dimensional groups, and this led us to the cohomology of the corresponding Lie algebras. A tentative calculation gave a surprising result: the second cohomology group of the Lie algebra of vector fields on the circle was neither trivial nor infinite-dimensional: it had dimension 1. Thus we discovered the central extension of the Lie algebra of vector fields on the circle which nowadays is known to all theoretical physicists under the name of the Virasoro algebra. During the next two years we made a lot of successful cohomology calculations. All this time it surprised me a great deal that Gelfand, who had a huge number of first-class results in mathematics, greeted every theorem as if it were his first success. But certainly this is the best attitude for scientific work.

The case of Esenin-Volpin. To finish my notes I want to record some events of Spring 1968 which later proved to be so important for Mekh-Mat.

The Brezhnev era began with several political trials, and the general atmosphere in the country was dreadful. But Mekh-Mat remained relatively free and happy thanks to the efforts of I. G. Petrovskiĭ, who was then Rector of MSU. This situation was very unstable, for its preservation required the acceptance of all kinds of compromises by members of Mekh-Mat's staff.

One of the important features of our political life in the late 1960s was "signing letters". After Sinyavski and Daniel's trial, some groups of people wrote collective letters to various power institutions with various kinds of protest (from rather mild to very strong). Of course, the authors of the letters were punished, but the letters kept on being written.

The letters concerned various things, mainly violations of the rights of various individuals and groups of people. To sign a letter became fashionable. There were so many letters that most of them caused no reaction from the authorities, and this fact encouraged various kinds of people (scientists, writers, artists, even schoolchildren) to write more and more letters. In order to stop the campaign, the authorities had to choose one of the letters for an exemplary punishment of its authors. And it seems that they chose the letter about Esenin-Volpin.

A. S. Esenin-Volpin was the son of the great Russian poet [Sergeiĭ Esenin], a good mathematical logician and a confirmed dissident. In January 1968 he was taken to a "*psikhushka*", a special psychiatric hospital for political deviants. Even now I do not know the real reason for this action, though I can guess that it was connected in some way with the approaching fifteenth anniversary of Stalin's death on March 5, 1968. After some unsuccessful private efforts to release him, a letter was written and signed by 99 mathematicians. Even now I do not dare list those of them whom I remember; I restrict myself to a neutral statement, that almost all well-known mathematicians who already had their Ph.D.'s in 1968 and who were seldom seen abroad later, were among those who signed the letter. The letter was very clever; it did not contain any general complaints or bitter accusations of the KGB (as was usual in those letters). It just stated the plain fact that there were some formal rules for putting people in asylums, and that these rules were definitely violated by the "emergency medical help" service which had taken Esenin-Volpin. The whole matter was briefly broadcast by the Voice of America. The violations were so evident that Esenin-Volpin was actually released immediately after the Voice of America broadcasting. (He was transferred to the Academy's hospital, where he was out of danger.) Later the authorities tried to prove that Esenin-Volpin had been released before the letter was written, and that the authors of the letter were aware of the fact. But that was too evidently wrong.

The signers began to be pursued. There were meetings at various places where the signers were vilified by their colleagues. But many of these meetings went wrong, partly because those who were to conduct them had only a vague knowledge of the events. Some signers were offered statements to sign such as "I certainly did not mean to say that... " or "I disapprove of the Voice of America's interfering... " or other such things. Some agreed but others refused.

Gelfand refused in the most persistent way. There were signs that the authorities wanted to represent him as the leader of the whole rebellion. There appeared nasty rumors that Gelfand had directed the whole campaign by phone and so on. It was definitely wrong. The Academy's authorities tried to make him sign a letter to some American mathematicians stating that everything was fine with Esenin-Volpin, and that the American newspapers that highlighted the events were wrong. Gelfand even agreed to sign the statement that everything was fine with Esenin-Volpin at the moment, but they insisted on the phrase "Your newspapers have deceived you", which Gelfand refused to include. He was in bad condition; his blood pressure went up, but he was resolute not to yield.

Nobody knew what might happen. And we did a wise thing. A small group of mathematicians (including Shafarevich, Arnold, Tyurina, myself, and some others) went to a remote skiing place in the Caucasus. All of us took our formal leaves. We had no connection with the outer world and did not want to know what was going on in Moscow. And when we returned everything had been settled.

There had been discussions in some high spheres, and it was decided to act without extreme measures. Thank God, nobody was arrested. Two people lost their primary jobs, and some ten people lost their secondary jobs. Many people had difficulties with promotions. No one was allowed to go abroad; for example, no one of the 99 went to the Nice Congress. (Even Novikov could not receive his Fields medal in person.)

But it was not personal matters that really counted. A year later Efimov was replaced as the Dean of Mekh-Mat by Ogibalov, who was one of the Heroes of the 1937–1938 purges. Every side of Mekh-Mat's life was terribly affected: entrance examinations, selection of graduate students and new members of staff, the possibility to organize seminars, etc. Only now has Russian mathematics begun to recover from the blow; we shall know before long whether it is still possible to recover.

In the Other Direction

A. B. SOSSINSKY

I hesitated a good deal before writing the text that follows: is a person of my age (early fifties) and of my modest mathematical achievements justified in inflicting his memoirs upon the international mathematical community? Three factors helped me overcome my hesitation. First, the fact that my scientific path, between two continents and three cultures, is rather unusual, going, as it were, the "wrong way"—from West to East, rather than from East to West. Secondly, my involvement in a number of activities in Moscow that are practically unknown in the West, although—as I strongly believe—they certainly deserve to be. And finally, will I ever have another opportunity to speak out like this?

It has been a great privilege, and a fascinating experience for me, to come in close contact with several great mathematicians of my time, including Kolmogorov, Gelfand, Maslov, Arnold, Novikov. However, I do not feel ready to concentrate my reminiscences on these (and other) personalities, and this account will mainly be about happenings and atmosphere. It is about people only insofar as they fit (or do not fit) into the atmosphere and participate in the events.

1. Personal Background

I was born in Paris in 1937, in a family of Russian emigrés. On my father's side, I come from Russian nobility that can be traced back to the sixteenth century; however, the family had lost their land by the turn of the century, and my paternal grandfather, Bronislav Sossinsky, was not landed gentry, but a highly qualified and well-known railroad engineer. In contrast, my maternal grandfather, V. M. Chernov, was of peasant stock (his father was born a serf), but emerged as a brilliantly educated politician and rose to become the only democratically elected president of Russia, only to be overthrown (less than 24 hours after his election by the short-lived Constituent Assembly) by Lenin and the Bolsheviks (1917). His wife, my grandmother O. E. Kolbasina-Chernova, came from a well-off literary family (her father was a close friend

A. B. Sossinsky

of Ivan Turgeniev), and was also a “professional revolutionary”.

My parents met in Paris. My father, Vladimir Sossinsky, settled there after stops in Constantinople, Shumen (Bulgaria) and Berlin following the defeat of the White Army by the Bolsheviks (he was a very brave cavalry officer). My mother got there via Prague, with her two sisters and her mother, after the latter had been freed from Bolshevik prison in 1923, thanks to Gorki’s intervention on her behalf. My parents’ main interest in life was Russian literature, hardly a lucrative activity in Paris at the time; my father held a variety of positions, eventually saving enough money to open a tiny printing shop of his own. Our circumstances were very modest, but my parents and their circle of friends cared more about the quality of poetry, often read out loud at their gatherings, than that of the food and wines served there.

When World War II began, my father volunteered for the French Foreign Legion, once again covering himself with glory on the losing side of the better cause, then spent time in the German stalags and (after being freed in 1943) fought in the French Resistance. In 1946 he was granted Soviet citizenship, but his attempt to return to Soviet Russia then was unsuccessful (fortunately so, because practically all Russian emigrés who returned at that time were soon sent to Stalin’s camps). In 1947 my father got a job as an international civil servant at the UN (he headed the Russian Verbatim Reports section for 13 years) and our family settled in Great Neck, Long Island.

Thus, a timid French-Russian boy of ten, bilingual in French and Russian as far back as he can remember, entered the fifth grade of PS 21 in Great Neck in the fall of 1948, not knowing a word of English. Everyone was very friendly and helpful (typical of the US), and I adapted very quickly to the "American way of life". Two years later, noticing that I was learning nothing in school (except the English language, which I had picked up in a few months), my parents opted for a French education, and I eventually (1954) graduated (with all kinds of honors) from the Lycée Français de New York.

My interest in (or should I say fascination with) mathematics began at age 13, when the French curriculum first introduces algebra and geometry. Geometry was my favourite subject, and I began "research" at age 14: I "proved" that Euclidean geometry is contradictory and "showed" that the universe is "closed" in the sense that straight lines "don't have two ends" but are "like very big circles". I was too shy to communicate my "results" to my teacher (or to other grown-ups), but wrote them up in great detail, in a calligraphic handwriting, and sealed them in an envelope, meant to be opened to the world at large later on, when I would be old enough to be taken seriously. (In my later life I have found that social and emotional timidity combined with intellectual conceit is typical of many mathematicians.) A year later I learned about non-Euclidean geometry, discovered the logical error in my "argument", and shamefully flushed the torn up shreds of my first mathematical paper down the toilet.

I should mention in passing how grateful I am for my French secondary-school mathematical education. The French curriculum then, the result of the pioneering educational ideas of Borel, Hadamard and Poincaré (and Felix Klein, although the French don't like to admit that), was very stimulating for creatively-inclined people. Not surprisingly, it led to the postwar Renaissance of French mathematics: Leray, Serre, H. Cartan, Grothendieck, Chevalley, Weil, Dixmier, Dieudonné, A. Borel, Douady, Deligne, Cartier are all products of the system, whereas 30 years of the Bourbakized curriculum seem to have produced no more than one or two mathematicians of the same caliber.

In 1954, having obtained my *bachot* (the French high-school diploma), I had no doubts that I would study mathematics. My parents could not afford a campus college, so that my basic choice was between Columbia and downtown NYU, the latter winning out (because of the Courant Institute). Unfortunately, I got only a year's worth of credits for my *bachot* and, what is worse, my faculty advisor at Washington Square College would not let me take Advanced Calculus and other serious math courses, because I didn't have credit for the prerequisites. I was stubborn too, and would not take any "beginner's math" that I felt I already knew. So in the first semester I took no math at all, and, besides some general courses, did some English Lit. I was interested in my studies and made the dean's list, but dropped out after one semester. This was 1955, Stalin was two years dead, the "Khrushchev spring" had set in and my father had been allowed to return to Russia for summer vacation. I was planning to continue my education in Europe, either in Moscow or in Paris, the following year.

Our two-month trip to Russia was quite a shock for the family: we saw what the standard of living there was truly like and obtained a first-hand account of the tragedy of Stalin's camps (which until then had been discounted by many leftist intellectuals in the West as "bourgeois propaganda"). It was clear to me personally (my parents did not press me one way or the other) that I needed time to make up my mind about where I would continue my education... and my life.

In the meantime, back in the US with my parents, I returned to NYU to continue my studies.

2. NYU: First Steps in Mathematics (1955–1957)

This time around I convinced my faculty advisor to let me take Advanced Calculus and Differential Equations without any prerequisites. It was a period of anxiety and doubt in my mind; I was undecided about everything, even about doing mathematics—for a while. (One of the options I seriously considered at the time was emigrating to... Iceland, and beginning a life of "isolation, meditation and study".) The person who got me back on track was John van Heijenoort (who was giving the calculus course I had signed up for), a great teacher and an extraordinary personality, whose varied achievements include a doctorate in Paris in functional analysis, fluent knowledge of many languages (including Russian), the design and construction of the first really operative high-fidelity stereo record player, work on radar systems with Shannon, Wiener and von Neumann during the war and (unbelievable but true) the position of Leon Trotsky's personal secretary at the time of the latter's assassination. Jean van H (as his graduate students liked to call him) showed me many of the most beautiful aspects of mathematics; he introduced me to algebraic topology (via Lipman Bers' superb mimeographed lectures, in the framework of a math honors course) and stimulated my interest in mathematical logic (his field of research at the time), an interest later revived by Kolmogorov, Markov and L. Levin.

I had no trouble adapting to the atmosphere at NYU (although my fairly leftist political views were not too well regarded by some of the faculty in that period of late McCarthyism), I was very successful in the role of American student, not only academically but socially and even athletically (I was cocaptain of the undefeated tennis team in my senior year; the other cocaptain, also a math major, was Herman Gluck, now a distinguished topologist at the University of Pennsylvania). Yet deep down I never really felt American; the strong European cultural heritage that was prevalent in school and family life never let me really adapt to American culture and American lifestyles, except for the superficial behavior patterns which I had easily assumed. I should say, however, that I never shared (and am still very irritated by) the snobbishly superior attitude of some Europeans to American culture, usually the result of their own narrow-mindedness. I got my B.S. from NYU in 1957 with the usual honors (Pi Mu Epsilon, Phi Beta Kappa, honors in Math and English Lit, *cum laude*); I would have made *summa cum laude* except for an extraordinarily stupid and biased course of American Government (which did not do justice to the remarkable institutions of the USA), where I was rewarded with a C for making fun of an incompetent instructor. This was a prelude of the problems I would later have with... another course, Communist Party History, in Moscow.

Overall, however, I look back with great pleasure to my two-and-a-half years at NYU, where I began to feel myself becoming a mathematician. An instructive episode that sticks in my mind is my interview with Lipman Bers (when I had been recommended by the math department for graduate study at the Courant Institute). After asking me lots of questions about non-Euclidean geometries and his own algebraic topology course (I had no trouble in answering them), Bers asked me what other interests I had in mathematics. I told him that I had read a great deal about complex projective spaces and had thought a good deal about certain questions... but he interrupted me with a lecture about projective geometry being a "finished science", that had reached a dead end at the turn of the century with the work of Veblen and his school, and that there was nothing to do there anymore. I was very impressed and felt ashamed of my foolishness. Only many years later did I realize that better people than I (Henkin, Gindikin, Penrose), apparently unmindful of similar advice, were about to reopen this remarkable field of modern mathematics and physics.

In the summer of 1957, after a two-month vacation near Moscow, my parents returned to New York, while I, after overcoming a lot of bureaucratic red tape, transferred from NYU to Moscow University (third year) and stayed on in Moscow. It was a tough decision to make; I was aware that I could not expect to leave Russian again in the foreseeable future; my parents were neither supportive nor opposed to my decision; I was clear about the difficulties that lay in store for me, although I did have naïve hopes that the Khrushchev thaw was only a beginning, that he would soon be replaced by a younger, more educated and more liberal man, that the Soviet Union would adopt a non-totalitarian regime, that some form of socialism with a human face would prevail...

3. Mekh-Mat: The Golden Era (1957–1968)

My career as a Soviet student at the Mechanics and Mathematics Department (Mekh-Mat) of Moscow University began with a very unfortunate interview with Professor Shirshov, then Deputy Dean of Studies. I politely explained to him that although I spoke fluent Russian, I had never done any mathematics in Russian before, so that I could be expected to have problems with terminology at first, and would he please bear this in mind when asking mathematical questions. Shirshov, however, did everything possible to ignore my request, wording his questions in typical Slavic words when synonyms with Latin roots were available in Russian (e.g., he asked for the definition of an "*opredelitel*" rather than that of a "determinant"). He concluded the interview by saying that although I had easily answered some difficult questions, I had certain significant lacunas in my mathematical education, I would not be able to follow third-year courses and should begin at the second-year level. This unfortunate decision was crucial to my mathematical life, as will be explained below.

I don't think that the story of how I adapted to Mekh-Mat life (generally quite well) is of much interest. However, I feel that the atmosphere of my undergraduate and postgraduate years there (1957–1964) deserves some description.

Those years, in the unanimous opinion of practically everyone who had the good fortune to be at Mekh-Mat then, were a period when mathematics and mathematicians flourished in a highly stimulating environment. Undoubtedly, the one person most responsible for this state of affairs was the Rector of Moscow University, I. G. Petrovsky. An outstanding mathematician (who headed the chair of differential equations for nearly two decades), Petrovsky will be remembered even more for his honesty, his personal courage and his remarkable ability as an administrator. He began his rectorship in the late Stalin years, managed to concentrate a great deal of power in his own hands ("he has more clout than many Central Committee members, although he's not even in the Party", a well-informed administrator once told me) and used it to expand and enrich the university in general, but especially the Mechanics and Mathematics Department, the apple of his eye.

As its name indicates, Mekh-Mat is divided into two sections (*otdeleniya*): mechanics and mathematics. The mathematics section, whose main administrative function is running the graduate math program, was then headed by the distinguished topologist P. S. Alexandrov, who was always fond of and helpful to talented students of mathematics. During his tenure as the head of the *otdelenie matematiki*, he did his best to ensure that scientific talent and achievement be the prevailing factors in the choice of graduate students, as well as in new appointments to the department. With the powerful help of I. G. Petrovsky, he was often successful in implementing this policy, getting his way in continuous struggles with party bosses and the rank-and-file, especially in the period when N. V. Efimov was the Mekh-Mat dean (1959–1969). An able administrator, a very popular and careful man, Efimov in fact did most of the infighting, in his friendly low-key style, with the party people, and, to my mind, is second only to Petrovsky as the individual most responsible for the golden era of Mekh-Mat.

It must be difficult for Western mathematicians to understand how, in a totalitarian society, scientific achievement as the main criterion of success in a scientific institution is something absolutely unusual. The usual criterion at the time in Russia, as in almost all other places of science, was politics or ideology, not scientific truth. The phrase *partiĭnost nauki,*[1] coined by a party functionary claiming to be a philosopher, was a guide to action in such *causes célèbres* as the Lysenko case in biology, the official banning of cybernetics (as a "bourgeois pseudo-science"), of psychoanalysis (as another "capitalist fraudulent science"), of mathematical methods in economics (as "inapplicable in principle") and of sociology (as such), as well as the attempts to denounce "Mach-inspired" physics (including all the work of Einstein, Plank, Bohr, etc.), stopped only by the development of the H-bomb. With this as background, it is remarkable that Mekh-Mat, until the end of 1968, was a unique place, an oasis, a haven where the objective value of one's research work was one's best asset. This was understood and accepted by most students and teachers; it was an essential feature of the atmosphere at Mekh-Mat at the time. To the list of those primarily responsible for this state of affairs (Petrovsky, Alexandrov, Efimov) mentioned above, I think the name of Kolmogorov should be added: although he held no important administrative position (except for a brief tenure as dean), he symbolized the total scientific involvement, the intellectual probity viewed by many of us as the ideal for a mathematician.

This being said, I would like to describe more specifically what was going on then in my own field—topology—at Mekh-Mat. The year 1957 was a great one for this topic, with a number of striking results demonstrating the effectiveness of algebraic methods in the classical geometric problems of topology and confirming the central role played by algebraic topology in all of mathematics. Yet at Moscow University, one of the world's leading mathematical centers, there wasn't a single working algebraic topologist of serious international stature! P. S. Alexandrov had moved into pure abstract topology, A. N. Kolmogorov's interest in the field had been short-lived, L. S. Pontryagin had left topology for optimal control theory, M. M. Postnikov had stopped doing or publishing original research work, V. A. Rokhlin was just getting settled in Leningrad. The only competent person teaching the subject was V. G. Boltyanski, but his brilliant lectures struck me as being somewhat superficial, spoiled as I was by Lipman Bers' fundamental lecture course.

What I didn't know then was that during this period some hard-working Moscow U. students had actually begun teaching each other algebraic topology (the hard way: from recent original research papers). I didn't know about this for a good reason—this was not an official course or seminar, no one supervised their activity, and while I was in my second year, they were

[1] This phrase is basically untranslatable in English; it means something like "political orientation of science" and implies that there is no "abstract scientific truth" and that science is socially biased, the only correct science being communist-inspired.

in their third—the year I should have been in, if it weren't for the disastrous interview with Shirshov...

The names of these people are now well known in the mathematical world: S. P. Novikov, G. N. Tyurina, D. B. Fuchs, A. M. Vinogradov were the most active; D. Anosov and V. I. Arnold, a bit older, were also frequent participants. I'm sure that I would have become a member of that exclusive circle had I known about it, and my mathematical life may have evolved along different lines. At least, that's what I like to think: it is always nice to have someone other than oneself to blame for one's missed opportunities.

As things actually evolved, I had enough intuition to feel that the illustrious head of the Moscow topological school, P. S. Alexandrov, who had noticed me and was apparently willing to be my scientific advisor, no longer really understood the best work being done in his field, and that his narrowing research interests were not mainstream mathematics any more. But I was not bright enough to understand on my own *what* topology I should be learning. I spent most of the 1957/58 academic year in the Mekh-Mat library: I knew most of the mathematics being taught in the second-year courses and could cut practically all the lectures, only attending the exercise groups and, of course, Communist Party History, a subject (as I had been warned) not to be taken lightly. My scientific adviser during that year was A. S. Parkhomenko, a dedicated (although totally blind) teacher and fairly knowledgeable point-set topologist (along the lines of the Polish school). Under his guidance, instead of teaching myself the really important topics (spectral sequences, homotopy theory, fiber spaces, etc.), I learned practically all there was to know about two-dimensional geometric topology, as well as a lot of three-dimensional topology (e.g., R. H. Bing's work), and wrote my first serious research paper, proving an old conjecture claiming that a certain class of continuous mappings cannot raise a continuum's dimension. Parkhomenko made me write and rewrite the proof until every detail was clear, so that he had no doubts about its correctness. However, when I reported the result at P. S. Alexandrov's seminar, it transpired that there was a counterexample to the theorem by R. D. Anderson. The reader can easily imagine my bewilderment and embarrassment, as well as my subsequent despair. (Actually, the argument in the proof was entirely correct but used an erroneous lemma due to Rozhanskaya, published without proof in the Doklady; I learned the hard way—don't use other people's lemmas unless you known how to prove them!) My consternation did not lead me to abandon the topic, however. A year later I proved a general theorem describing monotone open maps of plane continua, based on a technically very difficulty (but correct!) construction by L. V. Keldysh, who by then had become my scientific adviser.

Let me say a few words about the late Lyudmila Keldysh, my teacher, a person who to my mind is a striking example of dedication to science, courage and intellectual honesty. She was herself a pupil of N. N. Luzin, along with A. N. Kolmogorov, P. S. Alexandrov, N. K. Bari, M. A. Lavrentiev and P. S. Novikov (her husband). She was the only one of his pupils to remain true to him, even in the ominous year of 1937 when considerable pressure was applied on her to denounce Luzin publicly in the framework of an official campaign against him. (This fascinating topic is mostly *terra incognita*; it is unknown who was behind this campaign, or why it was aborted without developing into a political purge of mathematics and mathematicians, as could logically have been expected.[2])

L. V. Keldysh came from a large, close-knit family. Her father, Vsevolod Keldysh, was a military engineer who rose to the rank of general in the Tsarist years but succeeded in adapting to Bolshevik rule; her brother, Mstislav Keldysh, is known for his work in the theory of functions of a complex variable, but more for his long tenure as the President of the USSR Academy of Sciences; of her five children, four became scientists, two of them outstanding ones (Leonid Keldysh in semiconductor physics, and Serge Novikov).

At the time, she was in her late fifties and occupied a senior research position at the Steklov Mathematical Institute of the Academy of Sciences (MIAS); the image of her that most often comes to mind is that of Lyudmila Vsevolodna sitting behind her desk in her modest office at MIAS (where we

L. V. KELDYSH

[2]*Editor's note*: Cf. the articles by Demidov and Yushkevich in this volume.

met on Tuesday mornings, almost every week, for many years) commenting on something I would be writing on the blackboard...

Although L. V. Keldysh, a geometric topologist, was unable to keep up with the rapidly expanding field of algebraic topology, she encouraged her students—in contrast to P. S. Alexandrov—to learn a lot of algebra and algebraic topology, and in fact was most insistent about this. Her pupils at the time included A. V. Chernavskiĭ, M. E. Shtan′ko and L. V. Sandrakova. We did not take our teacher's instructions lightly, and in the late fifties and early sixties organized extremely intensive informal schools (usually at somebody's dacha out of town), where we taught each other a lot of algebra and algebraic topology, mainly under the leadership of Alexei Chernavskiĭ, who became a good friend.

Looking back, I must say that these years at Mekh-Mat were extremely rewarding, not only because of my youthful enthusiasm and naïve political expectations, but also because of the extraordinary feeling of kinship with people whose main interest in life was serving science, from the established older generation of mathematicians (Kolmogorov, Alexandrov, Petrovsky, Markov, Menshov, Gelfand) to my own (the generation of 1937, as I like to call it), whose talents flourished early (Anosov, Arnold, Kirillov, Fuchs, Tyurina, Sinaĭ, Manin, Novikov) in the stimulating atmosphere of the Mekh-Mat of the sixties.

Our love of mathematics was not, for most of us, only an escape from the tough realities of a totalitarian society, but part of a common outlook, characterized by anti-establishment political views and by great interest in the artistic and literary life of the times and in active sports (especially mountain hiking, camping, canoeing, cross-country and downhill skiing). My own involvement in these sports activities got me acquainted with many interesting Mekh-Mat people informally, in particular V. Arnold, Dmitry Fuchs (who became a lifelong friend), A. Kirillov, A. Kushnirenko, Maxim Khomyakov (my closest friend), Galya Tyurina (who became Fuchs' wife), N. Svetlova (who married Galya's brother Andrei, and later A. I. Solzhenitsin) and Marina Orlova (who became my wife).

My own academic career was proceeding successfully. By the time I graduated (1961) I had written three research papers and was recommended for post-graduate work in the topology section. There was a hitch, however, at the State examinations: in "Party History", I got a "4" (="B") instead of the "5" (="A") *de facto* required of applicants for post-graduate work; however, P. S. Alexandrov's influence and his high regard for my research work (although I was not "his" pupil) were enough to get me through. After three years of graduate work (and a thesis on multidimensional knots) I was offered an assistant professorship in Alexandrov's topology section. I was in a situation similar to tenure track in the US, at the best university in the USSR, at the time one of the best research centers in mathematics in the world. Nothing (or very little) forewarned of the troubles to come.

4. THE KOLMOGOROV SCHOOL

In 1962, when I was still a graduate student, P. S. Alexandrov recommended me to A. N. Kolmogorov as a possible teacher at the Moscow Specialized School No. 18, a boarding school for talented out-of-town students interested in mathematics and physics, that he had recently founded with the help of I. G. Petrovsky and Isaac Kikoin, the well-known H-bomb physicist. I had attended several of A. N. Kolmogorov's lectures (on the foundations of probability theory and self-reproducing automata) before then, and although he had the reputation of being an extremely abstruse and bewildering lecturer, I had had no trouble in following them. I had also seen him once at the famous "topological picnics" organized by P. S. Alexandrov, and recall that he was listening attentively while the latter was questioning me during an oral exam on homology theory.[3] But I had never spoken to the man before.

My first impression when I did speak to him for the first time (in his small office at School No. 18) confirmed the feeling that I had experienced during his lectures: that he and I were "on the same wavelength". Apparently, Kolmogorov was impressed by my taste for the synthetic approach, based on transformation groups, in teaching geometry which I had learned in the French lycée, and which he himself was developing in his own high-school geometry textbook. I was recruited to teach some exercise classes in calculus (following Kolmogorov's lectures on the subject) and to jointly[4] head an optional seminar with him (outside of regular class hours) in geometrical problems. That seminar was a fascinating teaching experience: the dozen students or so who stayed on to the end have all become research scientists since then; the most famous one (although not our best problem-solver) was Yu. Matyasevich.

As a lecturer, Kolmogorov had a strong tendency to overestimate the possibilities of his listeners and did not like to repeat anything (including the formulations of the main definitions and theorems), but the contents of his lectures were remarkably to the point and always bore the imprint of his original mind. His lectures for high-school students were easy for professional mathematicians (from the graduate level up) to follow, and thus students who had trouble understanding the material would later get a clear explanation from the instructors, who always sat in on the lectures and took notes. One of Kolmogorov's pedagogical principles (for teaching math to bright students)

[3] It is typical of the informal democratic atmosphere of Mekh-Mat of those times that there was nothing unusual about a serious examination being conducted in such a setting.

[4] Here again, there was nothing unusual about an established world-class scientist heading a seminar jointly with an obscure graduate student.

was that quite difficult material can be presented provided it is specific (concrete rather than abstract), related to our intuition of the physical world, and given motivation (e.g., its usefulness for solving real-life problems should be stressed). In proclaiming and implementing this principle, Kolmogorov was swimming against the current: Bourbakization, "new math", was in the process of flooding high-school curricula worldwide. I am stressing this point because, paradoxically, Kolmogorov was later accused (in particular in an underhand political campaign headed by L. S. Pontryagin that sought to denigrate his contribution to mathematical education) of ruining the secondary school curriculum by introducing abstract, set-theoretic and "Semitic mathematics" in place of traditional, applications-oriented "Russian math".

In the next semester, the topic of Kolmogorov's lecture course changed from calculus to algebra (examples of algebraic structures, polynomial algebra over a discrete field, including Galois theory); the next school year he lectured on "Discrete Mathematics", which in his interpretation included a lot of combinatorics, some probability (very little), a bit of logic and a unique cycle of lectures on the theory of algorithms culminating in the algorithmic proof of Gödel's incompleteness theorem (!).

The latter course, which was absolutely absorbing (but was never published), stimulated Kolmogorov to carry out his last great cycle of research (on the complexity of finite binary sequences and the foundations of probability theory). For me, it revived my interest in mathematical logic and was a unique teaching experience: I was not only stimulated from above by Kolmogorov, but also from below, by the strongest group of students I have ever taught. They included Leonid Levin (now at Boston University), A. Zvonkin and Z. Maimin (whose future mathematical careers were shaped by Kolmogorov's high school discrete math course); E. Poletski (now in the US), V. Sklerenko, T. Lukashenko, Yu. Osipov, A. Uglanov (the strongest of the lot, whom a freak accident prevented from realizing his full potential) and A. Enduraev. I still recall the exhilaration I felt before each of the lessons in this class; much later, some of its former students told me how much they also had looked forward to these mutually stimulating classes.

The Kolmogorov school, where I taught (first as a graduate student, then as an assistant and later associate professor at Mekh-Mat) was of course exceptional by the sheer talent and deep interest in physics and math typical of its students. Although my democratic family background has biased me against elite schools, I must admit that I would never had done so well as a teacher, learned so much, enjoyed it all so much if it were not for Kolmogorov and the carefully selected student body of School No. 18.

Another essential feature of the school was the great interest in general culture (music, the arts, literature, sports) that resulted from Kolmogorov's selection of teachers, each of whom, besides being a highly qualified mathematician, had his own *violin d'Ingres* that he would readily put at the disposal of his students. Optional courses in (or "evenings" related to) music, poetry and the arts given by mathematicians and physicists, camping trips and sports events opposing students to teachers, were a normal part of studies. At the school I met some very unusual and interesting people: D. Gordeev (a graduate student of Kolmogorov's who eventually gave up mathematics to become a successful avant-garde painter), E. Gaidukov (who gave up a promising career as a violinist to do mathematics under Arnold, was active in the dissidence movement, and when things got too hot moved to Khabarovsk), Yu. Kim (the now famous bard and playwright, who taught history and literature at School No. 18, organized very successful—and dangerously anti-establishment—musicals and was forced to resign after cosigning a political letter of protest), A. Zilberman (the best teacher of physics I have ever met, still a graduate student then), the late V. M. Alexeev (a talented research mathematician, a softspoken man of colossal classical erudition), V. A. Skvortsov (another unusual mathematician, for many years the president of the University English Club) and several others, who became my friends.

For several years, especially under the directorship of the late R. A. Ostraya, a well-educated historian, the Kolmogorov school was not only an elitist mathematical institution, but a school where high-level general culture and intellectual freedom were part of the curriculum.

5. Moscow Mekh-Mat: the destruction of a haven (1968–1974)

The year 1968 was the turning point of many lives, including my own. It was the year of the May barricades in Paris, of draft-card burning and riots on American campuses, of the Prague spring crushed by Russian tanks. For me it was the year that put an end to my hopes and illusions, the year of dramatic events that mark the end of the Mekh-Mat golden era.

These events are described elsewhere in this volume (see pp. 220–222 of Fuchs article).

Briefly, let me remind the reader that in March 1968, A. S. Esenin-Volpin, a mathematical logician and well-known human-rights activist, was forcibly and illegally placed in a psychiatric institution by the KGB. This shocked the Moscow mathematical community; uncharacteristically, it reacted by writing a (mild) letter of protest known as the "Letter of the 99", signed by many of the leading mathematicians of the day. The letter was almost immediately published in the West, against the wishes of its authors and cosigners; most observers agree that this was a calculated leak by the KGB, then in the need of a *causus belli*. In any case, the letter was the pretext[5] for a crackdown at the Moscow University math department: the administration at Mekh-Mat and

[5] A pretext, but not the cause: the Communist Party was strengthening its ideological authority, the "party line in science" principle was being widely reasserted, and any oasis for talented and honest scientists could not be tolerated by the totalitarian nature of the system.

the party leaders were all subsequently replaced by hard-liners. The cosigners were subjected to humiliating public disavowal procedures and severely reprimanded; some eventually lost their jobs.

Another significant development was the organization of systematic anti-Semitic practices at the Mekh-Mat entrance examinations; but much has already been written about this in the West and I will contribute only one episode, to give the reader an idea of what +1 in the statistics can stand for from a personal viewpoint.

In the summer of 1969 (or was it 1970?—I have kept no written records), still teaching at Boarding School No. 18 and training the students for entrance examinations, I had spoken privately to each of the Jewish students, described what lay in store for them at the exams, and explained how they should prepare themselves for these. The most talented one of the lot, a boy by the name of Kogan, a year younger than his classmates, did not take my advice seriously enough, and was easy prey for the young enthusiastic anti-Semitic thugs at the oral math exam. "I've learned my lesson," he told me later, "but I'll be back again next July." And he was (being too young for the draft that year), highly trained this time, matured and sobered by his first experience in the world at large, ready to fight any odds. His results—written math: 5, oral math: 5 (surviving four hours of olympiad-level questions), oral physics: 5 (where the two examiners were also out to get him). That left the Russian literature essay, where even a passing grade would be enough to get him in; it was clear to him, and to me, that Kogan had beaten the system. We were wrong; the philology department examiners failed him (Kogan had been an A-student in literature, there were no spelling, grammar or stylistic mistakes in the essay, but it was given a "2" (=F): for "not clarifying the topic"). I have never seen Kogan since. (Thank God. What would I have said to him?) For the first time I asked myself the question: what moral right did I have, as a teacher, to be, if not an accomplice to, then a passive observer of such practices?

The following years gave me several occasions to ask myself this question again: In 1971, when the Mekh-Mat Party Bureau decided to forbid me to teach at School No. 12; I was the second to go, after A. Zilberman, in a clean-up campaign that eliminated, one after another, the people Kolmogorov had handpicked as teachers for the school and who had contributed to creating its unique spirit; in 1972, when the most talented person I have ever had working under my personal guidance, a straight-A student named Isaac Lapitski, was deprived of doing graduate work by the Komsomol leaders, who would not give him a *positive kharakteristika*;[6] in 1973, when my teacher, Lyudmila Keldysh (a cosigner of the Letter of the 99) had been tricked into early retirement (she had agreed to retire so as to free her research position at the

[6] An untranslatable expression from the Bolshevik lexicon, standing for a kind of certificate of political and social docility.

Steklov Institute for A. V. Chernavskiĭ, but the second part of the bargain—hiring Chernavskiĭ—was never kept by the Institute's authorities); finally, in 1974, when two other students doing research under me had refused to take part in the so-called elections to the Supreme Soviet and had been thrown out of the University, a fact actually reported by the BBC; I had managed to convince them to act "reasonably", the incident had been smoothed over and they were almost immediately reinstated, but of course they were not allowed to do graduate work, although neither was Jewish and both were clearly Ph.D. material. In this episode I particularly felt the ambiguity of my own ethical position—was I right in counselling them to retreat from their stand "for their own good"?

Two other happenings, not directly related to mathematics, finally led to my cutting the Gordian knot. The first was a very strongly worded letter of protest about the forced exile of Solzhenitsin that I had written jointly with my friend Maxim Khomyakov and that we had circulated in *samizdat*; this made my chances for reelection to my associate professorship, due in 1975, pretty uncertain. The second concerned something called the "University of Marxism-Leninism", a two-year course in political indoctrination for faculty members. In my ten years at the Section of Topology, I had managed, under various pretexts, to avoid taking it, but it was made clear to me that this year was it: I must either take it, or leave the Section.

This was the fourth of March, 1974. I was very depressed and tired. My wife had been in the hospital for several months, my mother was very ill, I had two children to take care of, an overwhelming workload at the University (besides my own courses, I was standing in for M. M. Postnikov, in the hospital for heart troubles, in a new untested third-year course on Differentiable Manifolds). I was ready to capitulate, and I took the application form for the Marxism-Leninism University with the intention of filling it in. I read the mimeographed text: "Professor [blank] kindly requests the Party Bureau of the Department to recommend him for a two-year course at... " That proved to be too much: to write that it was I, I who was "kindly requesting to be recommended... " I tore up the application form, took out a blank piece of paper and wrote my resignation from the department.

A few days later, Yu. M. Smirnov, a professor at my former Section, tried to convince me to take the resignation back; I refused. Then the chairman, P. S. Alexandrov, invited me to his apartment to discuss the situation. This was a memorable talk, practically a monologue, that began with the following remarkable opening gambit:

"Alyosha," he said, using the informal diminutive of my name, "traditionally, we intellectuals of the Russian nobility have always placed our duty to our fatherland (he used the archaic *otchizna* rather than the typically Soviet *rodina* for homeland) above our personal interests and feelings. A Russian nobleman does not leave a sinking ship—he fights to keep it afloat. It is people like Kolmogorov, like you and me, who have made this department into the unique scientific oasis that you know. Even in the Stalin years, we have always done all we could, swallowing our pride if need be... "

I had known that Alexandrov's parents were small-landed gentry, but was not aware that he knew about my own origins, and had expected anything but an appeal to values that fifty years of Bolshevik rule were supposed to have eradicated, especially in a careful and successful establishment scientist. But the divulgence of this hidden facet of Alexandrov's personality and the implicit flattery in his monologue were not enough to make me change my mind. Although I did promise to give the remaining lectures that semester in the Differentiable Manifolds course, something no one else at the Topology Section appeared to be able to do, I confirmed my resignation.

And I became—as I then liked to joke when talking to friends—the only unemployed mathematician in the Soviet Union.

I recall the mixed feelings I experienced visiting the department, to work in the library or attend a seminar, in the months immediately following my resignation, the perverse pleasure I derived in refusing to shake hands with the most odious people there. And a strange form of solidarity from other faculty members: after furtively looking around, somebody (at times people whom I hardly knew) would walk up to me and give me a prolonged, silent handshake.

Looking back and trying to analyse how the Mekh-Mat we knew and loved was being destroyed, I think that the process (in which anti-Semitism was only one of the aspects) can be understood only in terms of Bettelheim's remarkable analysis of situations of stress in Hitler's camps. The basic premise—to achieve automatic obedience—can be made good by systematically denying people their sense of dignity, their personal identity, by teaching them never to "stick out". By humiliating a student or a professor, forcing him to dig potatoes out of the mud by hand (as "voluntary" help for a local collective farm), making him hypocritically repeat, in public, obvious political lies about the system, the system succeeds in making this person lose his sense of self-respect. Then he becomes manageable. Talented people—who tend to be unpredictable and more difficult to control—are eliminated. They are flunked at the entrance exams, not recommended for graduate work, not given positions at the department, unless their sense of self-respect is broken and they can prove their docility. What the hard-line administrators wanted were good, competent, solid, stolid, servile mathematicians. And that's what they now have—there are very few world-class mathematicians holding a full-time position at Mekh-Mat today, while there were dozens and dozens in 1968.

6. "Kvant" (1975–1987)

After my resignation from the University, I did not start looking for another position, feeling that I could make a decent living for my family and myself as a free-lance translator of mathematical books. However, my family and friends were worried about my future: they felt that the law about *tuneyadstvo* (a kind of vagrancy law), calling for internal exile of unemployed residents of Moscow and other large cities, most generously applied to dissenters, could be applied to me.

Several months later, I finally agreed to look for an official position, and after some unsuccessful attempts (which all followed the same scenario—I would be enthusiastically recommended by future colleagues, but at the last minute the personnel department would suddenly declare that the position in question was no longer available), I got a job as a mathematics editor at the popular science magazine *Kvant*. It was an underpaid and demanding position. A. N. Kolmogorov had strongly recommended me for it, and Isaac Kikoin had used all his political clout in the Academy of Sciences (which publishes the magazine) to get me in. I should perhaps mention that during subsequent years I made several attempts to find a better-paid and less energy-consuming position, and they failed, all along the lines of the scenario described above.

Kvant magazine was another accomplishment of the liberal sixties. It was founded in 1969 by the same tandem (Kolmogorov-Kikoin) as School No. 18, with the assistance of Petrovsky, in the framework of the Academy of Sciences. Like the Kolmogorov school, its aim was to provide gifted high-school students living in rural areas or in small towns with stimulating materials in physics and mathematics that they would not be likely to get from their teachers.[7] Its circulation, despite the high level of sophistication of most of the articles, started at 100,000, rose to the incredible level of 370,000 during the math-physics boom (early seventies), dropped to well under 200,000 by 1977 (when the boom was over), rose and levelled off at around that figure in the 1980s. Most of the articles are not written by high-school teachers, but by research mathematicians with an interest in education (many of them former Olympiad winners, later involved in the organization of Olympiads). Systematic contributors included Kolmogorov, Arnold, Kirillov, Fuchs, Gelfand, Rokhlin, Pontryagin, Migdal, Kikoin, Frank-Kamenetski. A very important feature of the magazine was the problem-solving section, a permanent contest where our high-school student readers competed in the number of correctly solved problems (five math problems and five in physics were proposed in each monthly issue). The mathematics problem section, from the very beginning, was headed by N. B. Vassiliev; it still is today, and Kolya, a soft-spoken, very musical person, still remains one of the champion "elementary problem solvers" of our times.

Two other persons I became friends with at *Kvant* were L. Makar-Limanov (a sharp mathematician and strong personality, now at Wayne State University) and Yu. Shikhanovich (a mathematical logician, under whom I had

---

[7] In the Soviet Union educational levels generally decrease proportionally as one moves away from urban centers.

taught some calculus part-time at the Structural Linguistics Department of Moscow University in my graduate-student days). The latter was always very active in the human-rights movement, came to *Kvant* after being released from a psychiatric ward (where he was forcibly placed for his political activities) and stopped working at *Kvant* only when he was arrested by the KGB in 1982. He got a heavy sentence at his "trial" (a farcical imitation of justice) and was only freed in the political amnesty that marked the beginning of *perestroika.* It is characteristic of the people at *Kvant* that the authorities were not able to get any one of us to testify against him.

Generally, during the drab and totalitarian years of the Brezhnev and post-Brezhnev era, *Kvant,* under the protection of I. Kikoin, remained a strange little islet of liberalism, where a closeknit group of underpaid physicists and mathematicians were doing an unheralded job of making first-rate math and physics accessible to thousands and thousands of high-school students. Fifteen or twenty years later, the yearly lists of prizewinners of *Kvant*'s math context read like a Who's Who in Soviet research mathematics. A number of these people have told me that if it weren't for *Kvant* and Math Olympiads they would never had gone in for the subject. Especially rewarding were the cases (not too frequent, to be sure) when the first clumsy, but promising, problem-solving efforts of a teenager from the middle of nowhere would gradually evolve in technique and sophistication, and suddenly yield a totally unexpected solution to one of our more difficult problems...

The Western reader can get an idea of what *Kvant* was like by looking at any issue of *Quantum,* a magazine based on back issues of *Kvant* and now published by the National Science Teacher Association in Washington, D. C.[8]

An important happening, which saved *Kvant* from the sad fate of many achievements of the sixties, was the failure of a takeover bid, undertaken by L. S. Pontryagin in 1980, when he tried to wrest control of the journal away from Kikoin and Kolmogorov. I was present at the decisive proceedings that took place at the Steklov Institute; I will always remember the tragic and odious figure of L. S. Pontryagin, not as the great mathematician that he once was, but as an old man nervously clicking the beads of his rosary and lashing out at Kikoin, Kolmogorov and even at me (he described an article about Conway numbers in *Kvant* that I had written jointly with A. Kirillov and I. Klumova as an extreme case of the "wrong kind of mathematics" that Kolmogorov and his entourage were inflicting on innocent school children). The takeover bid failed, because Pontryagin's cronies had not done their homework properly: the Mathematical Section of the Academy did not have any legal authority to control the magazine (which depends directly on the Academy's Presidium), and Pontryagin's virulent attacks (supported by the anti-Semitic remarks of I. M. Vinogradov) were simply ignored by Kikoin.

[8] *Editor's note*: *Kvant* is now published by Springer.

My eleven years at *Kvant* were not often as arresting—much of the editorial work was routine, and although I did a lot of creative rewriting, the job was very time-consuming and often depressing. I was doing less and less research of my own, not attending any seminars (except Sacha Vinogradov's seminar on symmetries of PDEs, an interesting topic then, about which, however, I published nothing). I felt rather isolated mathematically: L. V. Keldysh had died, A. V. Chernavskiĭ had moved away from mathematics to neurobiological applications, I had no contacts with the West and I had given up on my own work on the applications of the homology theory of Zeeman's tolerance spaces to approximate solutions of differential equations.

I. M. Gelfand, V. P. Maslov and—surprisingly—P. S. Alexandrov tried to get me a job where I could do more mathematics, but failed. I am particularly grateful to P. S. Alexandrov, who offered me A. S. Parkhomenko's associate professorship (which I probably would not have accepted anyway) after the latter died, but this position disappeared into thin air when my name was proposed for it.

At *Kvant*, at least I felt I was doing something useful and had no qualms about the underlying ethics of my position.

7. "Bella Muchnik University"

This short-lived unofficial institution has been known under various names, e.g. ironically as "People's University", or as "the Jewish seminar". I prefer to call it by the name of its founder, B. A. Muchnik, an alumnus of (and Ph.D. at) Mekh-Mat, who decided that bright students flunked at the Mekh-Mat entrance exams (for being Jewish, part Jewish, or too smart) should have a chance to get as good a mathematical education as the successful applicants. Another person involved in the organization was V. Senderov, a young dissident, who had taught at an elite mathematical school in Moscow (No. 2) before being dismissed, but was still politically very active.

I learned about the school from A. M. Vinogradov in 1978 and taught there from the very beginning with him and several of his pupils. Later the "staff" was joined by D. B. Fuchs, Andrei Zelevinsky, two graduate students (Alexander Shen and Arkady Vaintrop, both talented and dedicated teachers) and others. I taught the "Higher Algebra" course and (if I remember correctly) an introductory course on "Analysis on Manifolds", usually twice a week in the late afternoon.

The "enrollment" was typically some 50 or 60 people (about half as much by the end of each year), the classes, first held in Bella's cramped apartment, were later made almost official as optional seminars at the Gubkin Oil and Gas Institute on Leninsky Prospekt. The students were also learning math at the official institutions where they had been accepted, so our curriculum was not traditional—we tried to teach basic modern mathematics in a novel way. Apparently, some of the courses were attractive to more than just first-year students; even some from Mekh-Mat actually attended.

Overall, the enterprise was probably doomed to failure for pedagogical reasons—a small group of teachers could hardly hope to play the role of a real university. But we tried, during classes and in talks outside of class hours, to give some glamour and purpose to the study of mathematics to people deprived of a stimulating environment.

Some of our students did eventually go on to do research in the field, under Arnold and Fuchs in particular. Some of the names I remember are (unfortunately, I have not saved any written records): V. Ginzburg, B. Shapiro, S. Rogov, V. Etkin, B. Kanevski, F. Malikov, A. Lifshitz, A. Gokhberg, M. Fradkin; they have all become mathematicians, except for Fradkin who is a physicist.

Bella Muchnik University ended its existence under very dramatic circumstances, in the winter of 1982/83. V. Senderov was arrested, as well as two students (one was soon released). Several students were interviewed by the KGB. Classes ceased of their own accord. Soon afterwards, Bella Muchnik, returning home very late in the evening, was killed by a truck; the police never found the hit-and-run driver; few of us believed it was an accident. Classes never reopened.

8. Perestroika: Will it Destroy what the KGB Could Not?

And yet, the combined efforts of the KGB and the party bureaucrats at the Academy of Sciences, at Moscow University and in the Ministry of Education did not succeed in destroying the unique spirit of the mathematical community that had flourished at Mekh-Mat in the sixties. To be sure, mathematical schools were closed down, but new ones opened, where people like Shen, Vaintrop, B. M. Davidovich continued to teach. The All-Union Olympiad, taken over by bureaucrats from the Ministry of Education, had been transformed from a stimulating meeting place for talented students and dedicated research scientists into a training polygon for future performers in the politically prestigious International Olympiad, but the old olympiad traditions were preserved thanks to N. N. Konstantinov and his unofficial "Tournament of Towns". *Kvant* went on doing its remarkable job. School No. 18, no longer the unique place that I have described, still continued to produce future first-rate mathematicians. Leading mathematicians who did not have (or gave up) full-time positions at Mekh-Mat (e.g., Gelfand, Arnold, Manin, Sinaĭ, Novikov, Fuchs), kept their seminars there going, and the Russian tradition of research work gravitating around large, close-knit groups of mathematicians of various generations, each headed by its own *maestro*, is still very much alive.

Things changed radically after 1985 with the advent of *perestroika*. It has certainly changed my own life completely. I have been allowed to travel abroad (after 30 years!), I finally have the kind of job a mathematician can only hope for (I hold a research position at V. P. Maslov's Applied Mathematics Section at MIEM, the Moscow Electronic Machine Design Institute, where I also do a little teaching), I am happy in my family life (my second wife, E. N. Efimova, shares my world outlook and attitude about mathematics, being herself an alumnus of the Mekh-Mat of the late sixties), and I feel I can now make my contribution to mathematics and to the mathematical community without restricting myself to marginal activities, without having to avoid mainstream mathematical life.

My main regret, however, is not that all of this didn't happen sooner. It has to do with my concern, shared by many Russian mathematicians of my generation, and clearly expressed by my friends O. Ya. Viro and Yu. S. Ilyashenko at the ICM in Kyoto, that our failing economy, together with the freedom of movement granted by *perestroika*, will lead to an unprecedented brain drain and the drying up of the Russian mathematical school. When my generation leaves the active mathematical scene, there will be no one here to teach talented students, since all the best people of the next generations will have emigrated.

I can only hope that this discouraging prediction will prove to be as accurate as my naïve optimism in the late fifties or my hopeless pessimism of the late seventies.

A Brief Survey of the Literature on the Development of Mathematics in the USSR

S. S. DEMIDOV (MOSCOW)

1. Introduction

This short article will serve as an introduction to a bibliography of Soviet literature on the history of mathematics in the USSR. For the user's convenience, the article is organized in sections, in each of which the cited literature is alphabetized by authors' last names.

2. General works on the history of mathematics in Russia and in the USSR

Until recently, it has been customary for Soviet literature on the history of mathematics to give separate discussions of mathematical work in pre-Revolution Russia and in the USSR. Here the development of schools that are now known in the West as "Soviet Schools of Mathematics" was treated almost exclusively as consisting of scientific achievements of the Soviet regime. In point of fact, the progress of these schools was predetermined by the outstanding achievements of the national mathematicians of the last third of the 19th and early 20th centuries. These achievements were mainly connected with the activities of the famous Petersburg School; with the Moscow School, which on the eve of the Revolution was acquiring an extensive reputation; and with the concurrent development of other scientific centers like Warsaw (in the form of the Russian University at Warsaw, later relocated at Rostov-on-Don), Kiev, Khar′kov, Odessa, Kazan, and Derpt (Yur′ev, now Tartu).

For the history of mathematics in Russia up to 1917, one may consult the following general references: a brief book by B. V. Gnedenko (1946); the relevant sections of the general monograph "Development of Science in Russia," published in 1977 under the editorship of S. R. Mikulinskiĭ and A. P. Yushkevich (both of these were intended for a wide audience); the very substantial reference on the history of mathematics in pre-Revolution Russia by A. P. Yushkevich (1968); and finally, the "History of National Mathematics," published in 1966–1970 in five parts (four volumes) by a staff of authors supervised by I. Z. Shtokalo, A. N. Bogolyubov, and A. P. Yushkevich. The first volume covers the period from antiquity to the end of the 18th century; the second, 1801 to 1917, including a bibliography of the best-known pre-Revolution mathematics. The third volume, and the fourth (in two parts), contain surveys of the development of mathematical research in the separate regions of the USSR; accounts of mathematical institutes, conferences, publications, and international scientific meetings from 1917 to 1967; and finally a survey of research in the various directions cultivated by well-known specialists. The third volume includes a bibliography of work by prominent Soviet mathematicians; and the second part of the fourth volume contains a short bibliographical dictionary of the most prominent Soviet mathematicians, as well as information about mathematical periodicals and occasional publications issued in the USSR. Every section of the "History of National Mathematics" is accompanied by an extensive bibliography that includes its origin, a summary, and a history of the relevant mathematics.

The "History of National Mathematics" was continued as "Survey of the Development of Mathematics in the USSR", issued in 1983 under the same editorship.

Surveys of work in various mathematical directions in the USSR, as well as corresponding bibliographies, can be found in the substantial publications "Thirty Years of Mathematics in the USSR" (1948, edited by A. G. Kurosh, A. I. Markushevich and P. K. Rashevskiĭ); "Forty Years of Mathematics in the USSR: 1917–1957" (two volumes, 1959, edited by A. G. Kurosh); the latter originated from a short survey by D. F. Egorov on ten years of the development of mathematics in the USSR (1927), and "Fifteen Years of Mathematics in the USSR" (1932, edited by P. S. Aleksandrov, M. Ya. Vigodskiĭ, and V. I. Gnedenko). A bibliography of mathematical research during 1958–1967 is contained in the second volume (in two parts) of the book "Mathematics in the USSR, 1958–1967" (1969–1970, edited by S. V. Fomin and G. E. Shilov). The first volume, which was to contain a survey, has never appeared.

Short biographical notes on national mathematics can be found in the biographical dictionaries by A. N. Bogolyubov (1963) and A. I. Borodin and A. S. Bugaya (1979).

The journal "Research in the History of Mathematics" (Istoriko-Matematicheskie Issledovaniya) began in 1948 under the editorship of G. F. Ribkin and A. P. Yushkevich; later it was edited by Yushkevich alone. The double volume 32–33 should appear during 1990. This journal contains a variety of material on the history of mathematics in Russia and in the USSR. The same applies to a Ukrainian journal with a similar title that was published in Kiev in 1959–1963 (four numbers appeared), and to the mathematical issues of the collection "History and Methodology of the Natural

Sciences" published by Moscow University (10 issues have appeared since 1966). Articles on the history of mathematics in Russia and in the USSR also appear in "Mathematics in the School", in "Progress of the Mathematical Sciences," and in the collection "Studies in the History of Natural Sciences and Technology," published in Kiev.

In using Soviet works published before the beginning of "perestroika," one should always keep in mind their dates of publication and consequently the influence on their authors of ideological and censorial directives. For example, in books published during the Stalin era the reader will find no mention of the repression of mathematicians; in literature published in the eras of Khrushchev or Brezhnev, no mention of scientists who emigrated to the West. Generally speaking, in all Soviet publications on the history of mathematics in the USSR, except perhaps those published most recently, it is typical either that the social context of the development of mathematics in our country is ignored, or that it is deliberately distorted.

3. Mathematics in Russia from the end of the 19th to the beginning of the 20th century

The literature on the history of mathematics in this period is quite extensive. We have selected only the work that is most significant for understanding the origin of the "Soviet School of Mathematics," principally the literature on the activity of the St. Petersburg and Moscow schools, but also some of that of the provincial centers.

A general description can be obtained from the 1968 book by A. P. Yushkevich, mentioned above, and the second volume, "History of National Mathematics" (1967; edited by I. Z. Shtokalo and others).

An extensive literature has been devoted to the creation of the "Petersburg School" by P. L. Chebyshev. For the development of mathematical research at the Petersburg Mathematical Society, see an article by I. Ya. Depman (1960); for the University of Petersburg, see an article by I. Ya. Galchenkova (1961).

Among the investigations concerned with various aspects of the activities of the School and its outstanding ideas, we mention the following: a book by B. N. Delone (1947) on the Petersburg school of number theory; the scientific biographies of P. L. Chebyshev (V. E. Prudnikov, 1976), A. I. Korkin (E. P. Ozhigov, 1968), E. I. Zolotarev (E. P. Ozhigov, 1966), N. Ya. Sonin (A. I. Kropotov, 1967), A. A. Markov (S. Ya. Grodzenskiĭ, 1987), A. M. Lyapunov (A. S. Shibanov, 1985; A. L. Tsikalo, 1988); V. A. Steklov (V. S. Vladimirov and I. I. Markush, 1981), and finally articles devoted to activities of the School in various directions: probability theory (S. N. Bernshteĭn (1940)); various problems of mathematical analysis (N. M. Gyunter (1928), A. I. Markushevich (1950), F. A. Medvedev (1961), M. I. Markush (1974));

algebra (I. G. Bashmakova (1949); computing technology (L. E. Maĭstrov (1961, 1973, 1975)).

The development of mathematical research at Moscow is reflected in the articles by M. Ya. Vygodskiĭ (1948) on mathematical activity at Moscow University, and in articles by K. A. Rybnikov (1950), F. I. Frankl' (1951), I. Ya. Depman (1952), F. Ya. Shevelev (1966), S. S. Demidov (1985, 1986), F. A. Medvedev (1986), S. M. Polovinkin (1986), and in scientific biographies of N. E. Zhukovskiĭ (V. V. Golubev, 1947; A. A. Kosmodem′yanskiĭ, 1984), K. A. Andreev (D. Z. Gordevskiĭ, 1955), and B. K. Mlodzeevskiĭ (S. D. Rossinskiĭ, 1950).

Among articles on the development of mathematics in other mathematical centers in the days of the Russian empire, we mention the work of D. D. Mordukhaĭ-Boltovskiĭ (1941) and S. E. Belozerov (1953) on mathematics at the Warsaw (Rostov-on-Don) University; S. N. Kiro (1956 and 1961) and È. B. Leĭbman (1961) on mathematics at Odessa; M. F. Kravchuk (1935) on Kiev University; D. M. Sintsov (1908, 1938), M. N. Marchevskiĭ (1956), and A. K. Sushkevich (1956) on Khar′kov University; I. R. Depman (1955) on Derpt (Dorpat, now Tartu) University; N. D. Bespamyatnykh (1963) Vilnius (Wilno, Vilna) University; and also scientific biographies of V. P. Ermakov (V. A. Dobrovol′skiĭ, 1981), F. E. Molin (N. F. Kanunov, 1983), P. Bol′ (A. D. Myshkis and I. M. Rabinovich, 1965), and N. A. Vasil′ev (V. A. Bazhanov, 1988).

A special position in Russian mathematical life at the end of the 19th century is occupied by the creative activity of S. V. Kovalevskaya. Various aspects of her work are discussed in her scientific biography (P. Ya. Kochina, 1981), as well as in articles by I. Ya. Depman (1954), P. Ya. Polubarinova-Kochina (1954, 1957), and by G. K. Mikhaĭlov and S. Ya. Stepanov (1985).

A number of articles deal with research on particular directions in mathematics. Here we single out the monographs by E. P. Ozhigova (1986) on the development of number theory in Russia; N. I. Styazhkin and B. D. Silakov (1962) on the history of logic; A. K. Sushkevich (1951) on work in algebra; V. V. Gussov (1952, 1953) and M. B. Nalbandyan (1965, 1966) on the theory of special functions; A. Ya. Khinchin (1934), A. N. Kolmogorov (1947), and B. V. Gnedenko (1948) on probability theory; and F. A. Medvedev (1963) on set theory and function theory.

4. Mathematics in the USSR

Serious historical research on the process of development of mathematics in the USSR has only begun. In previous years, it was hindered, in the first place, by the lack of that historical distance with which the sequence of events becomes actual history rather than the news of the day; and, in the second place, by the impossibility of an impartial analysis of historical material that was conditioned by the power of ideology and the force of censorship. How can history of science be written if it is practically forbidden to do research freely in the domain of general history, if the organization of research in mathematical institutes is strictly prescribed, if the creative work of individual scholars is located in archives and special collections that are inaccessible to the scholar? Nobody would care to accommodate research in the history of mathematics to officially approved opinions on civic history of this period; and those who found anything worthwhile would, as a rule, not be able to write about it. The only useful part of this work on the history of Soviet mathematics deals with the development of ideas (only rarely at the level of historical research; usually mere surveys). One needs to look very critically at its treatment of questions about the history of mathematical institutes, organization of research, and the creative work of individuals and schools. This material is often inexact (sometimes deliberately distorted); as a rule, incomplete; and often incorrectly connected to events in general civil history.

The most general presentation of these questions can be found in volumes 3 and 4 of "History of National Mathematics" (1968–70, edited by I. Z. Shtokalo and others), and in "Fifteen Years of Mathematics in the USSR" (1932, edited by P. S. Aleksandrov and others), "Thirty Years of Mathematics in the USSR", (1948, edited by A. G. Kurosh and others), and "Forty years of Mathematics in the USSR" (edited by A. G. Kurosh).

There are a vast number of articles that deal with mathematical life in various parts of the country. Many of these contain useful data on the history of state or local scientific institutes (academies, universities, or associations) but quite often are merely eulogistic.

The majority of articles are devoted to the development of mathematics in Moscow or Leningrad, and to the establishment of Soviet mathematical schools. Mathematical research in the Academy of Sciences of the USSR has been the subject of articles by B. N. Delone, L. D. Kudryavtsev and M. M. Postnikov (1963), V. I. Smirnov and A. P. Yushkevich (1964), and E. P. Ozhigova (1966). Mathematics at Moscow University in the first half of this century is discussed by D. F. Egorov (1928), P. S. Aleksandrov, B. V. Gnedenko, and V. V. Stepanov (1948), and by I. G. Bashmakova and S. S. Petrova (1980); on the activity of the Moscow Mathematical Society, in articles by P. S. Aleksandrov and O. N. Golovin (1957) and by D. E. Men′shov (1965). A series of papers by A. F. Lapko and L. A. Lyusternik (1957, 1958, 1967) are devoted to various aspects of the development of Soviet mathematics; articles by V. V. Stepanov (1947), L. A. Lyusternik (1967), P. S. Aleksandrov (1980), and D. E. Men′shov (1983) covered the early history of the Moscow school of function theory (Egorov and Luzin); P. I. Kuznetsov (1971) covered the creative work of D. F. Egorov; N. A. Lebedev and S. M. Lozinskiĭ (1967) covered the life and work of V. I. Smirnov, and mathematical research in Leningrad University during 1917–1967.

From the literature on the development of mathematics in other scientific centers of Russia and other republics, we mention the following articles: on the development of mathematics at Kazan, N. G. Chebotarev (1941), B. M. Gagaev (1947), V. V. Morozov and P. P. Norden (1947), V. L. Laptev (1959); at Rostov-on-Don, S. E. Belozerov (1953); at Tomsk, N. N. Krulikovskiĭ (1967); in the Ukraine, N. I. Akhiezer (1956), B. V. Gnedenko and I. M. Gikhman (1956), B. V. Gnedenko and I. B. Pogrebysskiĭ (1956), G. E. Shilov (1956), I. Z. Shtokalo (1958), V. B. Gnedenko (1959), I. A. Naumov (1966), A. N. Bogolyubov and V. M. Urbanskiĭ (1985); in Georgia, L. G. Magnarazde and G. S. Chogoshvili (1957); in Uzbekistan, V. I. Romanovskiĭ (1942), T. A. Sarymsakov (1942), I. S. Arzhanykh (1953), S. Kh. Sirazhdinov (1967); in Lithuania, V. A. Statulyavichus (1965); in Kazakhstan, O. A. Zhautykov (1958), O. A. Zhautikov and M. V. Pentkovskiĭ (1960); in Moldavia, V. A. Andrunakievich and V. P. Bychkov (1965), V. P. Bychkov and K. S. Sibirskiĭ (1974); in Latvia, A. Ya. Luens (1950, 1958), A. Ya. Luens and others (1966); in Estonia, S. A. Baron, Ya. Gabovich, Yu. Ya. Kazik, and others (1964); in Azerbaijan, Z. I. Khalinnov (1950, 1957), A. I. Guseĭnov and G. N. Agaev (1964), M. P. Efendiev and M. A. Dzhavadov (1967); in Turkmenia, A. M. Akhundov and A. A. Karryev (1964).

A significant amount of research, which we cannot discuss in this brief survey, is devoted to work on the development of individual branches of mathematics. Here we mention only surveys contained in the publications that we have repeatedly mentioned on the development of mathematics in the USSR for 30 years (1948, edited by A. G.Kurosh and others); for 40 years (1959, edited by A. G. Kurosh); in volumes 3 and 4 of the "History of the Country's Mathematics" (1968–1970, edited by I. Z. Shtokalo and others) and "Outline of the History of Mathematics in the USSR" (1983, edited by I. Z. Shtokalo and others).

A large amount of factual material is contained in the series "Scientific Biographies," published by the Academy of Sciences of the USSR, on F. E. Molin (N. F. Kanunov, 1983); D. A. Grave (V. A. Dobrovol′skiĭ, 1978); A. P. Kotelnikov (T. V. Putyat and others, 1968); G. N. Nikoladze (A. N. Bogolyubov, 1973); N. E. Kochin (P. Ya. Kochina, 1979); Yu. D. Sokolov (B. N. Fradlin, 1984); and also in scientific biographies in other publications, on V. A. Steklov (V. S. Vladimirov and I. I. Markush, 1981); D. M. Sintsov (I. A. Naumov, 1955); V. F. Kagan (A. M. Lopshitz and P. K. Rashevskiĭ, 1969); S. A. Chaplygin (V. V. Golubev, 1951); N. M. Krylov (Yu. A. Mitropolskiĭ and A. N. Bogolyubov (1979), A. N. Bogolyubov and V. M. Urbanskiĭ (1987)); N. A. Glagolev (S. V. Bakhvalov, 1961); V. V. Stepanov (P. S. Aleksandrov and others, 1956); N. I. Muskhelishvili (I. N. Vekua, 1961). Brief biographical notes on scientists, and their bibliographies, can be found in the series "Material for Biobibliography of Scientists of the USSR." In this series, pamphlets have appeared on A. N. Krylov, in 1945; N. M. Krylov, in 1945; N. N. Luzin, in 1948; B. N. Delone, in 1967; I. M. Vinogradov, in 1978; N. I. Muskhelishvili, in 1967; M. A. Lavrent′ev, in 1972; V. I. Smirnov, in 1949;

N. E. Kochin, in 1948; I. G. Petrovskiĭ, in 1957; I. N. Vekua, in 1963; L. S. Pontryagin, in 1983; S. L. Sobolev, in 1949; N. N. Bogolyubov, in 1959; A. A. Dorodnitsin, in 1974; V. S. Vladimirov, in 1987; and G. I. Marchuk, in 1985. Scientific biographies of distinguished Soviet mathematicians can be found in articles on the occasions of their anniversaries, and in obituaries, which appear regularly in Russian Mathematical Surveys (Uspekhi Matematicheskii Nauk) and in "Mathematics in the School".

[Translator's note: the "Mathematical Encyclopedic Dictionary" (published by "Sovetskaya Entsiklopedia", Moscow, 1988) contains several hundred capsule biographies of Russian and Soviet mathematicians, as well as of particularly distinguished non-Soviet ones, with portraits.]

The publication of the collected works of Soviet mathematicians has become extremely important for research on the history of mathematics in the USSR. Many of these collections are accompanied by valuable commentaries, and, as a rule, by surveys of the authors' creative activities. The collected works of the following authors have been published: N. E. Zhukovskiĭ (1935–1938, 1948–1950), A. N. Krylov (1951–1956), S. A. Chaplygin (1933–1935), S. N. Bernshteĭn (1952–54), N. N. Luzin (1953–1959), Ts. Burstin (1932), N. G. Chebotarev (1949–1950), P. S. Uryson (1951); selected papers of A. A. Markov (1948, 1951), F. E. Molin (1985), D. A. Grave (1971), P. Bol′ (1961), D. F. Egorov (1970), S. A. Chaplygin (1954), N. M. Krylov (1949, 1961), V. I. Romanovskiĭ (1959, 1964), N. A. Vasil′ev (1988), E. E. Slutskiĭ (1960), N. N. Luzin (1951), V. I. Smirnov (1988), A. M. Razmadze (1952), I. M. Vinogradov (1952), O. Yu. Shmidt (1959), A. S. Novikov (1979), I. G. Petrovskiĭ (1986), A. N. Kolmogorov (1985, 1987), A. O. Gel′fond (1973), L. S. Pontryagin (1988), N. N. Bogolyubov (1969, 1971), A. I. Mal′tsev (1976), M. V. Keldysh (1985), Yu. V. Linnik (1979, 1982).

Memoirs are a traditional source of historical information. Activity in this genre, after a long period of quiescence in the Stalin years (one of the few exceptions was the already classical collection of reminiscences of A. N. Krylov (1945)), resumed in the sixties with reminiscences of L. A. Lyusternik (1967), and since that time has grown steadily, with memoirs of P. Ya. Kochina (1974), A. P. Yushkevich (1976), P. S. Aleksandrov (1977, 1979–1980), A. O. Gel′fond (1977), D. E. Menshov (1983), and A. N. Kolmogorov (1986).

Perhaps the most important activity in recent years in the domain of history of mathematics in the USSR has consisted of annotated editions of archival material. This work had already begun in the post-war years. However, at that time it was pursued at an extremely modest rate, in slender publications, sometimes disfigured by cuts, and with commentaries written in Aesopian style.

Among the best instances of archival documents from those times of slow publication, we call attention to L. E. Maĭstrov's publication in the eighth issue of "Studies in the History of Mathematics" of two documents for a biography of N. N. Luzin: his report in 1914 on his foreign study (N. N. Luzin,

1955), and the report in 1916 of the faculty commission on his promotion to a professorship, written by N. E. Zhukovskiĭ, L. K. Lakhtin and D. F. Egorov (1955). The 14th issue contains a note by N. G. Chebotarev (1961). The 16th issue contains a comment by Luzin, found in the archives, on the proposed publication of a letter from Euler to Goldbach (1965). These four items are all that appeared in the pages of "Studies in the History of Mathematics", up to the end of the seventies, from archival material on the history of mathematics during the Soviet period.

The situation began slowly to improve at the end of the 70's. In 1978, there appeared, in issue 23, a letter from N. N. Luzin to A. Denjoy (published by P. Dyugak); in issue 24, a letter from W. Sierpiński (1979) to N. N. Luzin (published by V. A. Volkov and F. A. Medvedev); in issue 25, a paper by L. K. Arboleda (1980) on the correspondence of P. S. Aleksandrov and P. S. Uryson, and also a letter from D. F. Egorov (1980) to Luzin (published by F. A. Medvedev with A. P. Yushkevich); in issue 27, letters from N. N. Luzin (1983) to M. Fréchet (published by A. P. Yushkevich), and from C. J. de la Vallée-Poussin (1983) to Luzin (published by F. A. Medvedev); in issue 28, letters from N. N. Luzin (1985) to O. Yu. Schmidt, and from D. F. Egorov (1985) to D. Hilbert (published by A. V. Dorofeeva), and also a note by A. P. Yushkevich (1985), reconstructed from archival material, on L. G. Shnirel′man's visit to Göttingen; in issue 29, reviews by N. N. Luzin (1985) and by Luzin and S. A. Chaplygin (1985) of the work of S. P. Finikov. Finally, issues 30–31 contain documents, from another of N. N. Luzin's archives, of the eminent Russian philosopher and theologian P. A. Florenskiĭ, who in 1900–1903 was a student in the mathematics department of Moscow University. These documents: papers relating to the activities of the students in the mathematical circle of the Moscow Mathematical Society in 1902–1903 (S. M. Polovinkin, 1986), Florenskiĭ's account of B. K. Mlodzeevskiĭ's course of lectures in 1902 on the theory of functions of a real variable (F. A. Medvedev, 1986), Luzin's notes (1986), a fragment of Florenskiĭ's candidate's dissertation (published by S. S. Demidov and A. N. Parshin, 1986), and finally the correspondence of Luzin and Florenskiĭ (published by Demidov and others): these provide important corrections to our understanding of the early history of the Moscow school of function theory. In issue 31, N. S. Ermolaev published the correspondence of N. N. Luzin and A. N. Krylov (1989); and on the basis of materials in the archives of the Paris Academy of Sciences, A. P. Yushkevich (1989) published an article on the election of S. N. Bernshteĭn, I. M. Vinogradov, and M. A. Lavrent′ev to the French Academy of Sciences.

Archival material has begun to appear frequently in other publications as well. As examples, we cite two publications that have appeared in the Kiev series "Studies in the History of Natural Science and Technology": a paper by V. M. Urbanskiĭ (1989) on the influence of V. I. Vernadskiĭ on the development of mathematics in the Ukrainian Academy of Science, and the publication by L. V. Koval′chuk (1989) of an account by M. F. Kravchuk (1989) of a visit to Moscow in 1915–1916.

The publication of archival material should lay the foundation for a serious study of the history of mathematics in our country. We already possess the first results of this work. As an example, we mention the article by A. P. Yushkevich (1989) on the famous "Case of Academician N. N. Luzin" in 1936; and the work of V. M. Tikhomirov on open A-sets, which should appear soon in the pages of "Istoriko-Matematicheskie Issledovaniya".

Bibliography

Names of journals are given in the abbreviations used by Mathematical Reviews, when available, except for four special abbreviations introduced by the bibliographer:

IMI=Istor. Mat. Issled.
VIET=Voprosy Istor. Estestvoznan. i Tekhn.
UMN=Uspekhi Mat. Nauk
IMEN=Istoriya i Metodologiya Estestvennykh Nauk (History and Methodology of the Natural Sciences)

Items are in Russian except as noted.

I. General works on the history of mathematics in Russia and in the USSR

P. S. Aleksandrov, M. Ya. Vygodskiĭ, and V. I. Glivenko, editors, 15 years of mathematics in the USSR, "Gostekhizdat," Moscow, 1932.

A. N. Bogolyubov, Mathematics and mechanics. Biographical reference book, "Naukova Dumka," Kiev, 1983.

A. I. Borodin and A. S. Bugaĭ, Biographical dictionary of workers in the field of mathematics, "Radyans′ka Shkola," Kiev, 1979.

D. F. Egorov, Progress of mathematics in the USSR, in Science and technology in the USSR (1917–1927), vol. 1, Rabotnik Prosveshcheniya, Moscow, pp. 223–234.

S. V. Fomin and G. E. Shilov, editors, 1969–1970. Mathematics in the USSR, 1958–1967, vol. 2, nos. 1-2.

B. V. Gnedenko, Outline of the history of mathematics in Russia, "Gostekhizdat," Moscow and Leningrad, 1946.

A. G. Kurosh, editor, Forty years of mathematics in the USSR, 1917–1957, vols. 1, 2, "Fizmatgiz," Moscow, 1959.

A. G. Kurosh, A. I. Markushevich, and P. K. Rashevskiĭ, editors, Thirty years of mathematics in the USSR, 1917–1947, "Gostekhizdat," Moscow and Leningrad, 1948.

S. P. Mikulinskiĭ and A. P. Yushkevich, editors, The development of the natural sciences in Russia (18th to early 20th century), "Nauka," Moscow, 1977.

I. Z. Shtokalo, A. N. Bogolyubov, and A. P. Yushkevich, editors, History of national mathematics, vols. 1, 2, 3, and 4 (books 1 and 2), "Naukova Dumka," Kiev, 1966–1970.

——, editors, Outline of the development of mathematics in the USSR, "Naukova Dumka," Kiev, 1983.

A. P. Yushkevich, History of mathematics in Russia up to 1917, "Nauka," Moscow, 1968.

II. Mathematics in Russia from late 19th to early 20th century

I. G. Bashmakova, Foundation of the theory of divisibility in the works of E. M. Zolotarev, IMI, no. 2 (1949), 233–351.

S. E. Belozërov, Mathematics at Rostov University, IMI, no. 6 (1953), 247–352.

S. N. Bernshteĭn, The Petersburg school of probability theory, Uchen. Zap. Leningrad. Gos. Univ., vol. 55, no. 10 (1940), 3–11.

N. D. Bespamyatnykh, Mathematics at Vilnius University (1803–1932), Transactions of the Karelian Pedagogical Institute, vol. 14 (1963), 49–69. [Ucheniye Zapiski Karel. Pedagog. Inst.]

B. N. Delone, The Petersburg school of number theory, Press of the Academy of Sciences of the USSR, Moscow and Leningrad, 1947.

S. S. Demidov, N. V. Bugaev and the beginnings of the Moscow school of the theory of functions of a real variable, IMI, no. 29 (1985), 113–124.

——, On the early history of the Moscow school of the theory of functions, IMI, no. 30 (1986), 124–130.

I. Ya. Depman, Karl Mikhaĭlovich Peterson and his candidate dissertation, IMI, no. 5 (1952), 134–164.

——, V. A. Steklov at Petersburg University, IMI, no. 6 (1953), 509–528.

——, Toward a biography of S. V. Kovalevskaya, IMI, no. 7 (1954), 713–715.

——, On the history of mathematics at Derpt (Yur′ev) University, Uchen. Zap. Leningrad. Ped. Inst., Fiz.-Mat. Fak., vol. 14, no. 1 (1955), 128–137.

——, The St. Petersburg Mathematical Society, IMI, no. 13 (1960), 11–106.

F. I. Frankl′, On the work of 19th century mathematicians on the theory of characteristics of partial differential equations, UMN, vol. 6, no. 2(42) (1951), 154–156.

R. I. Galchenkova, Mathematics in Leningrad (Petersburg) University in the 19th century, IMI, no. 14 (1961), 355–392.

B. V. Gnedenko, The development of probability theory in Russia, Akad. Nauk SSSR. Trudy Inst. Istoriĭ Estestvoznaniya, vol. 2 (1948), 390–425.

V. V. Gussov, The work of Russian scholars on the theory of the gamma-function, IMI, no. 5 (1952), 421–472.

——, The development of the theory of cylinder functions in Russia and the USSR, IMI, no. 6 (1953), 355–475.

N. M. Gyunter, On the scientific activities of V. A. Steklov, Obituary of V. A. Steklov, Press of the Academy of Sciences of the USSR, Leningrad, 1928.

A. Ya. Khinchin, The theory of probability in pre-revolution Russia and the Soviet Union, The science and technology front, no. 7 (1934), 36–46. [Front Nauki i Tekhniki]

S. M. Kiro, Mathematics at the University of Novorosiĭsk (Odessa) (in Ukrainian), Historical-Mathematical Collection, no. 2 (1961), 22–42. [Istoriko-Matematichniĭ Zbirnik]

——, Mathematics in the periodical publications of the University of Odessa (Novororossĭsk), 1865–1955, part 1 (1865–1933), Proceedins of Odessa University, Mathematics Series, vol. 146, no. 6 (1956), 89–109. [Trudi Odesskovo Univ. Ser. Mat.]

A. N. Kolmogorov, The role of Russian science in the development of the theory of probability, Scientific Transactions of Moscow State University, 91 (1947), 53–64. [Ucheniye Zapiski Moscov. Gos. Univ.]

M. F. Kravchuk, Mathematics and mathematicians in Kiev University during the hundred years 1834–1934 (in Ukrainian), in 100 Years of the Progress of Science at Kiev University, Kiev, Press of Kiev University, 1935, pp. 34–69.

E. B. Leĭbman, The mathematical section of the Novorossiisk Natural Science Society (1876–1928), IMI, no. 14 (1961), 393–440.

L. E. Maĭstrov, P. L. Chebyshev's first arithmometer, IMI, no. 14 (1961), 349–354.

——, An evaluation of P. L. Chebyshev's arithmometer, IMI, no. 18 (1973), 295–300.

____, The preservation of P. L. Chebyshev's mechanisms, IMI, no. 20 (1975), 309–318.

M. N. Marchevskiĭ, The history of the chair of mathematics at Kar'kov University up to the 150th year of its existence, Proceedings of the Mathematical Society, Khar'kov, vol. 24, no. 4 (1956), 70–79. [Zapiski Matem. Obshchestva, Khar'kov]

____, The Khar'kov Mathematical Society in the first 75 years of its existence (1879–1954), IMI, no. 9 (1956), 613–666.

I. I. Markush, On the question of the creation of the V. A. Steklov Petersburg-Leningrad school of mathematical physics, IMEN, no. 16 (1974), 141–153.

A. I. Markushevich, Yu. V. Sokhotskiĭ's contribution to the general theory of analytic functions, IMI, no. 3 (1950), 399–406.

F. A. Medvedev, A. M. Lyapunov's contribution to the theory of the Stieltjes integral, IMI, no. 14 (1961), 211–236.

____, Preparation for set-theoretic and function-theoretic research in Russia, Studies in the history of mathematics and mechanics, Moscow, Press of the Academy of Sciences of the SSSR, 1963, pp. 45–56.

____, A course of lectures by B. K. Mlodzeevskiĭ on the theory of functions of a real variable, delivered in the autumn of 1902 at Moscow University, IMI, no. 30 (1986), 130–147.

G. K. Mikhaĭlov and S. Ya. Stepanov, The history of the problem of rotation of a solid body around a fixed point in the Hesse and Kovalevskaya cases, and their geometric simulation, IMI, no. 28 (1985), 223–246.

D. D. Mordukhaĭ-Boltovskoĭ, Mathematics at Rostov University, Rostov-on-Don University, Anniversary Volume, 1915–1940, Rostov-izdat, Rostov-on-Don, 1941, pp. 46–52.

M. B. Nalbandyan, The theory of elliptic functions and its applications in the work of E. I. Zolotarev, IMI, no. 16 (1965), 191–206.

____, The theory of elliptic functions and its applications in the work of Russian mathematicians of the 19th and early 20th centuries, IMI, no. 17 (1966), 361–370.

E. P. Ozhigova, The development of number theory in Russia, "Nauka," Leningrad, 1972.

S. M. Polovinkin, On the student mathematical circle in the Moscow Mathematical Society in 1902–1903, IMI, no. 30 (1986), 148–158.

P. Ya. Polubarinova-Kochina, For a biography of S. V. Kovalevskaya (from material in her correspondence), IMI, no. 7 (1954), 666–712.

K. A. Rybnikov, Viktor Viktorovich Bobynin, IMI, no. 3 (1950), 343–357.

F. Ya. Shevelev, On the history of the Moscow Mathematical Society, IMEN, no. 5 (1966), 62–74.

D. M. Sintsov, The chair of pure and applied mathematics at Khar'kov University on its 100th anniversary (1805–1905), Khar'kov, 1908.

____, Fifty years of the Khar'kov Mathematical Society, Transactions of the First All-Union Mathematical Congress, ONTI, Moscow and Leningrad, 1936, pp. 97–105.

N. I. Styazhkin and V. D. Silakov, A brief survey of the history of general and mathematical logic in Russia, "Vysshaya Shkola," Moscow, 1962.

A. K. Sushkevich, Materials for the history of algebra in Russia in the 19th and early 20th centuries, IMI, no. 4 (1951), 237–451.

____, Dissertation on mathematics at Khar'kov University from 1805 to 1917, Transactions of the Khar'kov Mathematical Society, vol. 24 (1956), 91–115. [Zapiski Khar'kov. Matem. Obshchestva]

M. Ya. Vygodskiĭ, Mathematics and its activity in Moscow University up to the second half of the 19th century, IMI, no. 1 (1948), 141–183.

III. Mathematics in the USSR

N. I. Akhiezer, The Khar′kov Mathematical Society, Transactions of the Khar′kov Mathematical Society, vol. 24, no. 4, (1956), 31–39. [Zapiski Khar′kov. Matem. Obshchestva]

A. I. Akhundov and A. A. Karryev, From the history of the development of mathematics in Soviet Turkmenistan, Ashkhabad, Press of the Turkmenistan Polytechnic Institute, 1954.

P. S. Aleksandrov, Mathematics at Moscow University in the first half of the 20th century, IMI, no. 8 (1955), 9–54.

——, Topology at Moscow University, IMEN, no. 9 (1970), 3–7.

——, Recollections of Göttingen, IMI, no. 22 (1977), 242–245.

——, Pages of an autobiography, UMN, vol. 34, no. 6 (1979), 219–249; vol. 35, no. 3 (1980), 241–278.

——, The Luzin school of mathematics, Mathematical Science at Moscow State University, in News of Life, Science, and Technology, Series: Mathematics, Cybernetics, no. 4 (1980), 21–29, "Znanie," Moscow, 1980.

——, B. V. Gnedenko, and V. V. Stepanov, Mathematics at Moscow University in the 20th century, IMI, no. 1 (1948), 9–42.

P. S. Aleksandrov and O. N. Golovin, The Moscow Mathematical Society: 90 years of scientific activity, UMN 12, no. 6(78) (1957), 9–46.

V. A. Andrunakievich and V. P. Bychkov, Mathematical life in the Moldavian SSR, UMN 20, no. 2(122) (1965), 247–258.

L.K. Arboleva, The birth of the Soviet school of topology, Remarks on letters of P. S. Alexandrov and P. S. Uryson to Maurice Fréchet, IMI, no. 25 (1980), 281–302.

I. S. Arzhanykh, On the question of the development of mathematics and mechanics in Uzbekistan, Izv. Akad. Nauk. UzSSR, no. 6 (1953), 102–105.

S. A. Baron, Ya. A. Gabovich, Yu. Ya. Kaazik, and others, Mathematics in Soviet Estonia in the past 20 years, Scientific Transactions of Tartu University, Work on Mathematics and Mechanics, no. 150, vol. 4 (1964), 15–52. [Ucheniye Zapiski Tartu. Univ. Trudi po Matem. i Mekh.]

I. G. Bashmakova and S. S. Petrova, History of mathematics at Moscow University, Mathematical Science at Moscow State University, in the publication "News of Life, Science and Technology," Series: Mathematics, Cybernetics, 1980, no. 4, "Znanie," Moscow, pp. 45–63.

A. N. Bogolyubov and V. M. Urbanskiĭ, Organization and establishment of mathematical sections in the Academy of Sciences of the Ukrainian SSR, IMI, no. 28 (1985), 160–187.

V. P. Bychkov and K. S. Sibirskiĭ, The development of mathematics in the Moldavian SSR: Brief survey, "Shtiintsa," Kishinev, 1974.

N. G. Chebotarev, Work of Kazan mathematicians in the post-revolution period (1917–1939), Scientific Proceedings of the Kazan University, vol. 101, part 1 (1941), 57–78. [Ucheniye Zapiski Kazan. Univ.]

——, Some applications of ideal theory to algebra (Publication and notes by V. A. Dobrovol′skiĭ), IMI, no. 14 (1961), 539–550.

B. N. Delone, L. D. Kudryavtsev, and M. M. Postnikov, Outline of the history of the development of mathematics in the Academy of Sciences of the USSR in the Soviet period (1917–1960), in Outline of the history of mathematics and mechanics, Press of the Academy of Sciences of the USSR, Moscow, 1965, pp. 3–44.

A. V. Dorofeeva, On the letters of D. F. Egorov to D. Hilbert, IMI, no. 28 (1985), 270–278.

M. R. Efendiev and M. A. Dzhavadov, On the history of the development of higher mathematical education and mathematicians in Soviet Azerbaijan, Scientific Proceedings of Azerbaijan University, Series: Physical and Mathematical Sciences, no. 3 (1967), 3–15. [Uchen. Zap. Azerb. Univ. Ser. Fiz.-Mat. Nauk]

D. F. Egorov, Work of the Institute for Scientific Research in Mathematics and Mechanics during the five years from 1923 to 1928, Proceedings of the Association for Scientific Research of the Institutes of the Faculty of Physics and Mathematics of Moscow State University, vol. 1, nos. 3-4 (1928), 301–303. [Izv. Assots. Nauchno-Issled. Inst. Fiz.-Mat. Fak. Moscov. Gos. Univ.]

——, Letter to N. N. Luzin. (Foreword by P. S. Aleksandrov, Publication and notes by F. A. Medvedev with the assistance of A. P. Yushkevich), IMI, no. 25 (1980), 335–361.

——, Letter to D. Hilbert. (Translation and notes by A. V. Dorofeeva), IMI, no. 28 (1985), 266–270.

B. M. Gagaev, 25 years of the development of mathematics at Kazan, Scientific Proceedings of Kazan University, vol. 106, part 2 (1947), 3–13. [Ucheniye Zapiski Kazan. Univ.]

——, V. V. Morozov, and A. P. Norden, 30 years of the Kazan school of mathematics, UMN, vol. 2, no. 6(22) (1947), 3–20.

A. O. Gel′fond, Some impressions of a scientific visit to Germany in 1930, IMI, no. 22 (1977), 246–251.

B. V. Gnedenko, Research on the theory of probability and mathematical statistics in the organization of the Academy of Sciences of the Ukrainian SSR, Ukrainian Mathematical Journal, vol. 11, no. 2 (1950), 123–157.

B. V. Gnedenko and I. I. Gikhman, The development of probability theory in the Ukraine, IMI, no. 9 (1956), 477–536.

B. V. Gnedenko and I. B. Pogrebysskii, On the development of mathematics in the Ukraine, IMI, no. 9 (1956), 403–426.

A. I. Guseĭnov and G. N. Agaev, On the history of the development of mathematical research in Azerbaijan, Proceedings of the Academy of Sciences of the Azerbaijan SSR, Series of physical technology and mathematical sciences, no. 3 (1964), 3–17. [Izvestiya Akad. Nauk. AzSSR. Ser. Fiz.-Tekh. i Mat. Nauk.]

Z. I. Khalilov, The development of mathematical sciences in Azerbaijan, Proceedings of the Academy of Sciences of the Azerbaijan SSR, no. 7 (1950), 61–76. [Izvestiya Akad. Nauk. AzSSR]

——, The development of physico-mathematical sciences in Soviet Azerbaijan, Proceedings of the Academy of Sciences of the Azerbaijan SSR, no. 10 (1957), 25–28. [Izvestiya Akad. Nauk AzSSR]

A. N. Kolmogorov, Recollections of P. S. Aleksandrov, UMN 41, no. 6 (1986), 187–203.

P. Ya. Kochina, Recollections, "Nauka," Moscow, 1974.

M. F. Kravchuk, An account of the work of stipendary Mikhail Kravchuk from November 1915 to November 1916; Report of Honored Professor D. Grave on professional stipendiary M. F. Kravchuk's report for 1916; Publication and commentary by L. V. Koval′chuk, Studies in the History of Natural Science and Technology, no. 37, Kiev, "Naukova Dumka," 1989, pp. 86–93.

N. N. Krulikovskiĭ, History of the development of mathematics in Tomsk, Press of Tomsk State University, Tomsk, 1967.

A. N. Krylov, My reminiscences, Press of the Academy of Sciences of the USSR, Moscow, 1945.

P. I. Kuznetsov, Dmitriĭ Fedorovich Egorov, UMN, vol. 26, no. 5 (1971), 169–210.

A. F. Lapko and L. A. Lyusternik, Mathematical congresses and conferences in the USSR, UMN, vol. 12, no. 6(78) (1957), 47–130; vol. 13, no. 5(83) (1958), 121–166.

——, From the history of Soviet mathematics, UMN, vol. 22, no. 6(138) (1967), 13–140.

B. L. Laptev, Forty years (1917–1957) of mathematics at Kazan University, IMI, no. 12 (1959), 11–58.

N. A. Lebedev and S. M. Lozinskiĭ, Vladimir Ivanovich Smirnov and mathematics at Leningrad University during 1917–1967, Bulletin of Leningrad University, Mathematics, Mechanics, Astronomy, no. 7, Part 2 (1967), 7–18. [Vestnik Leningrad. Univ. Mat. Mekh. Astronom.]

N. N. Luzin, Report on foreign travel for scientific work... , IMI, no. 8 (1955), 57–70.

——, Foreword to a letter from L. Euler to C. Goldbach, IMI, no. 16 (1965), 129–144.

——, Letter to A. Denjoy (published by P. Dyugak), IMI, no. 23 (1978), 314–348.

——, Letter to M. Fréchet (published by A. P. Yushkevich), IMI, no. 27 (1983), 298–300.

——, Review of the scientific work of Professor Sergeĭ Pavlovich Finikov, IMI, no. 29 (1985), 293–316.

——, Review of the scientific work of S. P. Finikov in his final period, 1938–1946, IMI, no. 29 (1985), 316–318.

——, Letter to O. Yu. Schmidt (published by S. S. Demidov), IMI, no. 28 (1985), 278–287.

——, [On restrictors] (Published and edited by S. S. Demidov), IMI, no. 30 (1986), 177–181.

N. N. Luzin and A. N. Krylov, Correspondence; Publication, preface, and commentary by N. S. Ermolaeva, IMI, no. 31 (1989), 203–272.

N. N. Luzin and P. A. Florenskiĭ, Correspondence; Publication and notes by S. S. Demidov, A. N. Parshin, S. M. Polovinkin, and P. V. Florenskiĭ, IMI, no. 31 (1989), 125–190.

N. N. Luzin and S. A. Chaplygin, The scientific work of Sergeĭ Pavlovich Finikov, IMI, no. 29 (1985), 319–321.

A. Ya. Lusis, Ten years of the works of mathematicians of Soviet Latvia. Proceedings of the Academy of Sciences of the Latvian SSR, no. 11 (1950), 109–121. [Izvestiya Akad. Nauk. Latviĭ. SSR]

——, Development of mathematics in Soviet Latvia during the last decase, Scientific Transactions of the University of Latvia, Physical and Mathematical Sciences, vol. 20, no. 3 (1958), 5–20. [Ucheniye Zapiski Latviĭ. Univ. Fiz.-Mat. Nauki]

A. Ya. Lusis, L. E. Reizin′, and E. Ya. Riekstyn′sh, Mathematics in Soviet Latvia, UMN, vol. 21, no. 3 (1966), 248–259.

L. A. Lyusternik, The youth of the Moscow mathematical school, UMN, vol. 22 (1967), no. 1(133), 137–161; no. 2(134), 199–239; no. 4(136), 147–185.

L. G. Magnaradze and G. S. Chogoshvili, The development of the mathematical sciences in Georgia, Fortieth anniversary of the October Revolution, Tbilisi, 1957, pp. 485–611.

D. E. Men′shov, The Moscow Mathematical Society during the period 1910–1920, UMN, vol. 20, no. 3 (1965), 19–20.

——, Reminiscences of the early years and of the origins of the Moscow school of function theory, IMI, no. 27 (1983), 312–333.

I. A. Naumov, The work of Ukrainian mathematicians on nonholonomic differential geometry, IMEN, no. 5 (1966), 105–117.

E. P. Ozhigova, Mathematics in the Academy of Sciences in the early years of the Soviet regime, IMI, no. 17 (1966), 381–390.

V. I. Romanovskiĭ, Mathematics and mechanics; in 25 years of Soviet science in Uzbekistan, Press of the Uzbek Branch of the Academy of Sciences of the USSR, Tashkent, 1942, pp. 60–63.

T. A. Sarymsakov, Theory of probability and mathematical statistics in the work of the Tashkent School; in 25 years of Soviet science in Uzbekistan, Press of the Uzbek Branch of the Academy of Sciences of the USSR, Tashkent, 1942, pp. 64–72.

G. E. Shilov, On the history of the development of functional analysis in the Ukraine, IMI, no. 9 (1956), 427–476.

I. Z. Shtokalo, Outline of the development of mathemtics in the Ukraine during 40 years of Soviet power (in Ukrainian), Press of the Academy of Sciences of the Ukrainian SSR, Kiev, 1958.

W. Sierpiński, Letter to N. N. Luzin (Published by V. A. Volkov and F. A. Medvedev), IMI, no. 24 (1979), 366–373.

S. Kh. Sirazhdinov, 50 years of results of research in probability theory and mathematical statistics in Uzbekistan, Proceedings of the Academy of Sciences of the UzbekSSR, Series of Physical and Mathematical Sciences, no. 5 (1967), 21–26. [Izv. Akad. Nauk. UzSSR. Ser. Fiz.-Mat. Nauk.]

V. I. Smirnov and A. P. Yushkevich, Mathematics, in History of the Academy of Sciences of the USSR, vol. 2, Press of the Academy of Sciences of the USSR, 1964, pp. 34–51, 286–306, 473–483.

V. A. Statulyavichus, Mathematical research in the Lithuanian SSR, Collection of Lithuanian Mathematics, vol. 5, no. 3 (1965), 361–372. [Litovsk. Mat. Sb.]

V. V. Stepanov, The Moscow school of function theory, Scientific Proceedings of Moscow State University, vol. 91 (1947), 47–52. [Ucheniye Zapiski Moscov. Gos. Univ.]

V. M. Urbanskiĭ, V. I. Vernadskiĭ and the development of mathematical ideas in the Ukrainian Academy of Sciences, Outlines of the history of natural science and technology, Vol. 36, 1989, "Naukova Dumka," Kiev, pp. 21–29.

C. J. de la Vallée-Poussin, Letter to N. N. Luzin. (Translated and published by F. A. Medvedev), IMI, no. 27 (1983), 301–312.

O. A. Zhautykov, Mathematics in Kazakhstan in the Soviet period, Work of the Section of Mathematics and Mechanics of the Academy of Sciences of the Kazakhstan SSR, vol. 1 (1958), 5–24. [Trudi Sekt. Matem. i Mekh. Akad. Nauk. KazSSR]

O. A. Zhautykov and M. V. Pentkovskiĭ, Questions of the development of the mathematical sciences in Kazakhstan, Bulletin of the Academy of Sciences of the Kazakhstan SSR, no. 8 (1960), 9–19. [Vestnik Akad. Nauk KazSSR]

N. E. Zhukovskiĭ, L. K. Lakhtin, and D. F. Egorov, Report of the Faculty Committee . . . , IMI, no. 8 (1955), 70–76.

A. P. Yushkevich,Memories of student days; in Matematika: A collection of articles on the scientific method, "Vysshaya Shkola," Moscow, no. 6 (1976), pp. 99–112.

——, L. G. Shnirel'man at Göttingen, IMI, no. 28 (1985), 287–290.

——, "The case of Academician N. N. Luzin, "Bulletin of the Academy of Sciences of the USSR, no. 4 (1989), 102–113. [Vestnik Akad. Nauk. SSSR]

——, On the history of scientific relations between the mathematicians of the USSR and France (in connection with the election of S. N. Bernshteĭn, I. M. Vinogradov, and M. A. Lavren'tiev to membership in the Paris Academy of Sciences), IMI, no. 31 (1989), 203–272.

IV. Scientific biographies

P. S. Alexandrov and V. V. Nemytskiĭ, Vyacheslav Vasil′evich Stepanov, Moscow University Press, Moscow, 1956.

S. V. Bakhvalov, Nil Aleksandrovich Glagolev, Moscow University Press, Moscow, 1961.

V. A. Bazhanov, Nikolaĭ Aleksandrovich Vasil′ev, "Nauka," Moscow, 1988.

A. N. Bogolyubov, Georgiĭ Nikolaevich Nikoladze, "Nauka," Moscow, 1973.

A. N. Bogolyubov and V. M. Urbanskiĭ, Nikolaĭ Mitrofanovich Krylov, "Naukova Dumka," Kiev, 1987.

V. A. Dobrovol′skiĭ, Vasiliĭ Petrovich Ermakov, "Nauka," Moscow, 1981.

——, Dmitriĭ Aleksandrovich Grave, 1863–1939, "Nauka," Moscow, 1968.

B. N. Fradlin, Yuriĭ Dmitrievich Sokolov, "Nauka," Moscow, 1984.

V. V. Golubev, Nikolaĭ Egorovich Zhukovskiĭ, Moscow University Press, Moscow, 1947.

——, Sergeĭ Aleksandrovich Chaplygin, 1869–1942, Moscow University Press, 1951.

D. Z. Gordevskiĭ, K. A. Andreev, an eminent Russian geometer, Khar′kov University Press, Khar′kov, 1955.

S. Ya. Grodzenskiĭ, Andrei Andreevich Markov, "Nauka," Moscow, 1987.

N. F. Kanunov, Fedor Èduardovich Molin, "Nauka," Moscow, 1983.

P. Ya. Kochina, Nikolaĭ Evgrafovich Kochin, "Nauka," Moscow, 1979.

——, Sof′ya Vasil′evna Kovalevskaya, "Nauka," Moscow, 1981.

A. A. Kosmodem′yanskiĭ, Nikolaĭ Egorovich Zhukovskiĭ, "Nauka," Moscow, 1984.

A. I. Kropotov, Nikolaĭ Yakovlevich Sonin, 1849–1915, "Nauka," Leningrad, 1967.

B. L. Laptev, Petr Alekseevich Shirokov (1895–1944), Biographical and bibliographical account, Kazan University Press, 1955, pp. 3–20.

A. M. Lopshits and P. K. Rashevskiĭ, Veniamin Fedorovich Kagan, Moscow University Press, Moscow, 1969.

Yu. O. Mitropol′skiĭ and O. M. Bogolyubov, Mikola Mitrofanovich Krilov, "Naukova Dumka," Kiev, 1979.

A. D. Myshkis and I. M. Rabinovich, The Riga mathematician Pirs Bol′, "Zinatie," Riga, 1965.

I. A. Naumov, Dmitriĭ Matveevich Sintsov (outline of life and career in science and teaching), Khar′kov University Press, 1955.

E. P. Ozhigova, Egor Ivanovich Zolotarev, "Nauka," Moscow and Leningrad, 1966.

——, Aleksandr Nikolaevich Korkin, 1837–1908, "Nauka," Leningrad, 1968.

V. E. Prudnikov, Pafnutiĭ L′vovich Chebyshev, "Nauka," Leningrad, 1976.

T. V. Putyata, B. L. Laptev, B. A. Rozenfel′d, and B. N. Fradlin, Aleksandr Petrovich Kotel′nikov, 1865–1944, "Nauka," Moscow, 1968.

S. D. Rossinskiĭ, Boleslav Kornelievich Mlodzeevskiĭ, 1858–1923, Moscow State University Press, 1950.

A. S. Shibanov, Aleksandr Mikhaĭlovich Lyapunov, "Molodaya Gvardiya," Moscow, 1955.

A. L. Tsykalo, Aleksandr Mikhaĭlovich Lyapunov, "Nauka," Moscow, 1989.

I. N. Vekua, Academician Nikolaĭ Ivanovich Muskhelishvili, Press of the Academy of Sciences of the USSR, Novosibirsk, 1961.

V. S. Vladimirov and I. I. Markush, Vladimir Andreevich Steklov, scientist and organizer of science, "Nauka," Moscow, 1981.

V. Collected papers, selected work, separate editions, classical papers

P. S. Aleksandrov, Dimension theory and related problems, in Selected papers, "Nauka," Moscow, 1978.

——, Theory of functions of a real variable and the theory of topological spaces, in Selected papers, "Nauka," Moscow, 1978.

——, General homology theory, in Selected papers, "Nauka," Moscow, 1979.

S. N. Bernshteĭn, Collected works, vols. 1–4, Press of the Academy of Sciences of the USSR, Moscow, 1952–1964.

N. N. Bogolyubov, Selected papers, vols. 1–3, "Naukova Dumka," Kiev, 1969–1971.

P. G. Bol′, Selected papers, Press of the Latvian SSR, Riga, 1961.

Ts. Byurstin, Mathematical works, Press of the Academy of Sciences of the BSSR, Minsk, 1932.

S. A. Chaplygin, Complete collected works, vols. 1–3, Press of the Academy of Sciences of the USSR, Leningrad, 1933–1935.

——, Selected papers on mechanics and mathematics, "Gostekhizdat," Moscow, 1954.

N. G. Chebotarev, Collected works, vols. 1–3, Press of the Academy of Sciences of the USSR, Moscow and Leningrad, 1949–1950.

D. F. Egorov, Papers on differential geometry, "Nauka," Moscow, 1970.

A. O. Gel′fond, Selected papers, "Nauka," Moscow, 1973.

D. O. Grave, Selected papers, "Naukova Dumka," Kiev, 1971.

M. V. Keldysh, Mathematics: Selected papers, "Nauka," Moscow, 1985.

A. N. Kolgomorov, Selected papers: Mathematics and mechanics, "Nauka," Moscow, 1985.

——, Probability theory and mathematical statistics, "Nauka," Moscow, 1986.

——, Information theory and theory of algorithms, "Nauka," Moscow, 1987.

A. N. Krylov, Collected works, vols. 1–12, Press of the Academy of Sciences of the USSR, Moscow and Leningrad, 1951–1956.

N. M. Krylov, Selected papers, vols. 1–3, Press of the Academy of Sciences of the Ukrainian SSR, 1949–1961.

Yu. V. Linnik, Selected papers, vols. 1–2, "Nauka," Leningrad, 1979–1981.

——, Selected papers: Mathematical statistics, "Nauka," Leningrad, 1982.

N. N. Luzin, Integral and trigonometric series, "Gostekhizdat", Moscow and Leningrad, 1951.

——, Collected works, vols. 1–3, Press of the Academy of Sciences of the USSR, Moscow and Leningrad, 1953–1959.

A. I. Mal′tsev, Selected works, vols. 1–2, "Nauka," Moscow, 1976.

A. A. Markov, Selected papers on the theory of continued fractions, and the theory of functions deviating least from zero, "GTTI," Moscow and Leningrad, 1948.

——, Selected papers: Number theory, probability theory, Press of the Academy of Sciences of the USSR, Leningrad, 1951.

F. È. Molin, Number systems, "Nauka," Novosibirsk, 1985.

P. S. Novikov, Selected papers, "Nauka," Moscow, 1979.

I. G. Petrovskiĭ, Selected papers, "Nauka," Moscow, 1986.

L. S. Pontryagin, Selected scientific papers, vols. 1–3, "Nauka," Moscow, 1988.

A. M. Razmadze, Selected papers, Press of the Academy of Sciences of the Georgian SSR, Tbilisi, 1952.

V. I. Romanovskiĭ, Selected papers, vols. 1–2, Press of the Academy of Sciences of the Uzbek SSR, Tashkent, 1959–1964.

O. Yu. Shmidt, Selected papers: Mathematics, Press of the Academy of Sciences of the USSR, Moscow, 1959.

E. E. Slutskii, Selected papers: Theory of Probability, Mathematical Statistics, Press of the Academy of Sciences of the USSR, Moscow, 1960.

V. I. Smirnov, Selected papers, Leningrad University Press, Leningrad, 1988.

P. S. Uryson, Papers on topology and other branches of mathematics, vols. 1–2, "GITTL," Moscow, 1951.

N. A. Vasil′ev, Conceptual logic: Selected papers, "Nauka," Moscow, 1989.

B. A. Venkov, Collected papers: Research on number theory, "Nauka," Leningrad, 1981.

I. M. Vinogradov, Selected papers, Press of the Academy of Sciences of the USSR, Moscow, 1952.

N. E. Zhukovskiĭ, Complete collected works, vols. 1–9, "ONTI," Moscow and Leningrad, 1935–1938.

——, Collected works, vols. 1–7, "Gostekhizdat," Moscow and Leningrad, 1948–1950.

Translated by R. BOAS

Библиография

S. S. DEMIDOV (MOSCOW)

I. *Общие сочинения по истории математики в России и в СССР*

Александров П. С., Выгодский М. Я., Гливенко В. И., ред. 1932. Математика в СССР за 15 лет. М.: Гостехиздат.

Боголюбов А. Н. 1983. Математики и механики. Биографический справочник. Киев: Наукова думка.

Бородин А. И., Бугай А. С. 1979. Биографический словарь деятелей в области математики. Киев: Радянська школа.

Гнеденко Б. В. 1946. Очерки по истории математики в России. М.-Л.: Гостехиздат.

Егоров Д. Ф. Успехи математики в СССР//Наука и техника в СССР (1917–1927). Т.1.М.: Работник. просвещения. С. 223–234.

Курош А. Г., Ред. 1959. Математика в СССР за сорок лет. 1917–1957. Т.1–2. М.: Физматгиз.

Курош А. Г., Маркушевич А. И., Рашевский П. К., ред. 1948. Математика в СССР за тридцать лет. 1917–1947. М.-Л.: Гостехиздат.

Микулинский С. Р., Юшкевич А. П. ред. 1977. Развитие естествознания в России (XVIII–начало XX века). М.: Наука.

Фомин С. В., Шилов Г. Е., ред. 1969–1970. Математика в СССР. 1958–1967. Т.2. Вып. 1–2.

Штокало И. З., Боголюбов А. Н., Юшкевич А. П., ред. 1966–1970. История отечественной математики. Т. 1, 2, 3, 4 (кн. 1), 4 (кн. 2). Киев: Наукова Думка.

Штокало И. З., Боголюбов А. Н., Юшкевич А. П., ред. 1983. Очерки развития математики в СССР. Киев: Наукова Думка.

Юшкевич А. П. 1968. История математики в России до 1917 г. М.: Наука.

II. *Математика в России конца* XIX*–начала XX* века

Башмакова И. Г., 1949. Обоснование теории делимости в трудах Е. И. Золотарева//ИМИ. В. 2. С. 233–351.

Бернштейн С. Н. 1940. Петербургская школа теории вероятностей// Уч. записки ЛГУ. № 55. В. 10. С. 3–11.

Выгодский М. Я. 1948. Математика и ее деятели в Московском университете во второй половине XIX века//ИМИ. В. *1*. С. 141–188.

Галченкова Р. И. 1961. Математика в Ленинградском (Петербургском) университете в XIX в.//ИМИ. В. 14. С. 355–392.

Гнеденко Б. В. 1948. Развитие теории вероятностей в России//Труды ИИЕ АН СССР. Т. 2. С. 390–425.

Гуссов В. В. 1952. Работы русских ученых по теории гамма-функций// ИМИ. В. 5. С. 421–472.

Гуссов В. В. 1953. Развитие теории цилиндрических функций в России и СССР// ИМИ. В. *6.* С. 355–475.

Гюнтер Н. М. 1928. О научных достижениях В. А. Стеклова// Памяти В. А. Стеклова. Л.: Изд-во АН СССР.

Делоне Б. Н. 1947. Петербургская школа теории чисел. М.-Л.: Изд-во АН СССР.

Демидов С. С. 1985. Н. В. Бугаев и возникновение московской школы теории функций действительного переменного//ИМИ. В. 29. С. 113–124.

Демидов С. С. 1986. Из ранней истории Московской школы теории функций// ИМИ. В. 30. С. 124–130.

Депман И. Я 1952. Карл Михайлович Петерсон и его кандидацкая диссертация// ИМИ. В. 5. С. 134–164.

Депман И. Я. 1953. В. А. Стеклов в Петербургском университете// ИМИ. В. 6. С. 509–528.

Депман И. Я. 1954. К биографиии С. В. Ковалевской//ИМИ. В. 7. С. 713–715.

Депман И. Я. 1955. Из истории математики в Дерптском (Юрьевском) университете//Ученые записки Ленинградского педагогичекого института им. М. Н. Покровского. Физ.-мат. ф-т. Т. 14. В. 1. С. 128–137.

Депман И. Я. 1960. С.-Петербургское математическое общество//ИМИ. В. 13. С. 11–106.

Киро С. Н. 1956. Математика в перюодических изданиях Одесского (Новороссийского) университета. 1865–1955. Часть первая//Труды Одесского ун-та. Сер. мат. Т. 146. В. 6. С. 89–109.

Кіро С. М. 1961. Математика в Новоросійському *(*Одеському*)* університеті// Історико-математичний збірник. В. 2. С. 22–42.

Колмогоров А. Н. 1947. Роль русской науки в развитии теории вероятностей//Ученые записки МГУ. Т. 91. С. 53–64.

Кравчук М. Ф. 1935. Математика та математики в Київському университеті за сто років (1834–1934)// Розвиток науки в Київському університеті за сто років. 1835–1935. Києв: Вид. Київськ. ун-ту. С. 34–69.

Лейбман Э. Б. 1961. Математическое отделение Новороссийского общества естествоиспытателей (1876–1928)//ИМИ. В. 14. С. 393–440.

Майстров Л. Е. 1961. Первый арифмометр П. Л. Чебышева//ИМИ. В. 14. С. 349–354.

Майстров Л. Е. 1973. Об оценке арифмометра П. Л. Чебышева//ИМИ. В. 18. С. 295–300.

Майстров Л. Е. 1975. О сохранившихся механизмах П. Л. Чебышева// ИМИ. В. 20. С. 309–318.

Маркуш И. И. 1974. К вопросу о создании Петербургской-Ленинградской школы математической физики В. А. Стеклова//ИМЕН. В. 16. С. 141–153.

Маркушевич А. И. 1950. Вклад Ю. В. Сохоцкого в общую теорию аналитических функций//ИМИ. В. 3. С. 399–406.

Марчевский М. Н. 1956. История математических кафедр в Харьковском университете за 150 лет его существования// Записки математического общества. Харьков. Т. 24. В. 4. С. 70–79.

Марчевский М. Н. 1956. Харьковское математическое общество за первые 75 лет его существования (1879–1954)// ИМИ. В. 9. С. 613–666.

Медведев Ф. А. 1961. Вклад А.М. Ляпунова в теорию интеграла Стилтьеса// ИМИ. В. 14. С. 211–236.

Медведев Ф. А. 1963. Подготовка теоретико-множественных и теоретико-функциональных исследований в России//Очерки истории математики и механики. М.: Изд-во АН СССР. С. 45–66.

Медведев Ф. А. 1986. О курсе лекций Б. К. Млодзеевского по теории функций действительного переменного, прочитанных осенью 1902 г. в Московском университете//ИМИ. В. 30. С. 130–147.

Михайлов Г. К., Степанов С. Я. 1985. К истории задачи о вращении твердого тела вокруг неподвижной точки в случаях Гесса и Ковалевской их геометрического моделирования//ИМИ. В. 28. С. 223–246.

Мордухай-Болтовской Д. Д. 1941. Математика в Ростовском университете// Ростовский-на-Дону университет. Юбилейный сборник. 1915–1940. Ростов-на-Дону: Ростовиздат. С. 46–52.

Налбандян М. Б. 1965. Теория эллиптических функций и ее приложения в трудах Е. М. Золотарева//ИМИ. В. 16. С. 191–206.

Налбандян М. Б. 1966. Теория зллиптических функций и ее приложения в трудах русских математиков XIX и начала XX вв.//ИМИ. В. 17. С. 361–370.

Ожигова Е. П. 1972. Развитие теории чисел в России. Л.: Наука.

Половинкин С. М. 1986. О студенческом математическом кружке при Московском математическом обществе в 1902–1903 гг.// ИМИ. В. 30. С. 148–158.

Полубаринова-Кочина П. Я. 1954. К биографии С. В. Ковалевской (по материалам её переписки)//ИМИ. В. 7. С. 666–712.

Рыбников К. А. 1950. Виктор Викторович Бобынин//ИМИ. В. 3. С. 343–357.

Синцов Д. М. 1908. Кафедра математики чистой и прикладной в Харьковском университете за 100 лет его существования (1805–1905). Харьков.

Синцов Д. М. 1936. Харьковское математическое общество за 50 лет// Труды первого Всесоюзного съезда математиков. М.-Л.: ОНТИ. С. 97–105.

Стяжкин Н. И., Силаков В. Д. 1962. Краткий очерк истории общей и математической логики в России. М.: Высшая школа.

Сушкевич А. К. 1951. Материалы к истории алгебры в России в 19 в. и в начале 20 в.//ИМИ. В. 4. С. 237–451.

Сушкевич А. Е. 1956. Диссертации по математике в Харьковском университете за 1805–1917 годы//Записки Харьковского математического общества. Т. 24. С. 91–115.

Франкль Ф. И. 1951. О работах русских математиков XIX в. по теории характеристик уравнений в частных производыых// УМН. Т. 6. В. 2 (42). С. 154–156.

Хинчин А. Я. 1934. Теория вероятностей в дореволюционной России и Советском Союзе//Фронт науки и техники. № 7. С. 36–46.

Шевелев Ф. Я. 1966. К истории Московского математического общества// ИМЕН. В. 5. С. 62–74.

Ш. *Математика в СССР*

Александров П. С. 1955. Математика в Московском университете в первой половине XX века//ИМИ. В. 8. С. 9–54.

Александров П. С. 1970. Топология в Московском университете//ИМЕН. В. 9. С. 3–7.

Александров П. С. 1977. Воспоминания о Геттингене//ИМИ. В. 22. С. 242–245.

Александров П. С. 1979–1980. Страницы автобиографии//УМН. Т. 34. В. 6. С. 219–249; Т. 35. В. 3. С. 241–278.

Александров П. С. 1980. Лузинская математическая школа//Математическая наука в МГУ - в изд. Новое в жизни, науке и технике. Сер. - Математика, кибернетика. 1980. № 4. М.: Знание. С. 21–29.

Александров П. С. , Гнеденко Б. В. и Степанов В. В. 1948. Математика в Московском университете в XX в.//ИМИ. Вып. 1. С. 9–42.

Александров П. С., Головин О. Н. 1957. Московское математическое общество. К 90-летию научной деятельности//УМН Т. 12. *В.* 6 (78). С. 9–46.

Андрунакиевич В. А., Бычков В. П. 1965. Математическая жизнь в Молдавской ССР// УМН. Т. 20. В. 2 (122). С. 247–258.

Арболеда Л. К. 1980. Рождение совецкой топологической школы. Замечания о письмах П. С. Александрова и П. С. Урысона Морису Фреше// ИМИ. В. 25. С. 281–302.

Аржаных И. С. 1953. К вопросу о развитии математики и механики в Узбекистане// Изв. АН УзССР. № 6. С. 102–105.

Ахиезер Н. И. 1956. Харьковское математическое общество//Записки Харьковского математического общества. Т. 24. В. 4. С. 31–39.

Ахундов А. И., Каррыев А. А. 1964. Из истории математики в Советском Turkmenistane. Ashhabad: Изд. Туркм. политехн.ин-та.

Барон С. А., Габович Я. А., Каазик Ю. Я. и др. 1964. Математика в Совецкой Эстонии за последние двадцать лет//Ученые записки Тартусского университета. Труды по матем. и мех. В. 150. Т. 4. С. 12–52.

Башмакова И. Г., Петрова С. С. 1980. История математики в Московском университете//Математическая наука в МГУ - в изд. Новое в жизни, науке и технике. Сер. - Математика, кибернетика. 1980. № 4. М.: Знание. С. 45–63.

Боголюбов А. Н., Урбанский В. М. 1985. Организация и становление математических учреждений АН УССР//ИМИ. В. 28. С. 160–187.

Бычков В. П., Сибирский К. С. 1974. Развитие математики в Молдавской ССР: Краткий очерк. Кишинев: Штиинца.

Валле-Пуссен Ш. Ж де ла. 1983. Письма к Н. Н. Лузину. (Перевод и публикация Ф. А. Медведева)// ИМИ. В. 27. С. 301–312.

Гагаев Б. М. 1947. Развитие математики в Казани за 25 лет// Ученые записки Казанского ун-та. Т. 106. Кн. 2. С. 3–13.

Гагаев Б. М., Морозов В. В., Норден А. П. 1947. Казанская математическая школа за 30 лет//УМН. Т. 2. В. 6(22). С. 8–20.

Гельфонд А. О. 1977. Некоторые впечатления о научной поездке в Германию в 1930 г.// ИМИ. В. 22. С. 246–251.

Гнеденко Б. В. 1950. Исследования по теории вероятностей и математической статистике в системе АН УССР// Украинский математический журнал. Т. 11 № 2. С. 123–137.

Гнеденко Б. В., Гихман И. И. 1956. Развитие теории вероятностей на Украине// ИМИ. В. 9. С. 477–536.

Гнеденко Б. В., Погребысский И. Б. 1956. О развитии математики на Украине//ИМИ. В. 9. С. 403–426.

Гусейнов А. И., Агаев Г. Н. 1964. К истории развития математических исследований в Азербайджане//Известия АН АзССР. Сер. физ.-техн. и мат. наук. № 3. С. 3–17.

Делоне Б. Н., Кудрявцев Л. Д., Постников М. М. 1965. Очерк истории развития математики в Академии наук СССР за советский период (1917–

1960)//Очерки истории математики и механики. М.: Изд-во АН СССР. С. 3–44.

Дорофеева А. В. 1985. О письмах Д. Ф. Егорова к Д. Гильберту//ИМИ. В. 28. С. 270–278.

Егоров Д. Ф. 1928. Работа научно-исследовательского института математики и механики за пятилетие с 1923 по 1928 г.//Изв. Ассоц. научно-исслед. ин-тов при физ.-мат. фак-те МГУ. Т. 1. В. 3–4. С. 301–303.

Егоров Д. Ф. 1980. Письма к Н. Н. Лузину. (Предисловие П. С. Александрова. Публикация и примечания Ф. А. Медведева при участии А. П. Юшкевича)// ИМИ. В. 25. С. 335–361.

Егоров Д. Ф. 1985. Письма к Д. Гильберту. (Перевод и примечания А. В. Дорофеевой)//ИМИ. В. 28. С. 266–270.

Жаутыков О. А. 1958. Математика в Казахстане за совецкие годы//Труды сектора матем. и мех. АН КазССР. Т. 1. С. 5–24.

Жаутыков О. А., Пентковский М. В. 1960. Вопросы развития математической науки в Казахстане//Вестник АН КазССР. № 8. С. 9–19.

Жуковский Н. Е., Лахтин Л. Д., Егоров Д. Ф. 1955. Доклад факультетскойкомиссии ... //ИМИ. В. 8. С. 70–76.

Колмогоров А. Н. 1986. Воспоминания о П. С. Александрове//УМН. Т. 41. В. 6. С. 187–203.

Кочина П. Я. 1974. Воспоминания. М.: Наука.

Кравчук М. Ф. 1989. Отчет о занятиях стипендиата Михаила Кравчука с ноября 1915 года по ноябрь 1916 год. Отзыв заслуженного профессора Д. Граве на отчет проф. стипендиата М. Ф. Кравчука за 1916 год. Публикация и комментарии Л. В. Ковальчук//Очерки истории естествознания и техники. В. 37. Киев: Наукова думка. 1989. С. 86–93.

Круликовский Н. Н. 1967. История развития математики в Томске. Томск: Изд-во Томского ГУ.

Крылов А. Н. 1945. Мои воспоминания. М.: Изд-во АН СССР.

Кузнецов П. И. 1971. Дмитрий Федорович Егоров // УМН. Т. 26. В. 5. С. 169–210.

Лапко А. Ф., Люстерник Л. А. Математические съезды и конференции в СССР//УМН, 1957. Т. 12. В. 6(78). С. 47–130; 1958. Т. 13. В. 5(83). С. 121–166.

Лапко А. Ф., Люстерник Л. А. 1967. Из истории совецкой математики// УМН, 1967. Т. 22. В. 6(138). С. 13–140.

Лаптев Б. Л. 1959. Математика в Казанском университете за 40 лет (1917–1957)//ИМИ. В. 12. С. 11–58.

Лебедев Н. А., Лозинский С. М. 1967. Владимир Иванович Смирнов и математика в Ленинградском университете за 1917–1967 годы// Вестник Ленинградского ун-та. Мат., мех., астр. № 7. В. 2. С. 7–18.

Лузин Н. Н. 1955. Отчет о заграничной командировке для научных занятий ... //ИМИ. В. 8. С. 57–70.

Лузин Н. Н. 1965. Предисловие к письмам Л. Эйлера к Х. Гольдбаху// ИМИ. В. 16. С. 129–144.

Лузин Н. Н. 1978. Письма к А. Данжуа. (Публикация П. Дюгака)//ИМИ. В. 23. С. 314–348.

Лузин Н. Н. 1983. Письмо к М. Фреше. (Публикация А. П. Юшкевича)// ИМИ. В. 27. С. 298–300.

Лузин Н. Н. 1985. Отзыв о научных работах профессора Сергея Павловича Финикова//ИМИ. В. 29. С. 293–316.

Лузин Н. Н. 1985. Отзыв о научных работах С. П. Финикова последнего периода 1938–1946 гг. //ИМИ. В. 29. С. 316–318.

Лузин Н. Н. 1985. Письмо к О. Ю. Шмидту. (Публикация С. С. Демидова)//*ИМИ.* В. 28. С. 278–287.

Лузин Н. Н. 1986. [О рестрикторах]. (Публикация и примечания С. С. Демидова)//ИМИ. В. 30. С. 177–181.

Лузин Н. Н., Крылов А. Н. 1989. Переписка. Публикация, предисловие и комментарии Н. С. Ермолаевой//ИМИ. В. 31. С. 203–272.

Лузин Н. Н., Флоренский П. А. 1989. Переписка. Публикация и примечания С. С. Демидова, А. Н. Паршина, С. М. Половинкина и П. В. Флоренского//ИМИ. В. 31. С. 125–190.

Лузин Н. Н., Чаплыгин С. А. 1985. Научные работы Сергея Павловича Финикова//ИМИ. В. 29. С. 319–321.

Лусис А. Я. 1950. Работы математиков Совецкой Латвии за десять лет// Известия АН Латвийскоую ССР. № 11. С. 109–121.

Лусис А. Я. 1958. Развитие математики в Советской Латвии за последнее десятилетие//Ученые записки Латвийского университета. Физ.-мат. науки. Т. 20. В. 3. С. 5–20.

Лусис А. Я., Рейзинь Л. Э., Риекстыньш Э. Я. 1966. Математика в Совецкой Латвии//УМН. Т. 21. В. 2. С. 248–259.

Люстерник Л. А. 1967. Молодость московской математической школы// УМН. Т. 22. В. 1(133). С. 137–161; В. 2(134). С. 199–239; В. 4(136). С. 147–185.

Магнарадзе Л. Г., Чогошвили Г. С. 1957. Развитие матемаических наук в Грузии// Сорокалетие Октябрьской революции. Тбилиси. С. 485–611.

Меньшов Д. Е. 1965. Московское математическое общество в период 1910–1920//УМН. Т. 20. № 3. С. 19–20.

Меньшов Д. Е. 1983. Воспоминания о молодых годах и о возникновении Московской школы теории функций// ИМИ. В. 27. С. 312–333.

Наумов И. А. 1966. Работы математиков Украины по неголономной дифференциальной геометрии//ИМЕН. В. 5. С. 105–117.

Ожигова Е. П. 1966. Математика в Академии наук в первые годы Советской власти ИМИ. В. 17. С. 381–390.

Романовский В. И. 1942. Математика и механика// 25 лет советской науки в Узбекистане. Ташкент: Изд. Уз ФАН СССР. 1942. С. 60–63.

Сарымсаков Т. А. 1942. Теория вероятностей и математическая статистика в работах ташкентской школы// 25 лет совецкой науки в Узбекистане. Ташкент.: Изд-во Узбекского Филиала АН СССР. С. 64–72.

Серпинский В. 1979. Письма к Н. Н. Лузину. (Публикация В. А. Волкова и Ф. А. Медведева)//ИМИ. В. 24. С. 366–373.

Сиражединов С. Х. 1967. Итоги исследований в области теории вероыатностей и матемаической статистики в Узбекистанея 50 лет//Изв. АН УзССР. Сер. физ.-мат. наук. № 5. С. 21–26.

Смирнов В. И., Юшкевич А. П. 1964, Математика// История Академии наук СССР. Т. 2. М.: Изд-во АН СССР. С. 34–51, 286–306, 473–483.

Статулявичус В. А. 1965. О работах в области математики в Литовской ССР//Литовск. мат. об. Т. 5. № 3. С. 361–372.

Степанов В. В. 1947. Московская школа теории функций//Учёные записки МГУ. Т. 91 С. 47–52.

Урбанский В. М. 1989. В. И. Вернадский и развитие математической мысли в УАН//Очерки истории естествознания и техники. Киев: Наукова думка. В. 36. С. 21–29.

Халилов З. И. 1950. Развитие математических наук в Азербайджане//Известия АН Азербайджанской ССР. № 7. С. 61–76.

Халилов З. И. 1957. Развитие физико-математических наук в Советском Азербайджане//Изб. АН Азербайджанской ССР. № 10. С. 25–28.

Чеботарев Н. Г. 1941. Работы казанских математиков в послереволюционный период (1917–1939)//Ученые записки Казанского ун-та. Т. 101. Кн. 1. С. 57–78.

Чеботарев Н. Г. 1961. Несколько приложений теории идеалов к алгебре. (Публикация и примечания В. А. Добровольского)//ИМИ. В. 14. С. 539–550.

Шилов Г. Е. 1956. К истории развития функционального анализа на Украине// ИМИ. В. 9. С. 427–476.

Штокало И. З. 1958. Норис розвитку математики на Украіни за 40 років Радянськоі влади/ Киев: Вид-во АН УРСР.

Эфендиев М. Р., Джавадов М. А. 1967. К истории развития высшего математическоро образования и математики в Советском Азербайджане//Уч. зап. Азерб. ун-та. Сер. физ.-мат. наук. № 3. с. 3–15.

Юшкевич А. П. 1976. Из студенческих воспоминаний// Математика: Сб. научно-методических статей. М.: Высшая школа. № 6. С. 99–112.

Юшкевич А. П. 1985. Л. Г. Шнирельман в Геттингене//ИМИ. В. 28. С. 287–290.

Юшкевич А. П. 1989. "Дело академика Н. Н. Лузина"//Вестник АН СССР. № 4. С. 102–113.

Юшкевич А. П., 1989. К истории научных связей между математиками СССР и Франции (об избрании в Парижскую академию наук С. Н. Бернштейна, И. М. Виноградова и М. А. Лаврентьева)//ИМИ. В. 31. С. 203–272.

IV. *Научные биографии*

Александров П. С., Немыцкий В. В. 1956. Вячеслав Васильевич Степанов. М.: Изд-во Московского ун-та.

Бажанов В. А. 1988. Николай Александрович Васильев. М.: Наука.

Бахвалов С. В. 1961. Нил Александрович Глаголев. М.: Изд-во Московского университета.

Боголюбов А. Н. 1973. Георгий Николаевич Хиколадзе. М.: Наука.

Боголюбов А. Н., Урбанский В. М. 1987. Николай Митрофанович Крылов. Киев: Наукова Думка.

Векуа И. Н. 1961. Академик Николай Иванович Мусхелишвили. Новосибирск: Изд-во Академии наук СССР.

Владимиров В. С., Маркуш И. И. 1981. Владимир Андреевич Стеклов—ученый и организатор науки. М.: Наука.

Голубев В. В. 1947. Николай Егорович Жуковский. М.: Изд-во Московского ун-та.

Голубев В. В. 1951. Сергей Александрович Чаплыгин. 1869–1942. М.: Изд-во МГУ.

Гордевский Д. З. 1955. К. А. Андреев—выдающийся русский геометр. Харьков: Изд-во Харьковского ун-та.

Гродзенский С. Я. 1987. Андрей Андреевич Марков. М.: Наука.

Добровольский В. А. 1981. Василий Петрович Ермаков. М.: Наука.

Добровольский В. А. 1968. Дмитрий Александрович Граве. 1863–1939. М.: Наука.

Канунов Н. Ф. 1983. Федор Эдуардович Молин. М.: Наука.

Космодемьянский А. А. 1984. Николай Егорович Жуковский. М.: Наука.

Кочина П. Я. 1979. Николай Евграфович Кочин. М.: Наука.

Кочина П. Я. 1981. Софья Васильевна Ковалевская. М.: Наука.

Кропотов А. И. 1967. Николай Яковлевич Сонин. 1849–1915. Л.: Наука.

Лаптев Б. Л. 1955. Петр Алексеевич Широков// Петр Алексеевич Широков (1895–1944). Био-библиографический указатель. Казань: Изд-во Казанского ун-та. С. 3–20.

Лопшиц А. М., Рашевский П. К. 1969. Вениамин Федорович Каган. М.: Изд. Московского ун-та.

Митропольский Ю. О., Боголюбов О. М. 1979. Микола Митрофанович Крилов. Київ: Наукова думка.

Мышкис А. Д., Рабинович И. М. 1965. Математик Пирс Боль из Риги. Рига: Зинатне.

Наумов И. А. 1955. Дмитрий Матвеевич Синцов (очерк жизни и научно-педагогической деятельности). Харьков: Изд-во Харьковского ун-та.

Ожигова Е. П. 1966. Егор Иванович Золотарев. М.-Л.: Наука.

Ожигова Е. П. 1968. Александр Николаевич Коркин. 1837–1908. Л.: Наука.

Прудников В. Е. 1976. Пафнутий Львович Чебышев. Л.: Нуака.

Путята Т. В., Лаптев Б. Л., Розенфельд Б. А., Фрадлин Б. Н. 1968. Александр Петрович Котельников. 1865–1944. М.: Наука.

Россинский С. Д. 1950. Болеслав Корнелиевич Млодзеевский. 1858–1923. М.: Изд-во МГУ.

Фрадлин Б. Н. 1984. Юрий Дмитриевич Соколов. М.: Наука.

Цыкало А. Л. 1989. Александр Михайлович Ляпунов. М.: Наука.

Шибанов А. С. 1955. Александр Михайлович Ляпунов. М.: Молодая гвардия.

V. *Собрания сочинений, избранные труды, отдельные переиздания классических сочинений*

Александров П. С. 1978. Теория размерности и смежные вопросы. Избранные труды. М.: Наука.

Александров П. С. 1978. Теория функций действительного переменного и теория топологических пространств. Избранные труды. М.: Наука.

Александров П. С. 1979. Общая теория гомологий. Избранные труды. М.: Наука.

Бернштейн С. Н. 1952–1964. Собрание сочинений. Т. 1–4. М.: Изд-во АН СССР.

Боголюбов Н. Н. 1969–1971. Избранные труды. Т. 1–3. Киев: Наукова думка.

Боль П. Г. 1961. Избранные труды. Рига: Изд-во Латвийской ССР.

Бурстын Ц. 1932. Матэматичныя працы. Менск: Выд-во АН БССР.

Васильев Н. А. 1989. Воображаемая логика. Избранные труды. М.: Наука.

Венков Б. А. 1981. Избранные труды. Исследования по теории чисел. Л.: Наука.

Виноградов И. М. 1952. Избранные труды. М.: Изд-во АН СССР.

Гельфонд А. О. 1973. Избранные труды. М.: Наука.

Граве Д. О. 1971. Вибрані праці Киї в: Наукова думка.

Егоров Д. Ф. 1970. Работы по дифференциальной геометрии. М.: Наука.

Жуковский Н. Е. 1935–1938. Полное собрание сочинений. Т. 1–9. М.-Л.: ОНТИ.

Жуковский Н. Е. 1948–1950. Собрание сочинений. М.-Л.: Гостехиздат.

Келдыш М. В. 1985. Математика. Избранные труды. М.: Наука.

Колмогоров А. Н. 1985. Избранные труды. Математика и механика. М.: Наука.

Колмогоров А. Н. 1986. Теория вероятностей и математическая статистика. М.: Наука.

Колмогоров А. Н. 1987. Теория информации и теория алгоритмов. М.: Наука.

Крылов А. Н. 1951–1956. Собрание трудов. Т. 1–12. М.-Л.; Изд-во АН СССР.

Крылов Н. М. 1949–1961. Избранные труды. Т. 1–3. Киев: Изд-во АН УССР.

Линник Ю. В. 1979–1981. Избранные труды. Т. 1–2. Л.: Наука.

Линник Ю. В. 1982. Избранные труды. Математическая статистика. Л.: Наука.

Лузин Н. Н. 1951. Интеграл и тригонометрический ряд. М.-Л.: Гостехиздат.

Лузин Н. Н. 1953–1959. Собрание сочинений. Т. 1–3. М.-Л.: Изд-во АН СССР.

Мальцев А. И. 1976. Избранные труды. Т. 1–2. М.: Наука.

Марков А. А. 1948. Избранные труды по теории непрерывных дробей и теории функций, наименее уклоняющихся от нуля. М.-Л.: ГТТИ.

Марков А. А. 1951. Избранные труды. Теория чисел. Теория вероятностей. Л.: Изд-во АН СССР.

Молин Ф. Э. 1985. Числовые системы. Новосибирск: Наука.

Новиков П. С. 1979. Избранные труды. М.: Наука.

Петровский И. Г. 1986. Избранные труды. М.: Наука.

Понтрягин Л. С. 1988. Избранные научные труды. Т. 1–3. М.: Наука.

Размадзе А. М. 1952. Избранные труды. Тбилиси: Изд-во АН Грузинской ССР.

Романовский В. И. 1959–1964. Избранные труды. Т. 1–2. Ташкент: Изд-во АН Узбекской ССР.

Слуцкий Е. Е. 1960. Избранные труды. Теория вероятностей. Математическая статистика. М.: Изд-во АН СССР.

Смирнов В. И. 1988. Избранные труды. Л.: Изд-во ЛГУ.

Урысон П. С. 1951. Труды по топологии и другим областям математики. Т. 1–2. М.: ГИТТЛ.

Чаплыгин С. А. 1933–1935. Полное собрание сочинений. Т. 1–3. Л.: Изд-во АН СССР.

Чаплыгин С. А. 1954. Избранные труды по механике и математике. М.: Гостехиздат.

Чеботарев Н. Г. 1949–1950. Собрание сочинений. Т. 1–3. М.-Л.: Изд-во АН СССР.

Шмидт О. Ю. 1959. Избранные труды. Математика. М.: Изд-во АН СССР.

On Moscow Mathematics — Then and Now

V. M. TIKHOMIROV

Once, when giving a talk at the Moscow Mathematical Society (MMS), I compared the Mathematics Department of Moscow University (Mekh-Mat) with the Pushkin Lyceum[1] and the Leningrad Fiz-tekh.[2] In the history of education in our country, only the Pushkin Lyceum, the Leningrad Fiz-tekh of the 1920's, the Physical-mathematical Department in Moscow of the 1920's–1930's and Mekh-Mat of the 1950's–1960's can be placed on the same level according to the number of outstanding individuals who graduated from these institutions. This is the result of the particular spiritual atmosphere that was characteristic for the history of each of these educational establishments. Such an atmosphere was and still is considered by the majority of those who witnessed it as a miracle, as a particular gift of Fate. And that, in and of itself, deserves posterity's grateful remembrance.

But if Pushkin's incomparable light illuminates the whole historical space of his lifetime, things are much more complicated in the other two cases. In particular, Mekh-Mat's history can sink into oblivion if one does not take the trouble to write a truthful chronicle of 20th-century Moscow mathematics. Therefore, the publication and re-publication of *Golden Years of Moscow Mathematics* is priceless. In this article I want to describe briefly the landmarks of the development of the Moscow mathematical school against the background of the historical events of the 20th century in Russia.

First of all, a few words about this 20th century. A factor contributing to the birth of the Moscow mathematical school, which was formed within Moscow University, was the MMS. One of the organizers of the Society was Nikolai Dmitrievich Brashman (1796–1866), who came from Austro-Hungary to Russia, and remained there for the rest of his life. Brashman was one of the initiators for creating a publication of the MMS, the journal "Matematicheskii Sbornik".

The successive presidents of the MMS were mathematicians who played a major role in the history of mathematics in Russia: N. D. Brashman (1864–1866), A. Yu. Davydov (1866–1882), V. Ya. Tsinger (1889–1898), N. A. Bugaev (1898–1903), P. A. Nekrasov (1903–1905), N. E. Zhukovskii (1905–1921), and D. F. Egorov (1921–1931). Then P. S. Aleksandrov was president for many years, followed by A. N. Kolmogorov, I. M. Gelfand, I. R. Shafarevich, S. P. Novikov and V. I. Arnold, who is the current president.

---

[1]The name given nowadays to the 19th-century privileged school establishment known for its political liberalism and for several remarkable graduates, including the iconic poet Pushkin.

[2]Physical-technical Institute, headed by A. F. Ioffe, who raised a whole Pleiad of outstanding physicists, starting with the Nobel Prize winners P. L. Kapitsa and N. N. Semenov.

At the beginning of the 20th century, a rather narrow mathematical direction in differential geometry, surface deformations, was being developed in Moscow, cultivated by Karl Mikhailovich Peterson, a graduate of Derpt University and a student of F. F. Minding. At Moscow University, V. Ya. Tsinger, B. K. Mlodzievskii and others worked on this circle of problems. The greatest Moscow mathematician of this period—Dmitrii Fedorovich Egorov (1869–1931)—was also interested in differential geometry. Another active area in Moscow was classical mechanics. Brashman started its development, whereas Nikolai Egorovich Zhukovskii played a major role in its coming into being at Moscow University and the country in general.

At the beginning of the 1910's, there was only one scientific seminar at Moscow University, the Egorov seminar.

And it is against this background that the new mathematical school was born in the middle of the second decade of the 20th century. The phenomenon is striking.

The Moscow scientific community sharply changes its interests, and follows the road outlined by the French mathematicians E. Borel, R. Baire and H. Lebesgue. In the course of some seven years, a whole Pleiad of prominent researchers arises: P. S. Aleksandrov, N. K. Bari, A. N. Kolmogorov, M. A. Lavrentev, L. A. Lyusternik, D. E. Menshov, L. V. Keldysh, P. S. Novikov, I. G. Petrovskii, M. Ya. Suslin, P. S. Uryson, A. Ya. Khinchin, and L. G. Shnirelman. Some mathematicians from the older generation are also active: V. V. Golubev, I. I. Privalov, V. V. Stepanov, S. P. Finnikov, and others. Thus began the first phase of the golden period of Moscow mathematics. In the 1930's, the number of seminars rose to dozens (while in the 1950's they numbered more than one hundred!).

All of the above-mentioned young mathematicians later chose their own path, and began acquiring their own students: suffice it to name I. M. Gelfand, A. I. Maltsev and S. M. Nikolskii—students of Kolmogorov; L. S. Pontryagin and A. N. Tikhonov—students of P. S. Aleksandrov; and in the 1930's, the mathematical school of Moscow University (a school formed within the confines of a single university!) became comparable with the greatest mathematical schools in the *world*. It would not be a great exaggeration to say that it assumed a leading position in the whole mathematical world. Such a flowering of outstanding mathematicians was unique; the German mathematical school was devastated by fascism, the French was experiencing a change of generations, and the American was at the stage of initial development.

How can one explain such an unprecedented phenomenon? There were global reasons, as well as reasons connected with the activities of particular individuals. Both are worthy of particular analysis.

The revolution of February 1917 was embraced by many representatives of the Russian intelligentsia. The nobility's privileges were abolished immediately, as well as the laws and regulations affirming national (ethnic) and religious inequalities. A huge mass of people were called to active work, people who had hitherto been cut off from it. In the hope that the ideals of freedom, equality and fairness had been achieved, people were inspired and propelled toward creativity, in particular scientific creativity. This enthusiasm, this creativity, this passion remained for a while even after the October revolution.

Altogether other realities turned out to be connected with the October revolution. A totalitarian regime ensued, which was not conducive to the development of the humanities and to a series of branches of the natural sciences. Many creative young people started to choose physics, mathematics and technical professions. The Soviet regime of the 1930's contributed to the development of these fields. To be a scientist-mathematician was prestigious.

The social conditions in the Soviet Union outside of the main centers were difficult, and this contributed to the unprecedented concentration of the creative intelligentsia in several cities, and particularly in Moscow. Such were the social peculiarities in our country, peculiarities that promoted the development of the Moscow mathematical school.

A second reason for the take-off of Moscow mathematics is connected with two outstanding personalities: Egorov and his student Nikolai Nikolaevich Luzin.

Egorov was born in 1869 in Moscow. He graduated from Moscow University in 1891 and began teaching in 1893 at the University, becoming a Professor in 1903. As mentioned before, he was the president of the MMS from 1921 until 1931 (in 1930, he was arrested, but remained president of the Society). Egorov was a deeply religious man and one of exceptionally high moral principles. There is no evidence of his ever acting against his conscience; he always stood firmly by his opinions. I will present a few stories I heard personally.

In these years, rooms were scarce, and several institutions of higher learning held classes in vacant churches. Egorov, who also taught outside of Moscow University, refused to teach in God's temples. And, as strange as it may seem, the authorities did not insist.

Egorov realized full well the tragic consequences of a totalitarian ideology, and tried with all his might to resist. At the end of the 1920's, he was fiercely criticized for not submitting to the new regime. At one public meeting he was reproached for being a reactionary and a freedom suppressor. Egorov courageously countered to his accusers: "You are the suppressors of freedom, the freedom of speech, that is!"

Once he was called, among some others, to a meeting for re-educating the old professors by the revolutionary youth. All "re-educatees" swore allegiance to the new authority. Finally, it was Egorov's turn. He was asked, "What are your political views?" He answered, "I am not sure this audience will understand me, but I deem it unworthy of myself to conceal my thoughts. I am a proponent of constitutional monarchy." One can only imagine the reaction of the audience, triumphant from the abolition of Tsarism and ready for the World Revolution. The behaviour of Egorov was completely unprecedented, and the tragic ending was predetermined. In 1930, he was arrested and shortly thereafter exiled to Kazan', where he died on September 10, 1931.

Egorov was the traditional professor type. He was very restrained, exact in everything, serious and reserved. His lectures were always well thought out, delivered in a strict manner.

Luzin, a student of Egorov, was of a completely different mould. As a personality, he was a romantic and a prophet. He often improvised during his lectures, giving his fascinated audience the opportunity to witness his free-flying thoughts.

Luzin was born in 1883 in Tomsk. In high school, he did not like mathematics, so his parents had to hire a tutor. The tutor turned out to be a university student who had a passion for mathematics. This student accomplished a conversion in the soul of his pupil. Subsequently, Luzin wrote, "He opened in front of me the door of mathematics not as a system of mechanical theories, to be learned by heart, but as a science firing the imagination." Luzin was accepted at Moscow

University. He later would say, "The brilliant pure math lectures made a huge impression on me." Mathematics presented itself to him as a science full of tempting secrets. He became a student of Egorov, who introduced him to the rudiments of the theory of functions of a real variable. Luzin later deduced from Egorov's theorem the fundamental Property S of measurable functions. After graduating from the University, Luzin was retained "to be prepared for the title of Professor", and then Egorov sent him on a study trip abroad, first to France, and then to Germany. In 1905–06 and 1912-14, Luzin was in Paris, and in 1910–12 he was in Göttingen. In Paris, Luzin attended lectures by remarkable mathematicians: H. Poincaré, J. Hadamard, E. Picard, G. Darboux and many others. He had fruitful scientific and personal contacts with Borel and especially with Lebesgue, toward whom he cultivated feelings of reverential admiration. Upon his return to Moscow, Luzin sharply changed the style of Moscow mathematical life. How indeed did this new era of Moscow mathematics start?

Luzin introduced completely new methods of working with youth. Egorov would distribute literature at his seminar on series, the students would read it and then talk about what they had read. By contrast, Luzin would start from the outset by posing to his students, who were barely out of high school, problems of the highest level, problems that stymied the most eminent scholars.

This is how P. S. Aleksandrov described the first meeting with his teacher: "I first met him when I was a second-year student. The impression from this meeting was, simply put, staggering, and I have remembered it for life. After one of his lectures, when I asked his advice on how I was to work in mathematics, I was first of all struck by his attention and—I cannot think of another word—his respect vis-à-vis his interlocutor, as strange as this might sound when one thinks that it concerns a conversation between a renowned—even if young—scientist and an 18-year old student. When he heard me out, Luzin quickly managed, by asking skillful questions, to discern the character of my mathematical inclinations, and immediately sketched—in a form accessible to me—some directions that he could propose for me to work on; he very carefully steered me toward a choice of one of these directions, and as far as I can tell now, correctly so." Luzin then gave Aleksandrov the Continuum problem for Borel sets—a problem that interested Lebesgue himself and that had been attacked in vain by such great mathematicians as Young and Hausdorff.

Another great accomplishment is connected with Luzin's name. He united all his students in a remarkable, unique community, *Luzitania.* This is what Aleksandrov wrote in 1977: "Formed already in 1914–16, the group of older Luzin students consisted of D. E. Menshov, A. Ya. Khinchin, M. Ya. Suslin and myself. In 1919–1920, a series of young mathematicians joined: P. S. Uryson, L. A. Lyusternik, M. A. Lavrentev, N. K. Bari, and a few other Luzin students. A very friendly collective was thus formed, which was later joined by two remarkable mathematicians, Kolmogorov and Shnirel'man. From the mathematicians of the older generation, V. V. Stepanov also joined, as did later on P. S. Novikov and L. V. Keldysh. Thus arose the famous Luzitania, a collective of young mathematicians, united not just by ties of close friendship, but also by a passionate love of, and selfless interest in, mathematics. Luzitania immediately considered itself as a whole, and declared itself an 'order', with Luzin as 'Commander' and Egorov as 'Grandmaster'. Luzitania was a collective with great creative, working and emotional élan, reflecting in a 'microcosm' the huge and all-embracing élan of the Mathematics Department of Moscow University of the day. An essential element of Luzin's interaction with students was at the meetings and discussions that he had with very small groups (1–3 people) working on one subject. There was also the weekly general meeting day for all of his students. These weekly meetings of Luzin's students at his home were often also attended by Stepanov and I. I. Privalov. These evenings consisted of two parts: first there was the math part in Luzin's study... I'll never forget the conversations we used to have, full of the liveliest mathematics. These conversations went on past midnight, but whenever they ended, they were followed by a tea with the invariable walnut cake. During this tea—held not in the study, but in the dining room of the Luzins, the conversation had a different, nonmathematical character, and touched upon the most varied cultural topics." Kolmogorov once wrote that a "common heartbeat" was characteristic of Luzitania.

Let us stop to recapitulate.

In the 1910's, an unusually bright star appeared at Moscow University: N. N. Luzin. Exhibiting unusual, new methods of interaction with the young generation, Luzin laid out already before the revolution the fundamentals of the Moscow mathematical school of the theory of functions. The living conditions in Russia in the 1920's were conducive to the development of mathematics on the one hand, and to the concentration of intellectual forces in Moscow on the other. And shortly thereafter, many outstanding students of Luzin's began working at Moscow University. Luzin's example served as a model to be imitated for many of them, and from this school began to gemmate the topological school of Aleksandrov and Uryson, the schools of Kolmogorov and Khinchin on probability, of Stepanov and Petrovskii on differential equations, of Lavrent'ev on complex analysis, and of Lyusternik and Shnirel'man on nonlinear analysis. These were joined by the schools on number theory of A. O. Gelfond, on differential geometry of S. P. Finnikov and V. F. Kagan (which moved from Moscow to Odessa), on algebra of B. N. Delaunay and O. Yu. Schmidt (from which D. A. Grave was an alumnus), and I. M. Gelfand's school as well.

The above-mentioned mathematicians obtained many outstanding results in the 1920's and 1930's, so that this period can naturally be called the first golden period of Moscow mathematics.

This brilliant period for Moscow mathematics was marked in the history of our country by many tragic events. Our country went through collectivization at the end of the 1920's, through a horrible famine in the Ukraine at the beginning of the 1930's, and through a period of most cruel repression in the mid-1930's. Then war started. The country suffered frightful, irreplaceable losses. Those who lived and worked in the 1940's had to endure the war on cosmopolitanism,[3] the decrees concerning questions of art, the flowering of Lysenkovshchina and much, much more. Thinking people lived in fear and expectation of the worst. But in the 1950's a clearing-up occurred, and it brought forth a new take-off in mathematics.

The second golden period of Moscow mathematics has clear temporal limits: from 1953 until 1968. A new generation, which I shall call mine, entered science. In order to understand the reasons for this unusual take-off in mathematics in Moscow, one needs to describe not only the political atmosphere of these years, but also the

[3] Code word for "being Jewish".

new generation entering science, and the people who were most responsible for the development of science in this period.

The year 1953 is forever engraved in world history; this year is as important as 1914 and 1939—the years when the two world wars started—and 1917, the year of the two Russian revolutions. On March 5, 1953, Stalin's rule over an enormous territory of the world ended, and with it his tyranny. A period began of some respite from the looming freeze, the period referred to as the period of thaw. Warm winds of hope began blowing, and the iron curtain was slightly lifted. It became possible for some chosen, deserving individuals to see the world, to attend congresses and conferences. Through the narrow cracks under the iron curtain, we acquired access to many cultural and scientific activities. The Mekh-Mat entrance exams loosened their questionnaire restrictions.[4] Such were the social upheavals connected with the development of mathematics in our country.

Now about the generation. In order to more easily talk about my generation, I shall give the list of those who studied with me and with whom I have collaborated later on, with whom I was connected by ties of friendship and/or scientific work. This list contains the names[5] of those who were born between 1929 and 1938 and entered Mekh-Mat between 1946 and 1955; the year of their entering Mekh-Mat is given in parentheses.

S. K. Godunov, I. Y. Piatetski-Shapiro (1946), R. L. Dobrushin, V. A. Uspensky, N. N. Chentsov (1947), F. A. Berezin (1948), A. G. Vitushkin, A. A. Gonchar, M. L. Lidov, R. A. Minlos (1949), V. M. Alekseev, N. S. Bakhvalov, O. B. Lupanov (1950), Yu. K. Belyaev, L. D. Meshalkin, A. L. Onishchik (1951), I. V. Girsanov, Yu. A. Rozanov, Ya. G. Sinai, A. N. Shiryaev (1952), D. V. Anosov, A. D. Wentzell, Yu. I. Manin, B. S. Mityagin (1953), V. I. Arnold, E. B. Vinberg, Yu. V. Egorov, A. A. Kirillov, S. A. Kruzhkov, M. I. Freidlin (1954), A. M. Vinogradov, V. A. Zorich, V. P. Palamodov, S. P. Novikov, A. B. Sossinsky, D. B. Fuchs (1956). All of them were leaders of their classes (sometimes that became clear after they finished the University), and all of them studied (some not from the start, I shall talk about that later) at Moscow University.

The lot of those whose destiny it was to be born in our country in the 1920's and 1930's was harsh; they had to endure many hardships at the beginning of their lives. The tragic fate of their fathers was typical: some were victims of the repressions, and an uncountable number of them did not return from the war. It was not rare for the mothers to be separated from their children, who were then raised by their grandmothers or nannies. Those who were born between 1920 and 1926 went to war, and "few returned from the fields". The ones I will be talking about did not go to war, but almost all had to suffer hunger and cold, the burdens of the evacuation, poverty, and unsettled existence in communal apartments. Such was life for almost everybody.

I shall give only one example from my list. Vladimir Mikhailovich Alekseev was born in 1932. His father was a very sick man and died when his son was nine years old. His mother remarried, and Volodya was raised by his grandmother in the village of Bykovo near Moscow. They lived in a village house on the grandmother's minimal scanty pension. Clearly, they had no means to buy books, or go to the

[4]The implicit way of ensuring restrictions on admissions, based on ethnicity.

[5]The transliterations of the names here and below are chosen so as to agree with those that the authors have been publishing under recently.

theater, concerts or exhibitions. But I have never met a person with such a breadth of knowledge, cultural interests and spiritual needs. He combined spiritual wealth and outstanding accomplishments in science with an unusual tranquility. Where did that come from? The answer is, of course, that it came first of all from a fundamental inner strength, but for many, it also came from Mekh-Mat and from the personal contact with unusual personalities, such as A. N. Kolmogorov and I. G. Petrovskii.

Those who were lucky enough to attend Mekh-Mat in those years have forever kept in their hearts the memory of the remarkable reigning atmosphere of daring and camaraderie.

Our generation had amazing teachers. Mekh-Mat managed to retain remarkable creative potential at the beginning of the 1950's. P. S. Aleksandrov, A. O. Gelfond, V. F. Kagan, A. N. Kolmogorov, A. G. Kurosh, D. E. Menshov, I. G. Petrovskii and A. Ya. Khinchin were heads of sections (so-called cathedras).

During the period 1946–1960 the compulsory courses at the university were taught by direct students and collaborators of Luzin's, by mathematicians who belonged to the first generation of their students, and also by eminent scientists who came to Moscow and by their students. Such were P. S. Aleksandrov, N. K. Bari, I. M. Gelfand, A. O. Gelfond, B. N. Delaunay, E. B. Dynkin, V. F. Kagan, M. V. Keldysh, A. N. Kolmogorov, M. A. Kreines, A. G. Kurosh, M. A. Lavrentev, A. A. Lyapunov, A. A. Markov, A. I. Markushevich, D. E. Menshov, S. N. Mergelyan, P. S. Novikov, O. A. Oleinik, I. G. Petrovskii, L. S. Pontryagin, Yu. V. Prokhorov, P. K. Rashevskii, S. L. Sobolev, V. V. Stepanov, L. A. Tumarkin, S. P. Finnikov, I. R. Shafarevich, B. V. Shabat, G. E. Shilov, S. A. Yanovskaya and others.

"A Pleiad of great mathematicians, gathered in one department, was a completely exceptional event, and I have never encountered such a thing anywhere else", wrote V. I. Arnold, describing the atmosphere at Mekh-Mat during his years as a student.

The knowledge acquired in the compulsory courses was reinforced by the practical seminars, led by experienced lecturers with long tenure. Such were N. D. Aizenshtat, B. P. Demidovich, S. A. Galpern, Z. M. Kishkina, A. S. Parkhomenko and I. V. Proskuryakov.

The famous physicist and alumnus of Moscow University, M. L. Levin, wrote the following lines on the role of our "seminarists"[6]:

Когда прямой была дорога,
И нам Вселенной был мехмат,
Мы все учились понемногу
    У Кишкиной и Айзенштат.

Тем, у кого мечта сломалась,
И тем, которых вознесло,
Нам всем от этих двух досталось
    Святое наше ремесло.

(When our road was straight and narrow,
And Universe was just Mekh-Mat,
Our knowledge came, slipshod or thorough,
From Kishkina and Aizenshtat.

[6]The translator is grateful for D. Burns' and A. Sossinsky's help with these lines in particular.

Both those, whose dreams were shattered,
And those who saw them made,
It was from these two icons
We learned our sacred trade.)

One of the incomparable qualities of the Mekh-Mat of these days was the wealth of special courses and seminars. Their number was around one hundred. Among the seminars that played a role in the history of mathematics in our country, one should mention the topological seminar of P. S. Aleksandrov, which started with the friendship of Aleksandrov and Uryson, the seminar of Bari and Menshov, which was the continuation of the Luzin seminar, Kolmogorov and Khinchin's seminar on probability, Lavrentev and Markushevich' seminar on complex analysis, the seminar of Petrovskii, Sobolev and Tikhonov on PDE's, and Gelfand's seminar on all of mathematics.

After the 1960's, the basic courses began to be taught by mathematicians of my generation: V. M. Alekseev, V. I. Arnold, V. A. Zorich, A. A. Kirillov, Yu. I. Manin, Ya. G. Sinai and others. They also authored textbooks for revamped courses on mathematical analysis, algebra and geometry.

And finally, one must give credit to those who played a crucial role in the life of Mekh-Mat. I shall only mention four of them, though one could add many more: Ivan Georgievich Petrovskii, Andrei Nikolaevich Kolmogorov, Israel Moiseevich Gelfand and Nikolai Vladimirovich Efimov.

Petrovskii became Rector (=President) of Moscow University in 1951. His tenure in this position turned out to be a selfless service to Science and to those who had devoted themselves to Science. He had an uncanny vision in picking people he could ask advice from and rely upon. V. M. Alekseev and V. I. Arnold belonged to the few mathematicians of our generation who enjoyed Petrovskii's confidence, as did Yu. I. Manin and V. A. Uspenskii. Petrovskii made it an absolute principle of his life to take particular care of the best graduates of Moscow University. At the beginning of his tenure as Rector, hard and fast restrictions based on social and ethnic origins were still in place. For example, after graduating from Moscow University, Ilya Yosifovich Piatetski-Shapiro was assigned to teach in a secondary school in Orekhovo-Zuevo, and Feliks Aleksandrovich Berezin was assigned to teach in a vocational school in Moscow for working-class youngsters. Petrovskii took Piatetski-Shapiro on the staff of Mekh-Mat despite strong objections from those who insisted on keeping the old ways.

Kolmogorov and Efimov enjoyed particular patronage from Petrovskii. They were both Deans of Mekh-Mat, Kolmogorov from 1954–1968, and Efimov from 1962–1969. These were years when Mekh-Mat flourished, all of which was possible because Petrovskii was Rector of the University. In these years, I. M. Gelfand, Kolmogorov and Petrovskii's scientific influence, either direct or indirect via their former students E. B. Dynkin and O. A. Oleinik, on practically all of the above-mentioned students who entered Mekh-Mat in 1946–1954 was decisive. One should, of course, note that others shaped the mathematical school of that period. Suffice it to mention the names of L. S. Pontryagin and I. R. Shafarevich, who laid the foundations of schools on optimization theory and algebraic geometry, respectively.

The students who began their studies after 1955 belonged to a new, "post-Kolmogorov" wave. Here, Sergei Petrovich Novikov played a crucial role; he laid the foundations for the development of modern geometry and topology within Mekh-Mat. This "wave" was backed in a most active way by Efimov, a most noble man and a remarkable mathematician, pedagogue and organizer. I deem it a duty for his students, colleagues and those who simply studied during his tenure as a Dean to publish a book of reminiscences about this fine man.

The Golden Age came to an abrupt end in 1969–1970, following the publication in the West in 1968 of the "letter by 99 mathematicians" protesting the harsh conditions of detention in a psychiatric hospital of Aleksandr Sergeevich Esenin-Volpin, a Mekh-Mat graduate whose specialty was mathematical logic and who was the son of the famous poet Esenin. New procedures were established more or less immediately. In the 25 years between 1970 and 1995, Mekh-Mat suffered irreplaceable losses. All of our teachers, the founders of Mekh-Mat, were gone: Petrovskii passed away in 1973, Efimov's death was accelerated by the grave worries connected with the fate of his beloved Mekh-Mat, Kolmogorov had to endure much grief, many remarkable mathematicians of my generation passed away in their prime (V. M. Alekseev, F. A. Berezin, R. L. Dobrushin and others), and many Mekh-Mat professors started emigrating. From the sixty or so above-mentioned mathematicians of my generation, only a few remained by the 1990's. Here are the names of some who left: Yu. K. Belaev, B. R. Vainberg, A. D. Wentzell, I. M. Gelfand, S. I. Gelfand, S. G. Gindikin, E. B. Dynkin, B. A. Dubrovin, Yu. V. Egorov, A. A. Kirillov, N. V. Krylov, Yu. I. Manin, M. Melnikov, B. S. Mityagin, S. A. Molchanov, V. P. Palamodov, I. I. Piatetski-Shapiro, A. Panchishkin, A. N. Rudakov, Ya. G. Sinai, M. I. Freidlin, D. B. Fuchs, M. A. Shubin, and A. M. Yaglom—these are all close friends of mine. And the above is not a complete list by far of those who had taught or led an active scientific life at Mekh-Mat. V. I. Arnold, S. N. Artemov, E. A. Gorin, R. L. Dobrushin, R. A. Minlos, A. L. Onishchik, and A. B. Sossinsky were forced to leave the University, as were many others.

Many Mekh-Mat graduates, who later obtained outstanding results, were not hired by their alma-mater, e.g. G. Margulis, V. Drinfeld and M. Kontsevich, to name but three. Often, when meeting former students who had been my friends (for example Fedya Bogomolov, Vitya Lomonosov or Asik Khovanskii, all students of a graduating class I was particularly attached to), I find it hard to conceal my bitterness over the fact that they are no longer part of our department.

But I do not want to end on a sad note. In the 15 years since the collapse of the USSR, much has changed to the point of being unrecognizable. And as a result, a new Golden Age does not seem to be imminent. But note that world history has not seen golden ages like the one during which, within a single city, nay, within a single university, one saw a concentration of many dozens of world-renowned scientists. And if one does not abandon hope that a peaceful and prosperous existence for all humanity is achievable, then one should not even dream about golden periods such as the one in the history of Moscow mathematics. People should be free in their choice of where to live, but they should also have the freedom of contact with everyone on Earth.

About 15 years ago many of my friends and colleagues obtained the possibility to move from Moscow. I do not know of anyone who has regretted their choice. Not all had a warm hearth in the old country—some emigrants have never returned to even visit Moscow. But they are in the minority. And of course, with the passing of time, few are those who do not pay tribute to the blessings of their youth, their teachers and their university.

And what did happen during these last years here in Russia? Of course, much that was abominable and unimaginable. But the University has remained. With heavy losses, it has remained nevertheless and even grown. And the MMS has been working very productively.

Mekh-Mat has a hard time keeping up with the developments of the new, modern mathematics. And the difficult events of Russian history in the past 15 years are not the only reason. It is a law of nature that all living things are doomed to age. And maybe one should not make Utopian plans for the re-animation of the department; it can still serve a purpose by providing a decent general education. I would hope that a special department of "contemporary mathematics" be created either within the University or under its Aegis. It can be small, but it should have the possibility to invite scientists from various countries.

A prototype of such a department has been formed in Moscow—the Mathematical College of the Independent University of Moscow. It functions within the framework of the Moscow Center for Continuous Mathematical Education, in the creation of which N. N. Konstantinov played a fundamental role. This Center, headed by a mathematician who started working during the Petrovskii years, I. V. Yashchenko, carries out most varied and fruitful activities. It publishes a large number of books, booklets, the Almanac *Matematicheskoe Prosveshchenie* ("Mathematical Education"), a series of booklets *Biblioteka Matematicheskoe Prosveshchenie*, and the scientific journal *Moscow Mathematical Journal* with Yu. Iliyashenko and M. Tsfasman as editors and an Editorial Board consisting of D. Anosov, V. Arnold, S. Artemov, A. Belavin, V. Buchstaber, P. Cartier, B. Feigin, V. Ginzburg, J. Guckenheimer, S. Gusein-Zade, A. Katok, A. Khovanski, A. Kirillov, I. Krichever, R. MacPherson, G. Margulis, N. Nadirashvili, Yu. Neretin, S. Novikov, J. Palis, S. Shlossman, S. Smale, A. Sossinsky, V. Vassiliev, and S. Vleduts. This is a remarkable combination of Mekh-Mat graduates and respected scientists who have never studied at Mekh-Mat.

Within the Independent University, there is a seminar, called "Globus" on all of mathematics "as a single organic whole". Two books have already been published based on this seminar. In the preface, Tsfasman (quoted above) gives a full list of the speakers for the past three years. The list is very impressive. No explanation is needed about who is who on this list for those who have spent some time at Mekh-Mat around the middle of the 20th century, whereas for the uninitiated it would take too long to explain (and many of the names have already been mentioned in my story). So I give the list without comment: S. Alesker, V. I. Arnold, S. N. Artemov, V. V. Batyrev, A. A. Beilinson, A. A. Belavin, O. I. Bogoyavlenskii, A. A. Bolibruch, A. M. Borodin, V. M. Buchstaber (twice), V. A. Vassiliev, A. M. Vershik, A. M. Vinogradov, S. G. Vleduts, S. I. Gelfand, S. G. Gindikin, R. I. Grigorchuk, S. M. Gusein-Zade, V. I. Danilov, P. Deligne (three times), S. Yu. Dobrokhotov, E. B. Dynkin, Yu. G. Zarhin, A. V. Zelevinsky, V. Ya. Ivrii, Yu. S. Ilyashenko, M. E. Kazaryan, V. A. Kaloshin, S. B. Katok, B. Keller, M. L. Kontsevich, S. K. Lando, L. Lafforgue (three times), G. Lachaud, L. A. Levin, D. A. Leites, G. L. Litvinov, A. S. Losev, Yu. I. Manin (three times), V. P. Maslov, S. K. Nechaev, V. V. Nikulin, S. P. Novikov, G. A. Olshanski, S. Yu. Orevkov, A. N. Parshin, A. N. Rybko, V. Serganova, A. G. Sergeev, Ya. G. Sinai, A. B. Sossinsky, J. Stienstra,

B. L. Feigin, M. V. Finkelberg, J.-M. Fontaine, D. B. Fuchs, A. Ya. Helemskii, A. G. Khovanskii, B. V. Shekhtman, V. V. Schechtman, M. Schlichenmaier, S. B. Shlosman, and M. A. Shubin. If one looks at the titles of the reports, one sees that the former Mekh-Mat graduates continue to be fruitful in the development of modern mathematics.

Science does not belong to one city, one nation, or one country—it is the property of all of humankind, and one can only dream of a Golden Age of mathematics of all of humanity. But personally, I would like it if, among the important world centers of mathematics, along with Paris, Princeton and some others that have already appeared or that will at some point show their worth, there were a city with a most rich mathematical tradition, whose name is Moscow.

Translated by Smilka Zdravkovska [7]

[7]The translator gratefully acknowledges the help of A. Sossinsky, D. Burns and P. Duren. All the footnotes were added in translation.

ERRATA

p. ix, line 1, "IMU" should be "IUM"
p. 8, line 11 from below, "V1" should be "VI"
p. 9, line 7 from below, "it the" should be "it"
p. 11, line 21 from below, "Louis K. Arboleda" should be "Luis C. Arboleda"
p. 12, line 11, "appeared 1980" should be "appeared in 1980"
p. 26, line 2 from below, "Petrova-Vodkina" should be "Petrov-Vodkin"
p. 30, line 17, "Komarovka" should be "Komarovo"
p. 30, line 10 from below, "Komarov" should be "Komarovo"
p. 30, line 15 from below, "Yudina" should be "Yudin"
p. 31, line 5 from below, "Komarovka" should be "Komarovo"
p. 37, line 17 from below, "195?" should be "1950"
p. 40, line 7, "19??" should be "1996" (she died in the meantime)
p. 45, line 14, "[A13]" should be "[A13])"
p. 49, line 24, "again Lusin" should be "against Lusin"
p. 56, line 13 from below, "Ivernova" should be "Iveronova"
p. 58, line 17, "consisting" should be "consisted"
p. 60, line 12 from below, "did not more" should be "did no more"
p. 66, line 20 from below, "at the MMS" should be "of the MMS"
p. 67, line 8 from below, "grater" should be "greater"
p. 72, line 5 from below, "abut" should be "about"
p. 76, line 4, "vector diagrams and" should be "vector diagrams, and"
p. 86, line 16, "Nordan" should be "Norden"
p. 88, line 13 from below, "when it came" should be "when it came time"
p. 90, line 19, "of this" should be "of his"
p. 91, line 20, "Aleksandrovna" should be "Arsentievna"
p. 93, line 8, "only partically" should be "only partially"
p. 93, line 15, "HET" should be "IIET"
p. 93, line 3 from below, "Sacceri" should be "Saccheri"
p. 94, line 17 from below, "Arbic" should be "Arabic"
p. 98, line 4, "in this city" should be "to this city"
p. 101, in the footnote, Tikhomirov's name is misspelled
p. 104, line 2 from below, "Honoray" should be "Honorary"
p. 105, line 23, "may seen" should be "may seem"
p. 117, lines 24–25, "of the 1930" should be "of 1930"
p. 123, line 3, "where overtaken" should be "were overtaken"
p. 125, line 4 from below, "USSR" should be "UkrSSR"
p. 126, line 6, "Martin-Lev" should be "Martin-Löf"
p. 126, line 17, "academy" should be "Academy"

p. 133, line 4 of the footnote, "трехизмерителъные" should be "трехизмерительные"

p. 134, line 6 from below, "works Andrei" should be "works of Andrei"

p. 135, lines 15 and 16, the subject of the dissertation should be "On the homotopy groups of the group of diffeomorphisms of the sphere"

p. 137, line 4, "Martin-Lev" should be "Martin-Löf"

p. 140, line 1 from below, "invesigational" should be "investigational"

p. 140, line 7 from below in the footnote, "did not recall" should be "reminded me of"

p. 143, line 11, "new" should be "now"

p. 143, line 12 from below, "reading-room for" should be "colloquium for high-school"

p. 145, line 4, "official" should be "waiter"

p. 149, line 19, "hears" should be "years"

p. 149, line 26, "this" should be "his"

p. 149, line 16 from below, "this death" should be "his death"

p. 151, line 19, "his" should be "one"

p. 152, line 20, "but although" should be "but when" (the second event occurs 40 years later)

p. 152, line 27, "mountains." should be "mountains)."

p. 152, the footnote should read: "Editors' note: The chairman of any Department of the Academy is called Academic Secretary."

p. 195, line 1, "contacts."" should be "contacts."

p. 195, line 21, "precise."" should be "precise."

p. 196, line 4, "I goes" should be "It goes"

p. 201, line 1 from below, "certain of closed" should be "certain closed"

p. 215, line 9 from below, "bet" should be "best"

p. 220, line 13, "listed" should be "listened"

p. 228, line 6 from below, "1959" should be "1962"

p. 229, line 11, "Plank" should be "Planck"

p. 246, line 16 from below, "Vigodskiĭ" should be "Vygodskiĭ"

p. 246, line 6 from below, "Ribkin" should be "Rybkin"

p. 247, line 7 from below, "Ozhigov" should be "Ozhigova"

p. 250, line 14, "Luens" should be "Lusis"

p. 250, line 16, "Khalinnov" should be "Khalilov"

p. 250, line 14 from below, "Putyat" should be "Putyata"

p. 251, line 21 from below, "A. S. Novikov" should be "P. S. Novikov"

p. 252, line 13, "L. K. Arboleda" should be "L. C. Arboleda"

p. 252, line 10 from below, "Ermolaev" should be "Ermolaeva"

p. 256, line 23, "L.K. Arboleva" should be "L. C. Arboleda"

Index of Names